Second Over All Overview Part 2

ISBN-13: 978-1539044352
ISBN-10: 1539044351

All rights are reserved.
No part, parts or the entirety of this book may be reproduced by publishing, electronically copied, duplicated by whatever means that form reproduction or duplication, without the prior written consent of the copy rite owner.

Written by Peet (P.S.J.) Schutte

© KOSMOLOGIESE EN ASTRONOMIESE TEGNIKA

I do find much pride in my status as being Afrikaner and would like to have my names used by pronouncing it in the manner Afrikaans dictates…therefore I would sincerely appreciate the courtesy when readers will take note that my name and last name are pronounced in Afrikaans, which is originally from Dutch and must be pronounced that way. Peet one would pronounce "here" which is the closest English to the pronouncing of the "ee". The "Sch" in Schutte is pronounced exactly as school is where both actually are pronounced Skutte or "skool". By pronouncing my name in Afrikaans you do me the utmost courtesy any one can. Being an Afrikaner is what I am most proud of. I submit articles and books to well known physics magazines but my articles are rejected on the most unappeasable grounds and for the most outrageously ridiculous reasons the Newtonians can think of. I explain how gravity forms but I am rejected because they are of the opinion that my work does not meet standards they lay down. When you read this work you can judge the truth about that. One such an article I may use because I said I was going to use the material as an open letter I gladly show.

<u>This book was done with a $25 ⁰⁰ scanner and a $35 ⁰⁰ printer and the reason I explain inside. For the same reason this book was not edited or linguistically checked. I could not because that does not work because I am in the writing business and not the spelling business and while I check spelling the writing gets more and so does the spelling and grammar errors. I had a choice; doing the books with no funding or not doing it at all because while I rubbish Newtonian science and show it is the fake it is, they will never publish my work because I trash Newton. Not having funds and trying to fight science for the truth with the truth was a fight that physically broke my health and still I am not published except in this manner. I apologise for the spelling and language but in poverty that was the best I could do under the prevailing circumstances in which I find myself…. This book is a first in every sense… it unites science and religion because science and religion was separated by human stupidity.</u>

Please take note that I sell information and not words or books and therefore the information takes priority and not the spelling or words used to inform the readers. This represents the work of God and not the word of God and so there is no interpretations applying and versus you can learn and sound intellectual but only cold facts you will have to understand.

Investigating Kepler is also the understanding of Kepler and that no one did since Newton named gravity and defined what he saw gravity was but gravity is out and out Kepler's discovery. There is a Universe inside the sphere and there is another Universe outside the sphere. One has to see it from the point singularity holds $k^0 = a^3/T^2 k$ and $a^3 = T^2 k$. Both represent as view from singularity but the two are not the same and neither do they share a principle. That is the relevance bringing about the gravity Kepler discovered. What would Kepler discover that Newton did not discover. Newton named gravity but Newton also missed the chance to discover gravity because Kepler discovered gravity and with all the laws on motion that Newton introduced Newton failed to recognise the work of Kepler as that Kepler introduced gravity. Newton did not see what Kepler introduced as relevance between participating object in motion forming sequence. The gravity Newton saw and the gravity Kepler saw is not the same gravity but since Newton meddled in Kepler's work by changing Kepler's work Newton then being the master on motion should have seen that while we on Earth uphold motion in serving the time factor T^2, which produces the space such a space factor will compensate for this inconsequence in the formula $a^3 = T^2 k$. Kepler saw gravity forming a relation that becomes a square when the two bodies interact between the different positions the bodies hold in space that is relevant to the motion each structure performs. That too was what Titius and Bode both observed. The Titius Bode and the Roche limit is only a part of the complete gravity process where gravity forms with motion placing relevancies. Mass does not bring about gravity but gravity produces mass by performing as a resistance to motion. Saying that we better find a means to distinguish the concept we have about more or less particles per unit forming more or less mass, and mass forming the gravity contracting resistance that the unit upholds. In our case we are descending or falling which gives us a negative distance growth. Normally the Kepler's formula read that $a^3 = T^2 k$ therefore $k = a^3/T^2$. This is normal where the planet is orbiting the Sun in a regular patter and does not bring about a decline to the factor **k**.

From the gravity "building" space through motion the motion of the building leaves an imprint which is detectable in the sequence the Titius Bode law saw in the number arrangement of 3; 6; 12; 24; 48; 96 etc. The incorrect application of the Titus Bode law lies in subtracting the figure of 3 from 10 leaving 7. The other way of reasoning is to add four each time to the firs value of three starting with 3 and so on. One has to see the Titus Bode as two relevancies in the unit bringing across the building of space-time. The true significance of the Titus-Bode law is that it points directly to a circular growth of 7 in the sphere leaving the marks as it grew in stages. The 7 relating to 10 is a precise derogative of the Roche limit or the Roche limit is a precise derogative of the Titius Bode principle because he two systems interlink.

According to Newtonian science space is simply nothing with no qualities but gravity separate space and space does not mingle, as one would expect if space was nothing because space does form borders. Disasters of unprecedented magnitude arise from such borders. The Challenger disaster of February 2003 is much testimony to those borders that was powerful enough to break the aircraft into pieces while the explaining contributed by Mainstream science is evidence of a shocking lack of understanding about what took place as cosmic laws were breached. Let us now start to timely investigate cosmic laws as presented by the cosmos.

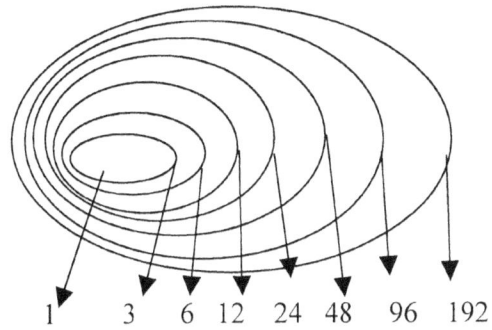

Planet	Mercury	Venus	Earth	Mars	Ceres	Jupiter	Saturn	Uranus
Bode's Law dist.	4	7	10	16	28	52	100	196
Actual dist.	3.9	7.2	10	15.2	28	52	95	192

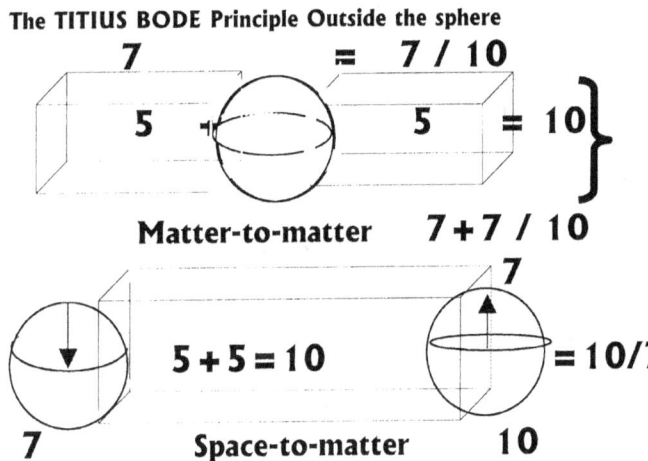

The TITIUS BODE Principle is gravity in space. It does no make sense that Mainstream physics denounced the TITIUS BODE law. Gravity is motion and motion in space comes about from The TITIUS BODE that is in principle gravity.

In the situation we are in and being captured by space-time control we have $1/k$ because we are within a reducing or growing smaller than the k the Earth introduce so we submit to the k of the Earth. Reduce k as it happens in the gravity Newton presented then k^{-1} would produce a smaller space a^3 if the time T^2 component of such motion relevancy is forced to be equal to both notwithstanding what differences there are in space a^3. Newton had available at the time the findings of Galileo and it was Galileo that showed by using a pendulum that swinging pendulum arm will reduce the space to compensate for the establishing of the time. By compromising space a^3 the pendulum can manage to uphold the time T^2 component because the pendulum is visual proof of $a^3 = T^2 k$. In our case the k attached to our position in space is in decline or negative which is the formula taken in another relevancy of $k^{-1} = T^2/a^3$. That is precisely what is happening. We are reducing the space between the Earth centre and us (our singularity) because we are travelling too slowly in relation to the Earth. Again Kepler showed himself being correct and not only that but he proves his brilliance.

We are all aware that the Titius Bode law is about positioning and aligning planets and yes that is true. I can explain that part in detail. However, the Titius Bode law holds much, much more in cosmology that planets finding relevance...

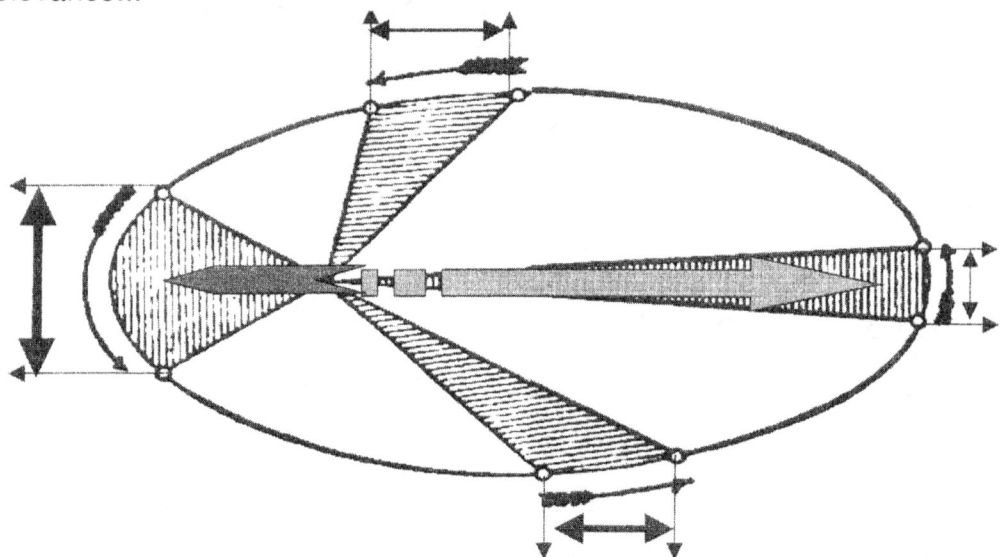

Newton saw in Kepler's work what all other people ever since Newton see, that gave Kepler's work the smallest attention and Newton was the only one that gave Kepler's work the smallest degree of attention. What all other people see in Kepler's work is the incorrectness I see in the work of Newton's idea about the

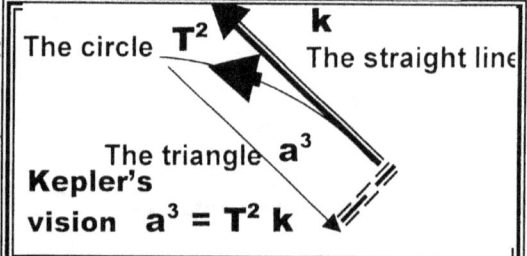

incorrectness in what Newton thought Kepler said. For that I have been dismissed for years. Kepler calculated the movement of the orbiting structure around a centre Sun. Newton did for starters not interpret this rotation as gravity. Secondly he saw what clearly is space, as motion. Therefore Newton was unable to see that Kepler said space in motion is gravity. I challenge who ever is of such an opinion that Newton made an in depth study of the integrate meaning of the work of Kepler, to prove that. Newton made a serious observation about the work of Kepler. Newton saw a circle forming by a planet that is in rotation, running around the Sun. To correct this Newton added $4\Pi^2$ on one side and on the other side he introduced his version of what he saw gravity to be. Newton as a person failed to see in Kepler's study any gravity applying. If Newton did recognise any gravity aspects then he, Newton would have concluded that forming another interpretation by adding is very unnecessary. Newton did not act incorrectly when he added $G (m + m_p) = 4 \pi^2$ but he merely duplicated what he then neutralised as a fact. He brought in his vision he had about gravity and that the two structures in motion is combining a time related value of $4 \pi^2$. He duplicated what Kepler said and by duplicating as well as neutralising the duplicating he brought misconception and he covered information, which I then uncovered. Through my uncovering, I came to the conclusion that what Newton said what Kepler saw is not what the cosmos told Kepler. This action Newton missed completely when he did not see that into the motion of the orbiting object the value of π^2 is built in and the circle will have to complete in $4\pi^2$ as a rotating action enforced by a centre. The centre is the measure holding π to enforce by motion π^2

From the orbiting structure (the planet) aligning with singularity only one structure, which is the very inside singularity applies as a position of reference and that is reference to the distance applied between points holding the governing singularity. From the Sun (governing singularity) the matter marker is 7/10 = 0.7 with the only one other forming a marker 7/10 = 0.7. The two form 1.4. The sectors provide individual singularity as a means in sustaining governing singularity by which provision comes through maintaining governing singularity the required spin in maintaining cooling. If this process did not apply, there would be no connecting individual singularity to major singularity.

The spherical positioning layout forming the Titius Bode Principle

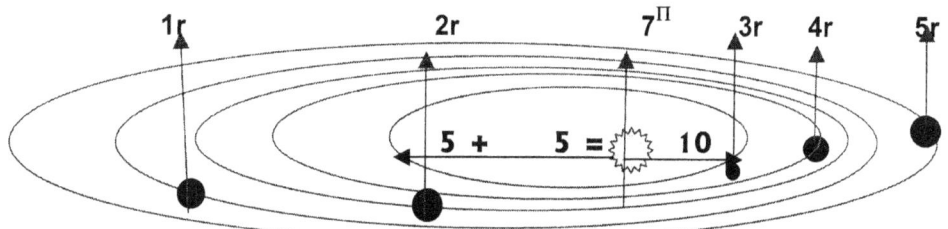

From the matter-to-matter relation in the Titius Bode configuration there are 7 / 10 + 7 / 10 = .7 + .7 = 1.4

From the space-to-matter relation in the Titius Bode configuration there is 10 / 7 = 1.42

(7 + 7) / 10 = 1.4

(7 + 7) / 10 = 1.4

(7 + 7) / 10 = 1.4

10 / 7 = 1.42

10 / 7 = 1.42

10 / 7 = 1.42
= .7 //\\ 1.42
= 1.4 //\\ 1.42

The 5 + 5 = 10 is a position of dimensions as space loses value to singularity. The 7 that matter diverts in points from singularity may seem, as coincidental but is valid. Still in accordance to our perception valuing the number in degrees, it seems coincidental but if it is coincidental, it is nevertheless a figure of diverting proven as accountable in all other calculations and plays a most dynamic role.

The Lagrangian 5 point system results as much from the Curvature of space-time as does the form the Black Hole holds. The Galactica is the opposing equivalent of the Black Hole and has identical but opposing similarities being the five points positioned to singularity. The galactica is generating space and the Black hole is degenerating space.

Because the space-to-matter is in the square at 10 placing the matter-to-matter at a square of .7 + .7 = 1.4 the space-to-matter forces the matter-to-matter to double the distance by number as structures are place father from the mainΠ^0 maintaining singularity.

Reasons why this does not fully apply to the solar system I give in book # 7.

The sectors provide individual singularity a means in sustaining a relation with the governing singularity by which provision comes through maintaining governing singularity the required spin in maintaining cooling. If this process did not apply, there would be no connecting individual singularity to major singularity.

SINGULARITY BY DIVIDING SPACE INTO MATTER AND MATTER INTO SPACE, ANG ALL OF THIS ACCORDING TO THE TITIUS BODE LAW OF 10 / 7 AND 7 / 10 IN CONJUNCTION WITH THE ROCHE PRINCIPLE OF $(\pi/2)^2$

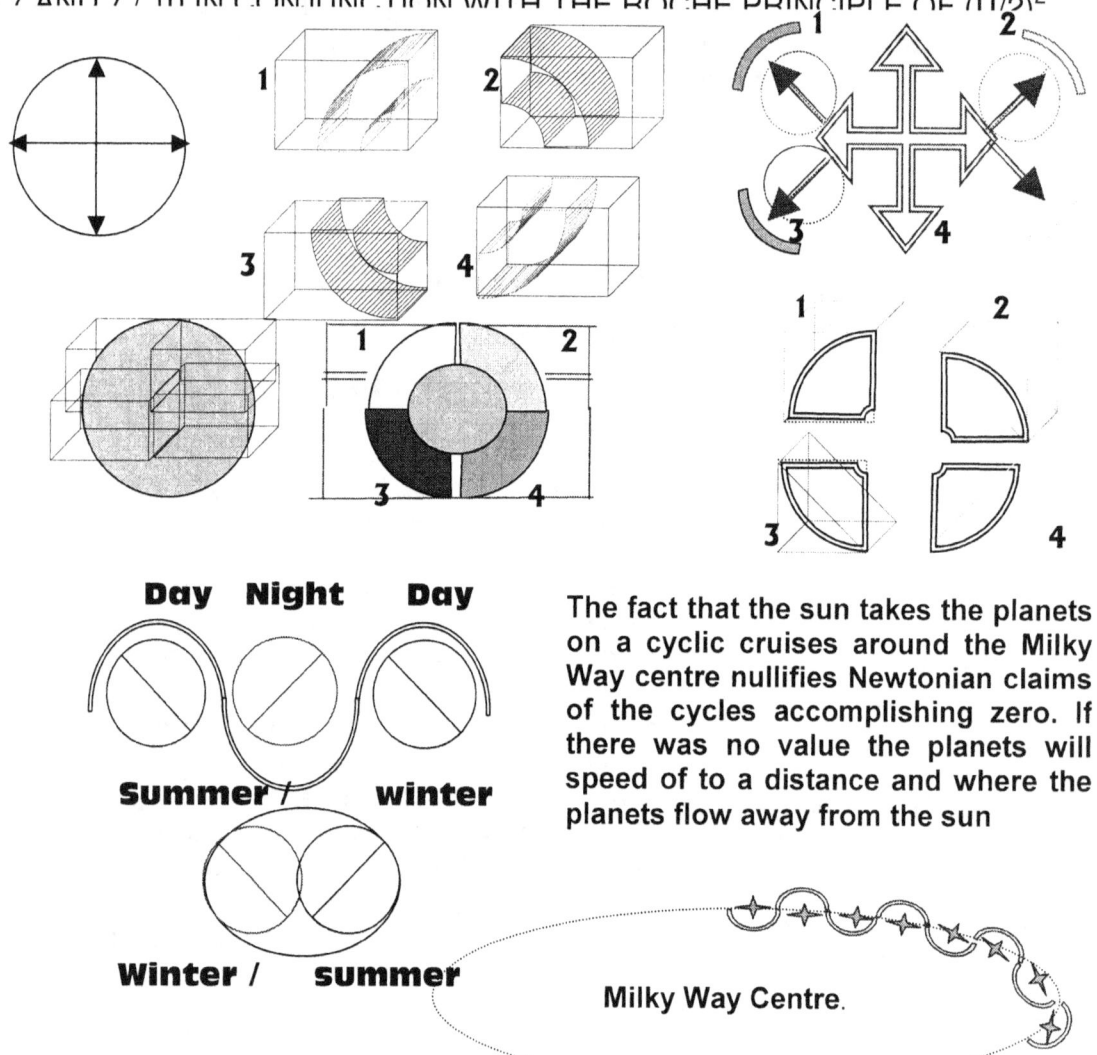

The fact that the sun takes the planets on a cyclic cruises around the Milky Way centre nullifies Newtonian claims of the cycles accomplishing zero. If there was no value the planets will speed of to a distance and where the planets flow away from the sun

Ice ages and droughts, flooding and Earthquakes are all alignment implications coming about as a result from alignments between the Earth and the sun and the sun and the Milky Way. It is all interlinked and zero is the last product one will ever find between the relevancies.

From the Sun (governing singularity) the outer planet forming the marker in search of position holds space in the square 10 / 7 = 1.42 in aligning with the 7 forming material of the Sun. Therefore there are two sevens relating to ten forming the material positioning of the structure in orbit and from the governing singularity all outside the Sun is the square of space (ten) aligning with one particle (seven) and not one of the other structure to the inside or the outside holds any value. Because .7 + .7 = 1.4 and 10 / 7 = 1.42 the distance doubles every time there is an aligning of three orbiting objects. In this there is definite proof of influences coming about between particles sharing gravity. But then again the entire Universe shares gravity and as such then all will influence everything. I do realise science do not recognise a relevancy between the rotation or orbit of the Earth and the position of the Sun as they choose to be keeping to what Newton claimed and I shall come to that in a brief time. Newton does not provide for this as a fact. On monitoring the rotation of the Earth to a graphic display one find that the Earth movement displaying in accordance to change in positional location does indicate a relevancy that imitates the flow of current to an almost exact. Seasonal change has all to do with the graphs influence derived from the cosmos and little to do with the position of the Sun and the Earth. It is the position singularity holds in relation to the Universe and the Milky Way forming currents and seasons moreover than the Sun shining brighter or not. The Sun in size over dominating the Earths in comparison disqualifies any positional influence that can alter the Earths heat standings. Through shear size the Sun can shine at the top and the bottom of the Earth simultaneously without effort from all normal possible angles. In **Seven Days Of Creation** I show a relation between singularity in different positions maintaining seasons and north/south polarity, not only as far as concerning the Earth but also outside influencing polarization. This has to do with the second position singularity holds in accordance to matter and space and is an "*electromagnetic*" (used for the lack of a better word) sustained positional opposing derived precisely from the graph in the manner when calculating electricity.

This results in $\Pi^2\Pi$ coming about By retaining the seven and carrying the seven on to the ten and into the three gravity produce Π^2 by duplicating Π in the form of 10 + 10 + 1 + {.9991 (motion reducing space)} / 7 = Π

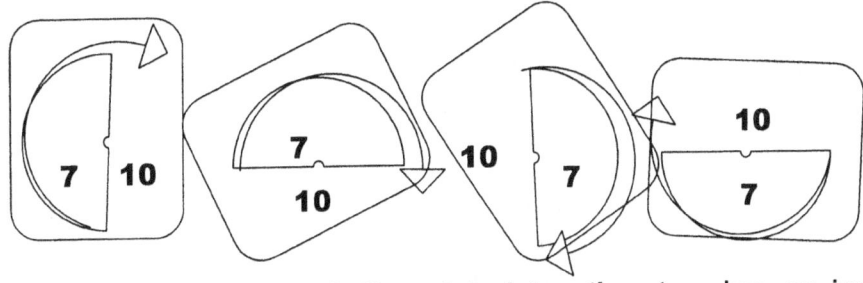

To give an example which I shall explain later, the stars has an inner core gravity charging value of **7/10 X 4(Π^2 + Π^2) = 55.27,** while outer space holds **10/ 7 X 4(Π^2 + Π^2) = 112. 8** as displacement

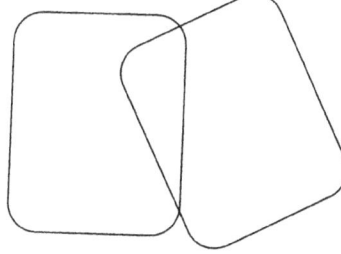

The motion duplicates space (10 + 10) by rolling over singularity (1) from the one side by reducing space (.09991) on the other side. All this motion of space 10 duplicating is directly related to matter 7 The total value established from this motion is the interaction of seven with ten producing Π^2 as the half of singularity $\Pi/2$ on the one side interact with half the singularity $\Pi/2$ on the other side and in this process Π the duplication of Π by means of matter and space produce gravity in the measure of Π^2

In this it is clear why the Titius Bode ([10 + 10 + 1 + .991] / 7) and the Lagrangian 5 \\ 7 systems part their ways when applying the different processes they hold. With all the differentiating, the observer must also consider the dual massage that light uses in travelling through the vastness of universal space. The thought of nothing is just what it is, a thought of nothing and although it is in the human mind common nature to present nothing as a value in the recalling of something, nothing is a presentation of the figment in the human mind. There can be no number such as nothing and that was (possibly) Newton's biggest error. Nothing represent non-existing and that is just what nothing is, it is non-existing. The Titius Bode influence in a manner that on the one side holds the matter-to-matter relation of 7+7/10 whilst on the other side during the same time holds the space-to-matter relation of 10/7 forming equal and opposing values. From this the orbits of cosmic structures are always oval favouring the singularity dynamics of the one structure at one point and switching the favouring to the other structure on the opposing side. Because the structures can never be equal in size (singularity will not permit that where the Roche principle will intervene) the shape is always "off centre" as well. This influences coming about as the Titius Bode principal manifest in other ways proving Kepler's time relation with space through distance from singularity controlling the factors.

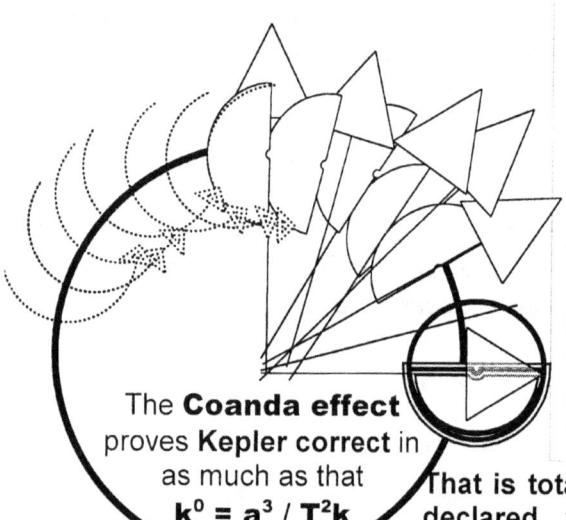

The **Coanda effect** proves **Kepler correct** in as much as that $k^0 = a^3 / T^2 k$

The first axis is the one every person knows. It is the straight line forming running from the sun to the planet and that has a symbol value of **k**. But then the is a second axes value coming about running in contrast to the first and that first line neutralising the second line forms another factor holding the square value. The planet is departing and stretching **k** as much as the sun is pulling back the planet and reducing **k**. The circle coming about from this motion produces the

That is totally the very opposite what Newton declared when he announced the circle motion forming a zero and therefore being neutralised as a circle. A circle came about from forming a double motion that is equal and opposing.

The understanding of this aspect is most crucial and all the condemning of my work in the past was a result from bias formed with my criticizing Newton. I feel I cannot overstate this condemning of my view just plainly because of favouritism about Newton. When a circle of cosmic proportions complete a rotation. The circle does not neutralise the combining effort by forming a rotation. Much rather is the circle the result of the square T^2 forming and with that reducing the extending line to one factor. Kepler did not find two particles in space pulling at each other and bringing on a force $F = G (M. m) / r^2$. By placing this factor on the other side of the changed concept duplicated what is a common fact but in doing that it neutralised all concepts forming about Kepler's research.

Every time the sun takes the planets on a route march around the Milky Way they travel through space-time and that does not come to nothing, since every aspect in the cosmos re-align totally in one rotation and in every rotation. What was never will be repeated in precise detail and what is in the past changes the future but never repeats the past. That can never add to zero accomplished because that is space-time. It is space in travel. It is gravity. After every rotation, everything in rotation grew by the Hubble parameter. How on Earth can that constitute to nothing...only a Newtonian will make sense of such double talk

Kepler already introduced the square in the time factor, but the square does not divide the gravity, it is the result of the combining of such relevancies doubling. As humans, it is very human to look at everything with a single-minded point of view. We have to see that where on object holds seven from one singularity, there is another singularity, which is concerned with 10. There is always a doubling of one where one in singularity holds seven, but the second part of singularity holds 10 whereby it is unified by relevancy. The doubling of the affect of the distance between the two cosmic structures is a result of the combining of the motions applied by both sectors bringing about the Titius Bode configuration, but it is the result of changing perspectives from both sides and that brings about gravity as a interacting of seven over ten and ten over seven on both sides of the relevancy ends.

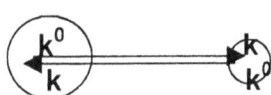 | The duplication of the relation **k** has forming gravity from one point to the other point is the gravity coming

This confirms that T^2 is gravity and is Π^2. The circle having four interacting relations with each other forms $4\Pi^2$ and by Newton's claim that the planets circling the sun is $G(m + m_p) = 4\Pi^2$ leaves no doubt that my argument about gravity and Kepler is correct.

SPACE = Π^2 = 9.8696 = Space and time in a dimensional implication.

The space to the outside of the gravity ring holds a dimensional relevancy of 10. The space to the inside holds a value of Π^2 in relation to the top. In one case the seven stands to ten through motion of reducing by spin and the reducing of ten then becomes Π^2 as the space reduces. It is an ongoing process carrying form from as wide as one wish to take it down to as small as it gets when entering singularity. The seven and the ten has very intermingling relevancies of relations swinging from top to bottom and from bottom to top.

The Titius Bode law is an extending dynamic deriving from the law of the gravity dimensional factor where the space factor in a square of ten relates to a matter factor in the square by half. Space makes four quarters in holding time or motion, which then places the next spot being the fifth into a located area being space. The fifth spot will be the half of the full value of the Universe on one side since nothing can be in two places in the Universe simultaneously of the matter factor of Π^{7+7} or the square of space (10) relate to the matter factor of 7. The Titius Bode is a relevancy of space reducing by implication of the square going to a circle.

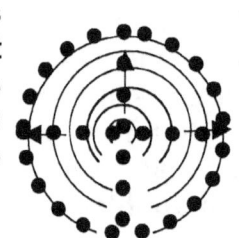

Every quarter provides a distinct value that indicates the progress of the flow of time from the one point Π to the next point Π.

Any changes occurring in Π will lead to an unequal triangle providing two different values to r and will alternate the link between r and Π^2 bringing about different form (Π) and time (Π^2).

Keeping these factors in mind it is clear that Π^2 are the choice of gravity and not r^2. What this says is that the gravity influence does not end on Earth. Although specific borders are in place in the atmosphere at various levels, the beginning and the end of such borders are in place in the atmosphere where the beginning and the end of such borders are as definite but also as vague as the gravity that forms them. If we go back to the spinning top might it might reveal some factors about motion that activates singularity that is forming borders by motion.

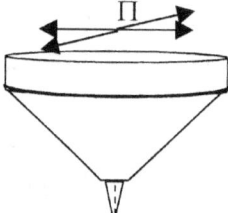

Every quarter provides a distinct value that indicates the progress of the flow of time from the one point Π to the next point Π.

Any changes occurring in Π will lead to an unequal triangle providing two different values to r and will alternate the link between r and Π^2 bringing about different form (Π) and time (Π^2).

It is motion that brings about 3D as much as it is motion bringing about gravity Π^2

Gravity comes because of interaction between spaces maintaining singularity of different independence. Space is created by motion thereof, forming dimensions which is involving Kepler's view totally. The material uses seven contact points to singularity and the material including space the material influences through the motion, is filling space a^3 or release of space filled a^3 and holds ten positions.

The figuration found in the planetary layout that has taken the name of the Bode or the Titius Bode system where such a system is the mould that remains in space as part of space. The form drives to form space after the concept establishes the relation to establish space through continuing with form. By providing space with forms, that relation of gravity is then detected by observation and growth leaving the mould in space as the departing of planets from a space concept in relation with the centre. It is a manifestation of the evidence what Kepler brought from his investigation of the cosmos. Gravity is where space vanishes and as such, space is in the centre of the sphere. The sphere controls not only the space included in the sphere but also the space excluded or on the outside of the sphere.

In the sphere or the including part of the sphere, there are seven points $k^0 = a^3/T^2k$, or the compliment in the dimensional forming of space $k^0 = k^{0=3} / k^{0+2} k^{0+1} = 7$ in relation with one another at all times, which totals seven all included. Then there are three more in space and three more at the same time coming from motion $a^3 = T^2k$, which then also is $k^{0+3} = k^{0+2} k^{0+1} = 6/2 = 3$. Adding the centre or sphere, the total is worth seven and any one side of the Universe you wish the sphere would move to a total of seven factors or dimensions fill with three more when the space is in motion filling the relevancy on one side with seven dimension characters and on the other side with ten dimensional relevancies. This role changes to a mirror image on the other side of the Universe where ten relates to seven.

Our judging what constitute of liquid or solids should not be tainted by a culture misconception. We must not presume the heavens as if the heavens were Earth. Looking at the cosmos we find hot and we find cold. Heat gives motion by space–time extending while cold contracts by motion declining space of value. There is no degree or quantifiable difference but in the minds of us small men. There are three positions in space –time that all space-time is accountable too. There is the electron state where heat brings about expansion. Then there is the neutron state giving unrestricted gravitational motion of space-time. Then there is the proton state of retaining or conserving. That means there is that which moves away, that which moves unrestricted in gravity and which moves towards and thirdly there is a pivot point aligning the motion to a fixed beacon. There is that which moves restricted as well as unrestricted in relation to that which does not move. When it moves it is liquid and when it does not move it is solid. The neutron is a solid too the electron and therefore the electron has mass. The proton is solid in relation to the Neutron therefore the proton has mass. The proton is a liquid in relation to the singularity and singularity by measure of Π is a liquid compared to singularity at measure of Π^0. It is about motion and the relevance thereof.

It is motion that brings about 3D as much as it is motion bringing about gravity Π^2

Gravity comes as a result of interaction between spaces maintaining singularity of different independence. Space is created by motion thereof, forming dimensions which is involving Kepler's view totally. The material uses seven contact points to singularity and the material including space the material influence through the motion it is filling space a^3 or release of space filled a^3 holds ten positions.

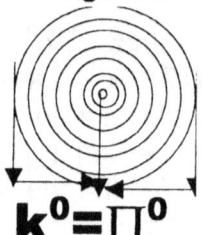

The figuration found in the planetary and has taken the name of the Bode or the Titius Bode system such a system is the mould that remains in space as part of space after the concept establishes the relation and as such is then detected by observation and growth leaving the mould in space as the departing of planets from a space concept in relation with the centre. It is a manifestation of the evidence what Kepler brought from his investigation of the cosmos. Gravity is where space vanishes and as such space is in the centre of the sphere. The sphere controls not only the space included in the sphere but also the space excluded or on the outside of the sphere.

In the sphere or the including part of the sphere, there are seven points $k^0 = a^3/T^2k$, or $k^0 = k^{0=3}/k^{0+2}k^{0+1} = 7$ in relation with one another at all times, which totals seven all included. Then there is three more in space and three more at the same time coming from motion $a^3 = T^2k$, which then also is $k^{0+3} = k^{0+2}k^{0+1=6/2=3}$. Adding the centre or sphere the total is worth seven and any one side of the Universe you wish the sphere would move too a total of seven factors or dimensions fill with three more when the space is in motion filling the relevancy on one side with seven dimension characters and on the other side with ten dimensional relevancies. This role changes to a mirror image on the other side of the Universe where ten relates to seven.

If gravity is motion T^2 the process may sound deception silly and simple for Masters such as the Academics of distinction fighting to create fusion and Space Whirls and lots of other very impressive goodies but it is very complicated to rein act. By motion T^2 space comes about forming a^3. But the motion T^2 will mean a crossing to the other side of the Universe since singularity divide the Universe into sectors. Mass is the frustration of the motion and mass can only become part as a fact when an object refuses to relinquish that object individuality by compromising the space it claim in favour of becoming part of the bigger space. No atom will do such relinquishing without an enormous increasing of time and that is impossible outside a massive star and in the massive star we call this process fusion.

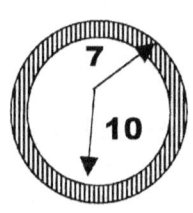

Material produces the dismissing or the concentration of space by applying the motion. Surrounding all elements is a layer we call the atmosphere and even Pluto and the moon must have the atmosphere because they have gravity. In this, the relevancy of ten to seven forms this layer and it results in forming a circle, because the combining of the motion duplicates the singularity factor Π, forming from that gravity as Π^2

THE PROCESS PARTED USING THE ROCHE PRINCIPLE

By establishing motion and creating motion singularity quadruples to 4Π in rotation. But since the rotation is motion duplicating the space established as four times the value of singularity the motion divide the space coming about by halving such space by the dimension, which is putting a square root over the quadrupling of space. In this comes about the direction gravity takes the Universe. The expansion is always double the square root but the square root is neutralising the expansion and that brings about that the neutralising of the expansion creates a contraction that seems dominant to us but it is not. The contraction is doubling the expansion by halving the effort of the expansion.

Space duplicates as space moves. The motion that is becoming another duplication is the result of singularity being unable to move. Since singularity is eternally motionless singularity cannot shift in a straight line. The only way singularity may achieve a motion is by re-directing the continuous duplication of space by re-alignment. That is what the Coanda effect proves. The singularity is unmovable since it is motionless but it does shift position as we see in the manner that the top spins. Our position we have in the Earth does not allow the spinning an everymen in which the top can spin. To the spinning top there are no other options to be within but to use the space the top claims from the Earth for rotating purposes. The duplication by employing a singularity allows the rotating or the continuing of time to be flowing in a fluid like manner but we also know that there are other functions involved. When a train and a fly go in a direct head on collision, it is obvious that at some point both objects must come to a standstill. This is because there is a re-arrangement of particles layout at structure re-arranging, especially as far as the fly's interest in the event is concerned. To find such re-establishing of material positioning there had to be some stopping and applying different motion to different particles to create the re-arranging of the structure of the fly's body. We know that when two object go in a head on collision between say a train and a fly at the moment of impact the fly stops the train as much as the train stops the stops the fly. Only in the very next motion does the rearranging of construction begin. This is because singularity has to demolish space-time reposition the centre forming singularity that every atom will reapply the next position they will hold to in relation of the motion that determines the previous and the following position of singularity relevancy.

After installing singularity at the next position the singularity then re-institutes the space-time in relation to what the space-time was just before. Only if there is an interruption by another singularity of more dominance in proportional displacement such singularity then will it stop the singularity on route to fill that position in the direction where it is heading. As that position is then already filled will it force the atom to place the singularity of lesser displacing qualities to seek another position to occupy in relation to all the surrounding singularity it forms a relation with. This all is the result of singularity being immovable the singularity must attach to a new focus point whereto the focus of the atom having the singularity will shift. In that it is not the shifting of the singularity because such shifting is impossible. It is the motion of the space-time generating another alliance with the same singularity (singularity everywhere is the same singularity) but finding in motion of space a new point, which the motion generates, excitement. That is why there can only be space-time anywhere. The motion or the momentum it has and that momentum continuously creates the Universe or creating of space is the value of time. After focussing a new point to install singularity in motion the space-time has to duplicate in relation to the new position that came about by the motion bringing new points on which to focus. One may never exclude the thought that it is one singularity having many points where each holds an individual Universe. In the event where the point has been filled and cannot find a vacant spot in time the singularity will not match the time component and will turn the space-time it then cannot duplicate into heat, which will then form a crack or a tear or show some form of material deformation. By duplicating space-time in relation to a new allocation the singularity had to diminish the space to time being motion then shift the point of focus and then reinstall the space from motion into the new found position that may or may not bring new relevancies about where every atom indirectly form a new alliance with every singularity in the Universe since all the singularity maintain a relevance but all remain the very same.

That is the same action we detect in the Pulsar where the space breaks down relocate then re-established and floods out into outer space. But the pulsar is only getting in the single atom stage while us observing the Pulsar find our position is one other Universe away from the pulsar. We can never reach the pulsar as much as the pulsar can never influence us directly. The controlling singularity remains to the most specific sequence there can be and the duplicating of singularity that is applying motion by producing space to move or time in releasing space in order to use time to reinvent space from inside the proton that is inside the atom. It is like letting a strobe light flash $((1836)^3)^2$ times per second and watch the content of the Earth move in that strobe light. Because of the fact that the motion will introduce a new space coming about such a space has to be within limits. The maximum limits or minimum limits depending on where you place the relevancy will be half the value of singularity dividing singularity as much as duplicating singularity. Gravity is about factor holding singularity and placing relevancies by putting relevancies in place. The motion creates new space $a^3 = T^2k$, but the motion duplicates the space created as well the motion producing space $T^2k = a^3$ where the singularity stands overall related to singularity changing into space and into motion $k^0 = T^2k = a^3$. The duplicating of space by dividing of space reduces space by half after it quadrupled on both sides of the divide. The pulsar swings between turning matter into heat as light and allowing that to manifest as the duplicating while the contracting takes the star to a point of singularity outside the realm of the Universe. The Pulsar is at the point where time catches heat by time delay returning to eternal while the heat is still cyclic providing space with the liquid that once formed a Universe. The pulsar is a neutron star relinquishing the last connection the star still holds with the formed Universe.

The Titius Bode principle is a relation where space is the ten factors and material is the seven factors. By space being diminished by material one relation comes about and where material dismisses space another relation of seven to ten comes about. Gravity is the motion where space conservation is applied by motion control. By applying motion in sequence to space movement Universal harmony is installed. By applying motion at a faster pace than space conserve motion the motion, which is also gravity, such gravity controls the space by motion duplicating space at an even rate. In that event gravity is applied in the manner we recognise the working of gravity and space.

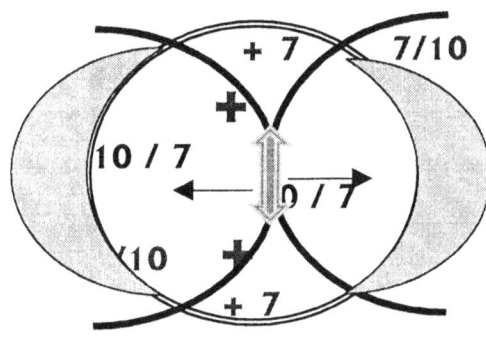

Gravity is motion in space and motion through space. Gravity is a relevancy between space travelling and the time it takes the space to travel. Gravity produces or reduces space during a certain period of synchronized spin of material in motion. Gravity is $a^3 = k\,T^2$ where it then becomes $k = a^3 / T^2$ In the light of this all other explaining fails the test of accuracy. It is no force because mass depends on gravity and gravity does not depend on mass.

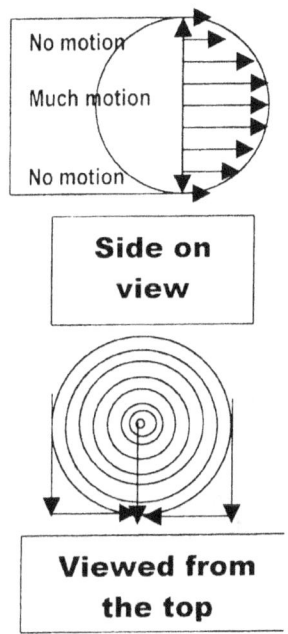

The motion produces a relevance by which the sphere in seven forms a working relation with the ten surrounding the sphere. What makes this so truly important is that the atom uses it my forming a relation between the electron and the neutron / proton. There is always space surrounding the space holding the material forming the solid. The sphere holding the solid is time driven forming the four that acts as the proton. $(\Pi^2+\Pi^2)$. This is the time part forming four positions and marks the material. Then there is the liquid or fluid or flowing part. To move it has three positions in Π and is the connection that bonds the motion to the solid. It is what serves gravity by establishing gravity and it is the neutron and not the proton that supplies the gravity. The neutron$(\Pi^2\Pi)$ produces motion which produce gravity and in that it serves the proton with gravity to which the atom is served by the proton the gravity that keeps the atom in a unit. In stars when the neutron falls away the star immediately thereafter start to collapse in space-time. The 3 one may think of as being the parcel wrapper that serves the atom by providing a paint layer or a veneer of protection.

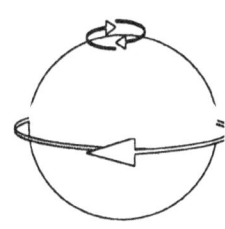

On the inside, there are the seven markers of which singularity is the focus point in the centre of the centre. The markers are representing one aspect of space, which for argument's sake let us call it cold. Then there are three more markers on either side being part of the space but not captured in the space. It is space in motion by the influence of the motion of the Earth.

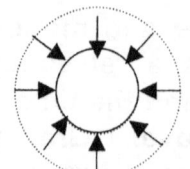 **Gravity is about reducing space**

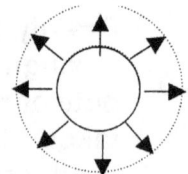 **Expanding is all about heating. Heating takes up more space and gravity reduces space.**

Remember there is no Universe towards the outside because there is no outside but image. The Universe runs towards the inside or away from the inside and every singularity is an individual Universe, only one Universe away from the next Universe. Since the Universe starts and ends in infinity and that end all definitive value big and small is merely human appreciation of what cannot be. It is a relevance of what came when and that is all. Everything past singularity is space created time driven temporarily substituted by the unreal. There is was and will never be one fixed solid Universe one can touch and smell, but the Universe is timely created space by motion of duplication in time delay. Once motion stops, time stands still and space falls into a black, Black Hole of eternal space less motionless reality where all the created concepts of space and time are contained in reality of eternity. That also is not religion but is physics. Time can only stand still in the Black Hole of empty space.

There was the spot that became lots of dots. The dot had no borders therefore there was no separation and still we know there were more than one in a group of one. When Π^0 moved to form Π the evidence of this move is very present in the cosmos at present and one can find such evidence all around us. The overall picture resulted in a ring or circle due to the release of from motion by all parts and all rings hold Π to secure the form. The only form that existed then was Π and therefore even today the borders use Π to indicate positions. But in the single dimension such definitions were far from clear and the only distinctions came from securing singularity in preserving the position of singularity to apply gravity and thereby absorb all anti -gravity. But anti-gravity could not control expansion by counter acting contraction through gravity so the overheating continued forming non-existing borders. The borders appeared in some material that was infinitely solid just as Einstein predicted because this took place before light came about and therefore before the speed of light. That which we refer to at this point even pre-dates light and therefore light at that point was excluded as being part of the cosmos. The cosmos formed a partnership with one side overheating forming antigravity by expanding into space through the applying of the overheating. In the relevance which the Universe is all about there is another side and the other side formed gravity or contracting of the expanding space

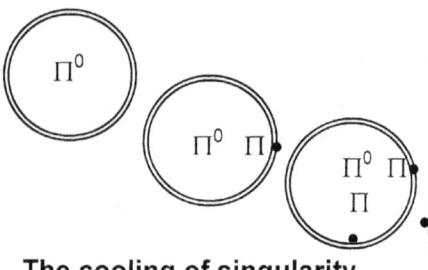

In the very beginning the overheating heat brought expanding into space producing extending singularity and the expanding brought along cooling to singularity

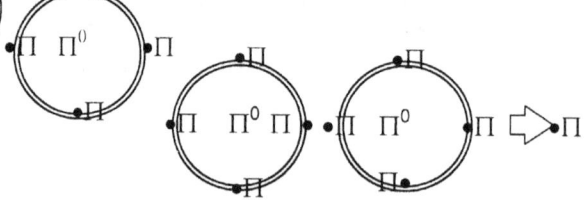

The cooling of singularity brought about cooling and the cooling brought with it the contracting

The five points we associate with the Lagrangian system find its routes as an outflow of the time zone. It is the first in the line where time was retarded beyond time control. But it also was before the atom in seven to ten could form therefore the four forming a dimension with singularity manifested in another forming Π^0 to Π. It had eventually developed towards forming the atomic bonding of the Coanda effect but the Lagrangian system is merely a bridge from one dimension in singularity connecting to another dimension as it becomes a point of the seven in relation to ten, which forms the full atomic gravity atomic structure

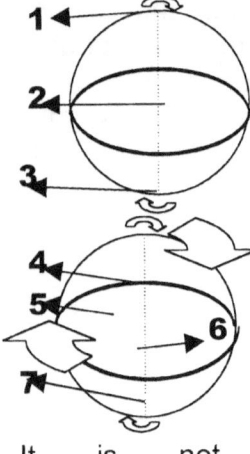

It is not possible for these points to have motion or to produce

By not having motion the lines also have no space as the space extends to form space forms space and the line includes serving the three points to the outside. Where there is no motion, there is no space and where there is little motion, there is little space. The only space the line may relate to can be a point that is on the border of the sphere that is crossing singularity and connecting the two edges on either side of the sphere that is forming the sphere. That means the line from one point holding singularity to another point holding singularity that line will cross the centre line which gives the line in singularity valid space-time to control. Singularity does not have the ability of motion therefore singularity does not hold space. Singularity is also eternally indifferent to motion and motion can excite singularity but singularity cannot be shifted by motion. Three points form the line.

spin since spin requires the refurnishing of singularity transfer or displacing space-time from one point holding singularity to another point holding singularity. Where space-time or space by motion is active, there the motion carrying space has to be transferred from singularity without motion and space to singularity being without motion or space. As the space transfers from one locality to the next locality there some energy has to be transferred with the space that the motion generates to a new location and such relocation is going about with the transfer of space-time. However, the motion excites singularity but does not shift singularity. Time shifts singularity because it is the exciting of singularity to get singularity to hold space for a period that forms time. Time is the flow of heat through space in space. **Space-time is conducted from one point to the next point just as if the flow of electricity is conducted. The conducting medium however is not similar. Should motion be required, there has to be a transfer of space-time from one location to the next location. This transfer is not that immediately instated but is a process that has to be developed. It takes time to break down space-time, relocate the position generating the space-time and then reinstate the energy that will produce space-time. The Newtonian view is that mass shifts but the truth is incomprehensibly more complicated than purely shifting mass.**

There is a breakdown of singularity attachment and then there is the motion transfer of space-time followed by the rebuilding and re-instituting of space-time that the motion generates, just like the motion in the Coanda effect is responsible for generating the pulsing of electricity. Generating electricity is in fact a process of creating and re-establishing space-time but only at the level of C by creating gravity at the level of C. Every point holding space is singularity being charged with receiving as much as distributing space-time at points that receive or let go of the next space-time it is charged with. The space-time, the point is charged with becomes the space-time it is displacing the charging. It is four points in receiving as much as the four points are in distributing

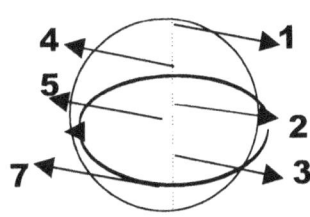 Every time motion takes place, the centre line holding 1,2 and 3 stands still. There is little generating going on and reducing to a point where there is no generating going on. On the outer edge of the rim however there are four points that do shift. The points shift from one location to the next location by generating space-time.

Those four points are the measure of conducting time by redistributing heat as material or in a state to receive material. We live in a Universe of time being from **10/7 $(4((\Pi^2 + \Pi^2)$** down to **7/ 10 $(4((\Pi^2 + \Pi^2)$**. The use of the four in front of the proton relevancy indicates the time aspect. The Involvement is the invert square law in the equation. Space-time involves the incorporating of motion in relation to singularity by distributing space from one sector to another sector. There is no wild theorising necessary of antimatter devouring matter when nature presents the most adequate explaining in the simplest form. By using mathematics, one can never reach the true origin of the Universe because such beginnings were even before mathematics. How do we know this? Because that part of mathematics are still with us and present in mathematics we currently use.

On the one-side space forms and in the forming process of space two relevancies come about. The seven departs in the direction putting the ten it is moving to in relation to the seven it is coming from. **7/10** Then by the same token and in a separate action which from our 3 D point of view is the same action the ten is losing the seven as the seven is departing from the ten. **10 /7.** All this is happening while the crossing is all concerning singularity moving from one sector of singularity to the other sector of singularity which is **(Π/2) X (Π/2).** Then there is the shift in space-time in the centre sphere component that holds four in orbit around the centre singularity line. As part of the motion is the singularity of the lesser-developed sphere, which is also part of the generated space-time in relation to the centre sphere because the singularity line came about at the same developing phase as the directly linked space-time of four points came about. In that it is the dominating sphere holding four orbiting and as well as the orbiting singularity that is related to the dominating sphere as merely space-time that carries three. Albeit that the orbiting singularity line is singularity, it is related to the dominating singularity line in the same manner as the same as the space-time in orbit forming the better developed sphere.

This then puts the fourteen in relation to ten, which gives **1.4**. From the other side there are the seven orbiting which includes the three reflected bringing the total to ten in orbit. That ten stands related to the centre sphere of seven and as it stands, related it is a division of **10 / 7 = 1.42**

The calculation of this 1.4 / 1.42 multiplied b y the square of space brings the value of gravity Π^2=9.8

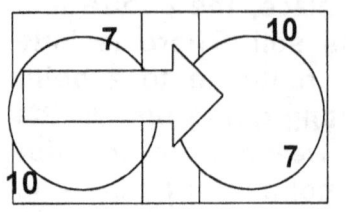

7/10 (Π/2) X (Π/2) 10/7

On the one-side space forms and in the forming process of space two relevancies come about. The seven departs in the direction putting the ten it is moving to in relation to the seven it is coming from. **7/10** Then by the same token and in a separate action which forms our 3 D point of view is the same action the ten is losing the seven as the seven is departing from the ten. **10/7**. All this is happening while the crossing is all concerning singularity moving from one sector of singularity to the other sector of singularity which is **(Π/2) X (Π/2)**.

The immovable

The spinning four

The improvised three   The spinning four

The rotating three

The relevancy forms part of the duplicating and dismissing displacement of space-time we call gravity. In that, we are looking at relevancies and no precise specifics. However, the Universe was built block by block in this manner. As it was but is no longer only form that applies in the Universe but concrete measurements also come into play therefore even the relevancies may apply in different relations as they switch over to compensate for other factors alternating as they are coming into prominence. The lesser developing sphere orbits the dominating sphere and between them, there are definitive relevancies. The centre circle singularity line of three is unaffected by spin which I shall call the immovable three. However, the immovable three holds such a stout position as far as the centre sphere is concerned. In relation to the orbiting circle the centre line is part of the building and destructing process that manifests as duplication as the centre singularity maintain domination and control over the orbiting structure. In addition, it has a major part in the motion building of the sphere in orbit and moreover building by generating the singularity line that generates the lesser and the orbiting sphere. In that relation the centre sphere reflects the centre line to serve the orbiting sphere by supplying the reference needed to establish motion in the orbiting sphere as one Unit. In that there is an undisputable reference of seven orbiting the centre and the centre providing three as a reflection of the seven which in all accounts for ten relating to the four which also is spinning as time and in total forms the seven taken in relation from the orbiting ten.

The form singularity insists on is rooted in seven relating to ten. That is what forms motion or gravity at $Π^2$ where space in singularity forms the axis around which time rotates by four.

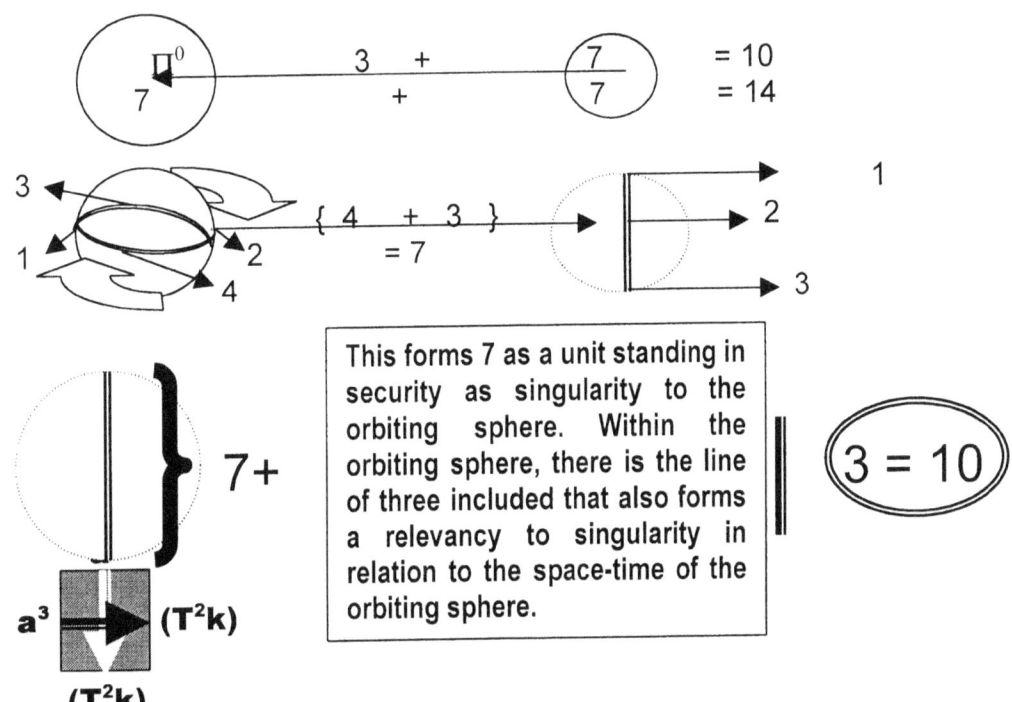

Realising that singularity k^0 produces $a^3 / (T^2k)$ in Kepler's Universal findings of $k^0 = a^3 / (T^2k)$ then one can see why gravity applies a sphere when form of free choice is an option. This statement has very and many far-reaching implications in how our Universe came about as it chose forms when it developed.

$T^2 > a^3 /k$	$T^2 < a^3 /k$
k reduces Gravity contraction comes into play	k increases Antigravity brings about expansion and "growth" comes about

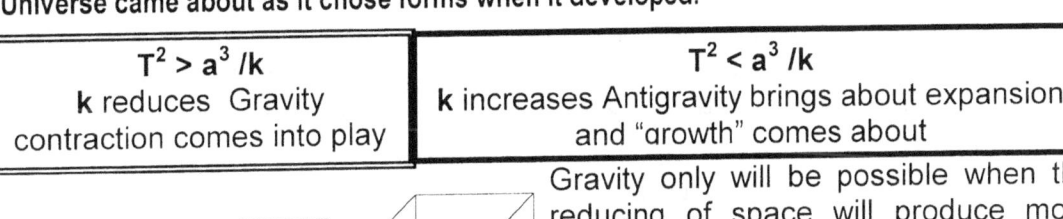

Gravity only will be possible when the reducing of space will produce more space for heat to ensure the reducing of heat and prevent overheating which will contribute to expansion. There will forever be growth.

$T^2 = a^3 /k$ $T^2 > a^3 /k$ $T^2 < a^3 /k$

The growth wills either be by expanding to secure more dense particles or the increase of space will produce the increase of space by reducing density. If the time effort $T^2 = a^3 /k$ can duplicate space from singularity no increase in space will come about. If that was the case, then the status quo in space-time will be in place and time will duplicate space to a precise eternal value. When time reduces space to a lesser value $T^2 > a^3 /k$ gravity applies reduction and space through movement becomes smaller than it previously was. That is the way the Earth perceives gravity. When **space produces larger expansion** than what time can duplicate $T^2 < a^3 /k$ by the extending of k, the growth of space will be larger than what time can manage and space will increase. This applies when a spacecraft launches into outer space and astronauts become taller. All three systems are in place and Universal growth comes about in such a manner.

Gravity is in place when space is in motion producing a duplication of lesser space than time will form a sequence. In the other scenario overheating or space expansion or antigravity will come about when gravity cannot sustain space duplication in the preventing of duplicating to bring about reduction. With space remaining at an even duplication without adding heat space will become lesser dense and such thinning will also increase space as antigravity, which is the same thing as overheating, and space growth comes about. The motion secures, prevents or support space in motion by forming harmony to frequency to motion duplicating space. One side of space is the duplicating of the other side because time is eternal at singularity. It is the double motion applying that performs the gravity between the objects in rotation. But the way Kepler introduced it diverts drastically from the way Newton introduced gravity. The moving away in relation to the coming towards forms the square that we see as the rotation. Te rotation does not bring about an accomplishment of zero as Newton suggested but it brings about a square, which Kepler introduced.

We must be clear on the subject of motion. In the beginning before the Big Bang commenced, the motion was in relevance because singularity even if activated still is beyond motion. Looking at singularity where we at present can trace singularity we find singularity still being without motion as singularity was then. But the number of singularity confirmed alliance since it had all the same value and the points created had the same relevance to other points forming a group. Time moving and not space moving brought on the motion.

When space started moving time was already so retarded, it could compact into forming a dense solid material we know as the atom. The units became inclusive by measure of $\mathbf{k^{-1} = T^2 / a^3}$ as well as $k = a^3 / T^2$ and if we release the time delay causing the atom we find the heat release to be almost totally destructive to the surrounding space-time. The atomic release of time by nuclear action releases devastating amount of heat that turns to space as soon as the release comes about. Such release is the product of the extending of $\mathbf{k}$ and by extending $\mathbf{k}$ the time constrain on the atom reduces from the speed of light which the electron has to the not containable $\mathbf{k}$ that release nuclear winds powerful enough to blow houses away. Those winds represent space that was previously contained by a number of atoms. In order to find where it started we must start where space parted from time.

In a short while we will start determining where space is, what space is and what time is. The space does not part from the space by any other way than having a different relation with heat. We also realise that space is heat and heat is space. In differentiating we then have to call one cold, which we know is incorrect and the other hot, which also by the same margin is as incorrect as the previous name given. There is not hot or cold. Yet for the sake of being humanly inferior we are forced to be incorrect in order to assert our incorrectness. The one limit has t be the furthest position away from the next and opposing limit.

Let us refer to this as cold. There is no possible manner to separate hot from cold and yet gravity does just that by concentrating heat into cold space. The unseen cold and therefore unnamed, unrecognised and unappreciated space, which forms the basis of the space, has the ability to move independent from the hot part, which is heat in space. We cannot separate the two forms of space and yet with the aid of gravity the Earth turns the shorter **k** which represents a lesser (volumetric) space by gravity in reproducing a smaller area where the very end limit that was in place one moment before the Big Bang, are to be a bitter cold. On Earth we named the coldest regions the Artic Pole and the Ant Arctic Pole since we have nothing better to call it. Our way of reasoning the difference there is between the hot equator and the cold poles has to be the Sun shining on the equator and not shining on the poles. The argument is that it is the Sun that is shining at such and such an angle that makes the Earth middle sector hot and as the Sun cannot reach the poles and that, not reaching the arctic regions, makes the pole regions cold. What a lot of baloney that lot is because the Sun is far too big to have the Earth roundness restrain the heat distribution of the Sun from the centre the top and the bottom of the Earth.

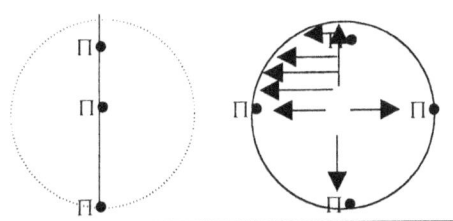
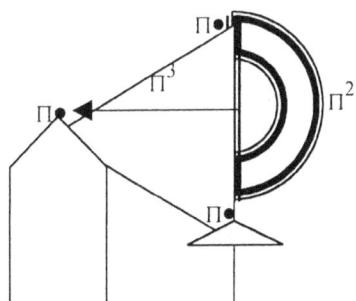

Points seen from the side form a line that never moves and is a line in singularity holding three points

The seven can never totally separate from the ten, but by singularity being the same but being on the other side it is withdrawing space-time altogether. See it as seven (let us think of that as the cold basis of space) spinning or turning in the ten (which then will represent the hot part in the cold basis) and the ten is part of the seven but the seven is not part of the ten. The third factor is the axis around which hot as well as cold will turn. Therefore when reading the next page, please envisage a cold base turning in a hot and cold space. The purpose of this is not to define whether the argument is correct or not but it is to help the reader gain understanding of the process principles involved. But motion also converts space to relate to space by changing relevancies through motion matter is in relation (part of) to the total dimension of space but is not the total dimension of space.

Space-time is allocated in progress from the line singularity offer. That space-time consists of heat and can therefore store more heat, which strands in contrast to the cold than singularity presents. By providing, more space-time at the equator there is more space-time to allocate to heat being there

It was thought that it is a mere coincidence that planets in our solar system should align by using the form as they do. However the truth is that it is one of the four pillars the Universe was founded on.

This is not possible to have any strobe with such sequence I know, but it serves as an indication of what applies because if it was possible we would have been able to see photons move about where as now we can see the photons flicker, while in reality the motion is so solid while flickering it sustains the Universe as a solid. In all of this we find gravity as the controlling factor holding the lot together as follows:

Matter in relation (part of) to the total dimension of space.
(10 / 7) \ (7/ 10) = 2.04
1.4285 / 0.7 = 2.04
Taking from both orbiting influences

SPACE DIVIDED INTO TIME
(7/10) / (10/7) = 0.49
.7 / 1.4285 = 0.49 Taking from both orbiting influences

SPACE MULTIPLIED WITH TIME
  7/10 / 7/10 = 1 and   10 / 7 X 7/10 =1 Therefore not influencing change

THE PROCESS PARTED USING THE ROCHE PRINCIPLE

10 / 7	$(\Pi/2)^2$ The Roche influence on Titius Bode
7/10	2.04 x $(\Pi/2)^2$ = 5.033
$(\Pi/2)^2$	2.04 x $(\Pi/2)^2$ = 5.033
10 / 7	5.033 +5.033 = 10.066 from both objects

7/10

7/10 / 10 / 7= 0.49 + 0.49   on both sides of the divide
10 / 7
0.49 X 2 =.98

$\dfrac{10/7}{7/10}=.49$ $\dfrac{10/7}{7/10}=.49$

  .49   +   .49   = .98
  .98 X 10.066 = 9.8 =Π^2 TIME SPACE = Π^2 = 9.869 TIME

Those factors being equal and therefore equal to one is not influencing change. The space that the motion establishes creates a relevancy of seven factors in space while the direction of motion involves another three dimensions or points, which is incorporating the other singularity in the unit. While the motion is at the same time moving out of ten points in relevancy and only occupying seven points the very opposite comes about through the same action being duplicated. The motion turns ten points to seven by moving from ten and filling seven points through the motion ending before the next cycle starts.

There is a lot more to say on this issue and I do just that in other books.. With the size that the Sun has it will fry the Earth rite through the very core the Earth has. The discrepancy we find in the equator regions compared to the poles is all about reproducing space as cold and reproducing heat in cold, which at the poles are much less to bring about through motion. If it was all about centrifugal forces flinging mass and thereby bulging the Earth as I have read some Academics announce is the case, we should also have all the massive particles such as the iron, copper, lead and so on swinging at the centre on the outside of the Earth core. It is about duplicating the cold, which is space without much heat, and by turning that it requires a fluid such as that which heat is to be within the space required to turn in. The seven can never totally separate from the ten but by singularity being the same but being on the other side it is withdrawing space-time altogether. See it as seven (let us think of that as the cold basis of space) spinning or turning in the ten (which then will represent the hot part in the cold basis) and the ten is part of the seven but the seven is not part of the ten. The third factor is the axis around which hot as well as cold will turn. Therefore when reading the next page please envisage a cold base turning in a hot and cold space. The purpose of this is not to define whether the argument is correct or not but it is to help **the reader** gain understanding of the process principles involved. But motion also converts space to relate to space by changing relevancies through motion matter is in relation (part of) to the total dimension of space but is not the total dimension of space.

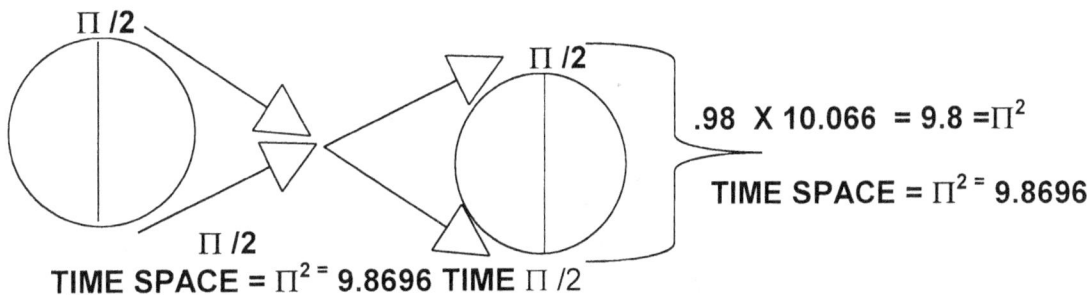

We cannot see any motion of any object as a secluded affair. There is forever the seven standing by three in relation to time holding ten circling around singularity.

Everything is space-time by confirming space in establishing time

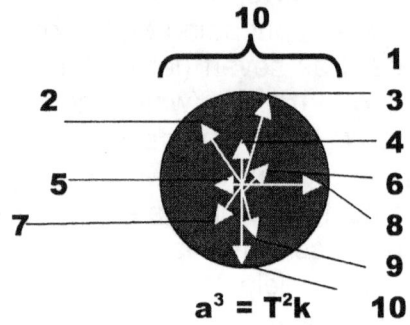

$a^3 = T^2k$

The value of ten to seven must not exclusively attach to our three dimensions we use. Before space parted from time the Universe went into a heat surge where the contraction reduced which brought cold in relevance

When the Universe was in the beginning with the entire cosmos still in a single dimension there were no limits as we know limits to form in the Universe we use and no borders indicating limits because after all it is the single dimension where there is only one dimension holding so much diversity. The borders were part of development because we can witness the legacy of such borders in the present day holding the 3D in place. But the actual value in the relevancy of ten coming into a relation with seven began long before our 3D Universe came into place.

while the motion distributed heat and the motion contained the heat surge. Once anything is part of the Universe it remains part of the Universe since there is nowhere to go. By us recognising this fact and use our recognising to track down cosmic laws there is a lot we will learn in the process. Such interaction of movement in what we know as outer space time provided space a means to move within there has to be an alliance forming. This alliance if it is must be gravity because the name gravity does attaché more to pulling and attraction than to what gravity truly represent nevertheless we have to change the concept about gravity if it is impossible to change the name addiction to gravity.

The concept of Π^2 is representative of the value gravity holds and is the square that motion represents as time. Changing that idea will prove not much gained and a lot lost so sticking to gravity by changing the definition of gravity might be more prudent.

Second Over All Overview

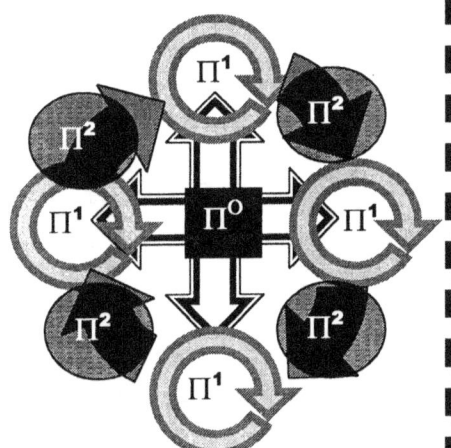

The motion is in relevance much more than it is in actual motion and this runs along since the first singularity produced the relation. At the very beginning singularity came about from overheating and in the process the heat that was too much within the original singularity spilled over to form three points in space singularity while the original singularity maintained the time reference to the motion to

> All rotating object has to be round to rotate. From the ends of the Unit rotating there will run a line running horizontally which turns with the top and as it turns with the top, the horizontal line is crossing another line that is running vertical but is not running at all because the three points cannot turn. The picture on the side is a picture showing the rotating object from the top as one would look down onto the line in singularity that does not turn. That line in singularity is representing the Universe because it is representing eternity as much as it is eternity. It will never disappear because it is never there to begin with. The line cannot stop turning because the line can never start turning. The line is absent because it can never hold space, yet the line is always there because any motion may charge the line into presenting the centre of the Universe. To find the centre of the Universe is to reduce the line because Kepler said $k^0 = T^2 / a^3$.

be. However the motion was a relation and not an actual moving. Since there is a law that once something is part of the Universe such a law can never become unattached to the Universe, this principle of motion by four standing related to space in three which then combines to be motion in the set time position of three is what produces gravity. It is the flickering as directional relations change that produces the act, which can be seen as motion. Only much later when time delay produced space did actual motion became part of the Universe. That process they named the Big Bang event.

The interaction forming gravity where gravity is the flow of space towards as well as the flow of space away from a specific centre holding singularity is directly contained by the time it takes such space to duplicate the space in the motion of expanding as well as contracting. However in the flow of this action the flow is centred by a rotating action about a centre in singularity. Singularity's square is halved by the half thereof during the time it takes the motion to establish the rotation as well as the four newly developed positions in singularity. While the overheating caused expansion time went on to establish four positions that divided the motion coming from Π^0 forming Π but since there was no space Π rotated by establishing Π^2 and this rotating about its position became a time issue of four in motion which then became singularity moving to form a square that was during the event when the time it took to rotate about its axis was the same time it took to establish the four positions which is time. Anything closer than $(\Pi/2)^2$ singularity would consider as fodder.

The crossing of the divide will implicate singularity on both sides of the divide bringing about the Roche factor
10 / 7 $(\Pi/2)^2$ The Roche influence on Titius Bode   7/10   2.04 x $(\Pi/2)^2$
SPACE DIVIDES INTO TIME

On the one side of the Universe
7/10 /10 / 7 = 0.49
on the other side of the Universe
7/10 /10 / 7 = 0.49
And on both sides of the Universe
.49 + .49 = .98
(10 / 7) \ (7/ 10) = 2.04
.98 X 10.066 = 9.8 = Π^2

The value science use for gravity is 9.81, which they measured to much detail but the moon has a lot to do with the recorded difference coming about.

This is the prime element; the state where everything started. It started at the time when mathematics was still to be invented by nature and only singularity had a value of Π and a reference of 180^0 in all directions in relations to other positions singularity established. It is at this point where everything other than **singularity was $\Pi^0\Pi$ becoming Π going on to be Π^2 through motion forming Π^3.** One must take into account that gravity is different motion of particles forming a relevancy about duplicating space and dismissing space in relation to the effort of the particular and specific element. The motion differences in the motion between two particles bring about relevancies. It is a seven factor standing in relation in motion to a ten factor of which the seven then is included and part of the ten factor. The four time factors applying gravity as time is on the edge of the sphere in relation to the centre line forming singularity in the sphere.

A boat sailing uses wind to power the sailing. Such sailing is done on the principle of wind pushing canvas and canvas pushes boat while boat moves in a direction secured by rudder...or so it seems from a distance. The wind blowing onto the canvas can never quite reach the canvas and that inability to reach the canvas is what brings motion across. Clinging onto the canvass of the sail is a part of material, which forms the time component to the canvass. The canvass is never bare but is covered by a layer of heat (time-space) and when the wind pushes onto such a canvas it cannot touch the canvas but it does cause friction as the wind try to tear the layer of heat surrounding that canvas from the canvas. It is the measure by which the canvas secures the layer and by which the layer protect the canvas that cause the friction to drive the sailboat. That is why a sail can have as much force driving it when slanted sideways as what is has when the wind hits the canvass straight on. Under the water the sailboat hull has a veneer of water that it secures as private property and this layer it drags along as the hull cleaves though the water. We call this friction but friction it is not because it is there mainly to protect against friction.

In outer space where such layer is absent the surface corrode gingerly as there is no protective time cover. Recognising this fact will take one some distance in understanding the process that leads to the sound barrier. Around every particle is a bubble of time-space secured to that particle.

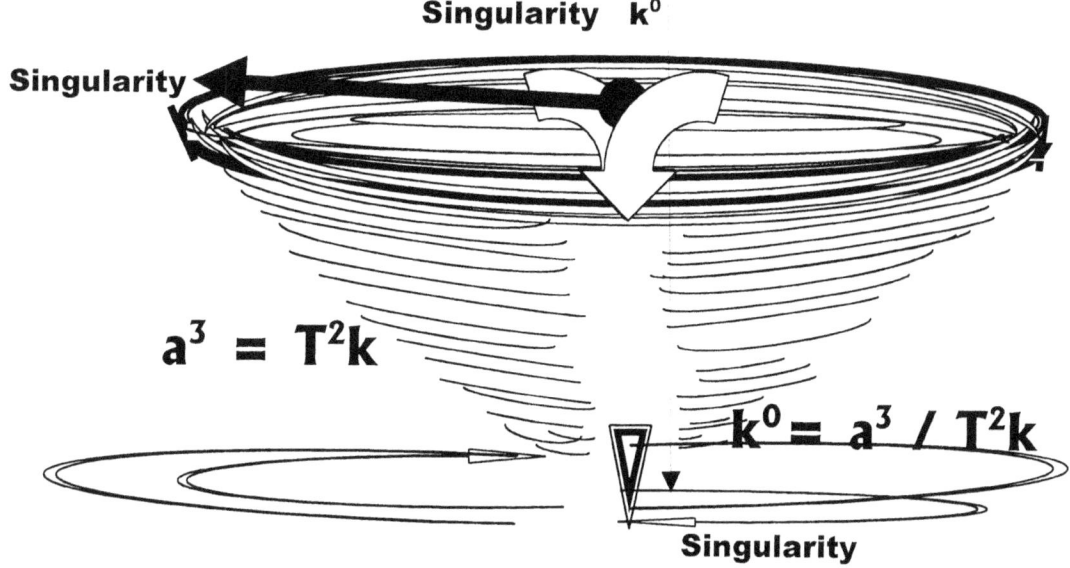

When the top rotates it forms this layer around the material, which is part of the Coanda gravity process. The spinning top initiates a newly established $k^0 = a^3 / (T^2k)$ that has independence from the governing singularity simply by the motion that generates the layer ant therefore the stance the top holds.

Within the circle incorporating a sphere as the formula $k^0 = a^3 / (T^2k)$ shows it holds gravity centred in the precise middle of the circle. By using mathematics in the way Kepler used it, shows that those rules and laws, when used correctly in the investigating of the formula that Kepler introduced is a testament of the cosmos. When using the Kepler formula it shows a space-time relation that must then form the basis of cosmology. Also when doing the investigating as we now are doing, such intense investigation then must be without Newton interfering and telling Kepler what he (Kepler) should have found. The formula should be applied directly instead of Newton incorrectly correcting Kepler whereas instead Newton should have been looking at what he (Kepler) found because only then he (Newton) could have seen what gravity is. He (Kepler) said that the cosmos said that gravity is $a^3 = T^2 k$. The space is held in check by motion from a centre and that is gravity. It becomes more than clear that space a^3 is time by dimension T^2 and time is space a^3 without dimension k Gravity is a^3 / k but k is an addition of motion T^2.

The major issue is to find where the cosmos started by using Kepler. Let's go back once more and reduce the line by half every time. By repeating the process some other aspect comes to light every time. What happens to one line is also applicable to all lines. It is cutting every distance there is in the Universe by half every time. The process we refer to at this point I might add is not in the space we use and know as outer space but in the inner, inner space where the proton is and we cannot use, so please do not confuse the space with outer space. Then repeat the reducing process until it can repeat no more. The reducing of the line by half every time will get to a point where all the ends land on the same position without any possibility if halving the two ends further. There all possible points share one position and moving the points in any direction will lead to an immediate increase of extending the line once more.

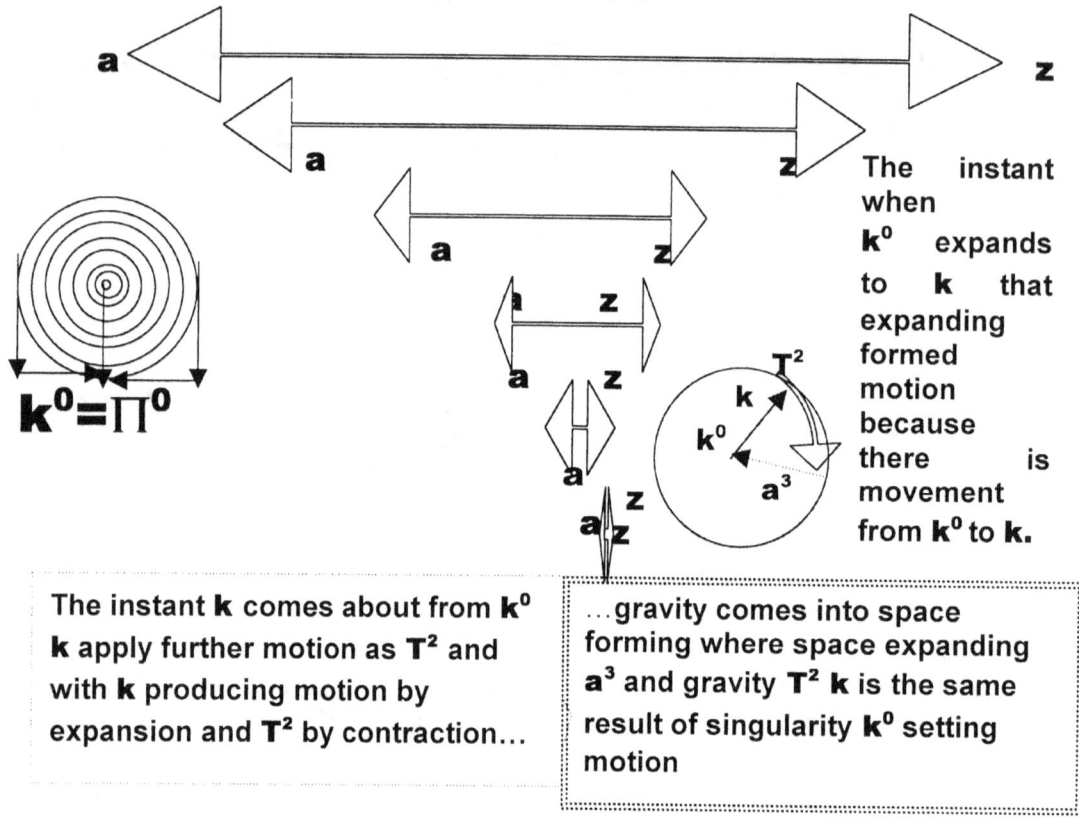

In the sketch I made shown below of lines reducing I left space between the two ends of the line that is symbolising the end of the reducing will share one location even by having one single line. There is no chance that I can present any sketch reducing the line to a point where the points are sharing one location in the single dimension. The points are there and with the points being present they may not be dismissed as nothing. From there no reducing in a natural manner can lead to nothing without changing the rules of mathematics in such reducing. Any further effort of reducing must bring about extending because every point possible share space with every other possible point at the point of singularity where all points share common space.

From where all possible points landed on the same spot any further moving of any point must then bring about an increase of space. No matter what motion comes to that point when moving such an act will result in the increase of space between such a point and singularity. At that specific point further reducing will bring redundancy of the line so further reducing is no longer an option. Since all the reducing that did not bring about the redundancy of the natural line that there are covers everything that is going to develop from the line, the fact of zero that is not included is a proven fact. With the reducing zero was never encountered and zero therefore cannot then enter the process because the process fill without the need for zero. This also applies to the circle because the circle uses a line to indicate size, such reducing of the line and by reducing the circle will end up in the same position. It is this fact of the moving of any point from that spot holding singularity that such motion will introduce space as the space exceed the previous limits of singularity. This process is a natural normal occurrence everywhere in nature without any person ever noticing. We see this so clearly in the spinning top. ⇐When the top is spinning such spinning of the top creates a centre and the lines start reducing space in the direction of the centre of the spin. The centre establish a balance in space-time where at a point it finds partial independence from the dominance of the Earth's gravitational motion that is depressing the tops movement to a standstill.

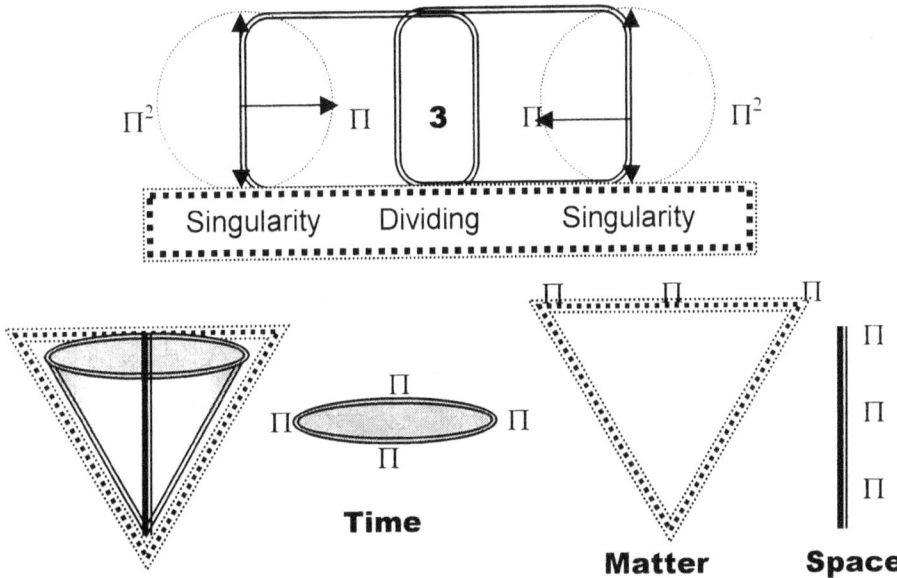

When the top is spinning there is an obvious centre point established. This point is also the application of the Coanda effect. It is establishing a "non-existing" but also eternally circling centre that is allocated but cannot be located. That point secures a centre by motion, because the motion establishes a centre $k^0 = a^3/T^2 k$, which means that centre establishes individual singularity. The space created a^3 must be equal to the motion created around singularity $a^3 = T^2 k$. The only part is that no such point exists in our human reality, but such a point can only be realised as much as such a point can only be understood. That point where directions change while the spin continues, is without space a^3 which is k^0 but on the condition of motion $k^0 / k T^2$.

One must draw this statement of motion back to the point where singularity is getting sides. After all, that is where creation started. When there is singularity there can be no sides. It is 1 (one) position away from all angles there can be. That one fills no space but all space flows from there. The space we allocate to that spot does not really exist in the manner we humans see space to exist. It is a spot that is there without being there. It does not visually exist because it is not filling any substance and it cannot be recognised. It is clearly holding a position in a location where the location where such a locality is found is beyond this Universe. Once one accepts the fact of singularity, that accepting of singularity then is contradicting all the things we know by not being any of the things we can recognise. In singularity there is no space.

There can be no motion because there can be no space to have the motion within. It is a line that is so small it is not there and the only reason why we know it is there is because of the results it left as an imprint of its not being there. It is not recognised by its absence because it is never absent. It cannot be absent because it is eternally present. It cannot go absent but it can never be there where it should be if any person wishes to locate it. If it was absent then it was zero or nothing but since it is there it is not there and that makes it present. The centre spot that we cannot see and that we cannot detect has no sides to any side and has no place it fills because it fills all the places we cannot detect.

The only way such a spot can fill space is by doubling the space it fills to become more than one place to fill. But the very instant that happens it halves the space it fills because it then cuts the space it has into two parts. That brings about that the point of not being is doubling the not being and by doubling the not being into being it also cut the not being that became present into half. We have to find this spot as we find religion. It is something that we can only know is there because we cannot disprove it is there but we can never prove it to be there either. It is something seen through intellect and not through the eyes or light transfer. It is a point far beyond light. It is in our being and not in our vision. Most important about this is to confirm what liquids are and what solids are in space. Because of control of heat, the confirming of the control of heat or the lack of control of heat we find the state matter occurs in.

Flames are a liquid heat. Smoke is a liquid heat. Vapour forming compressed space is a liquid heat. Rocket thrust is liquid heat. Air and atmospheric space is liquid heat.
The Coanda effect is the manifestation between solids, liquids, heat and motion. The fact of being solid or being liquid is locked in a relevancy that brings across a comparison. Hydrogen in Jupiter may just be ore solid than iron is on Earth. However, hydrogen on Jupiter will always be a liquid compared to iron on Jupiter. The cosmos puts a huge difference between solids and liquids. In fact the Universe divide on that perception. Both are extensions of singularity but it is the extension that brings across the distinction because in the one case gravity forms part of the inside of an atom and in the next extension that extension allow gravity to bring about the motion of space which is gravity.

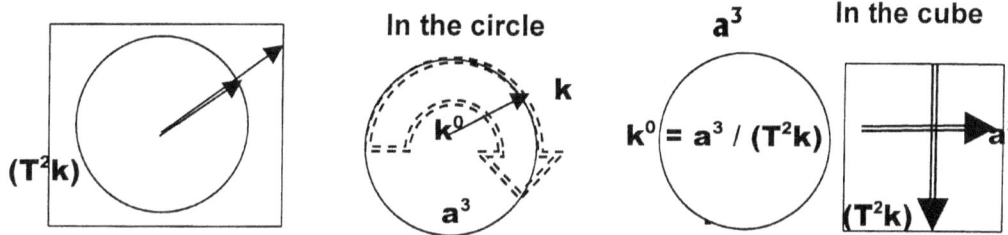

The big proof coming from the Coanda effect as explained above is that space flows through time as a product of the creation by motion op space-time. The opposing of **a³ by (T²k)** on the "other side" produces the **six sides** we now came to use. However, that means gravity is where space disappears within spherical or circular structures and not in outer space.

We think of the Universe as a solid structure filling an empty part where both have space. What makes the solid structure solid is the rotating of the motion of the particles as the particles drive a linear motion that forms a circle motion around a singularity holding more prominence in time. It is the motion that secures the solid state because the solid state is heat or time retarded to almost a standstill. I repeat the part about almost standing still because once it stands still, the solidness collapse into a vacant ness. This process of duplication secures the solidness of material. The front secures the back while the back support the front as the motion pulls and drags a rhythmic flow of material in a line in time.

$a^3 = T^2k$ therefore $1 = a^3 \backslash T^2k$ and is the same as $k^0 = a^3 / T^2k$ which means that
$k^0 = a^3 / a^3$ or $k^0 = T^2k / T^2k$

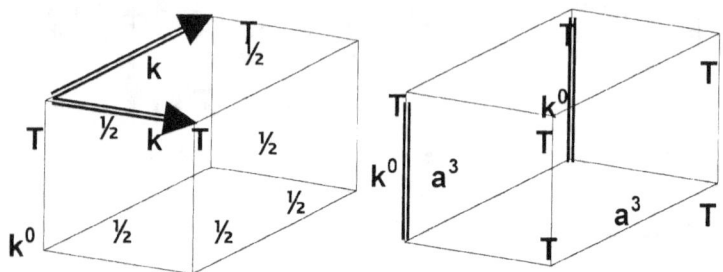

Gravity (cooling by contracting) and anti gravity (heat bringing expansion) is space created by the motion of the space in a line running between two points. Space in the cube a^3 is the space in the square T^2 by motion through a straight-line k. That is mathematically what Kepler says. Space a^3 is formed by the square T^2 going double in creating a straight-line k. That means space a^3 is the doubling through motion T^2 by a straight line k and time T^2 is creating space a^3 by implementing motion T^2 using straight- line k. By the motion T^2 of singularity k^0 space a^3 comes about in a straight-line k. Applying motion T^2 is the forming of gravity or anti gravity. Gravity is much more about one particle following the other particle as motion displaces their position complementing one another in a following aligning. The motion of displacement by duplication as well as replication is more involved than what the human brain can manage. Gravity is the one atom following the other atom by spinning into the position of the atom in front and emptying the space for the one to follow.

In a solid structure the structure select a point of forming a centre that will refer the function of such a centre onto a centre which all the atoms sharing the unity of the space of the group holds which by motion separated the isolated space as a separate Universe from the rest of the Universe. That is solid structures and stars form an excluding unit with one singularity at the centre which dominate as much as control all singularity within the unit. In such a unit an object loses independence when the object does not bring about independent motion to confirm the singularity as being apart holding individual space-time outside the units space-time confirmed by every atom in the Unit. All motion of the unit will presume and any motion will be affective above and beyond the motion of the unit operating as the larger structure. When the dominated body finds any ability of producing motion such motion will serve as an attachment of the larger singularity in motion by spin. By it having individual spin, such accumulating of the group effort is secured by the larger objects being the accumulated unit's motion and find security in that motion. But since the structure is finding security in motion by adopting the combining unit it has to adapt to the combining unit space-time by establishing a measure of duplication of space in line with that which the combining unit produce to sustain the combining unit gravity. This connection comes about since the object does not have an independent created singularity it can sustain.

All singularity points are one singularity point and singularity does not apply motion because singularity is without motion. It is the space-time that transmits space as motion but since all singularity is identical and one, the transmitting of the space-time is not extending as far as the singularity but only as far as charging singularity. Again I have to insist that when one think of space-time and space motion one has to exclude any comparing that with life. Singularity is charged by motion and it is motion that sets the space-time. By charging singularity, division comes about while singularity remains one spot with unmeasurable dots, but it is motion charging the singularity into dots. Singularity is as much a dot as it is a line.

The only fact that the | line is there is because the ⋮ line is not there. By duplicating the ‖ line not being there the line establish the fact that it is there but by ‖‖ duplicating the line the line at the same instant becomes half of what it is at the time it is duplicating what it was not. This is coming straight from the horse's mouth. The Cosmos told Kepler that $a^3 = (T^2k)$. That is it. That is gravity. That is how the cosmos unveiled the cosmos. The cosmos said that from singularity k^0 came space a^3 with the motion establishing time T^2k. If it was this simple why make it so terribly complicated just to please the playing of the games by the mathematicians. The moment space realised a^3 such realising presented itself as motion. Translating Kepler's statement from the mathematical equation of $a^3 = (T^2k)$ to verbally spoken English can only be translated as "space moves" and "space is motion" and "space becomes motion". It is space filling as motion or motion will complete space. There is no other translation one can draw from this except by altering the concept through incorrect adding of facts the Universe never entered or stated in the formula the Universe unveiled to Kepler.

Space is either filled or waiting to be filled whereby when waiting to be filled, time validates the existence of such space. The one side is space not filled and on the other side it is space only filled by the space not filling one point duplicating such a point and then only filling the point through the motion that excludes the point from being on either the one or the other side of the Universe. Without the one side that is going to fill the other side and without the other side that is going to halve the one side not filled into two sectors of half the standing of one not one side is possible. Even if the following space seems to us as filled, to the space in motion every particle finds the space in which it is, as being filled and the space it is moving to, as unfilled or empty. To every singularity the Universe is only filled by what that specific singularity fills the Universe with. All the rest is empty space.

Wherever we are and whatever we see is a line we see placing what we are in relation to where we are. The line is the Universe as **k** determines the direction of development. The factors came about at the Big Bang mainly because at the Big Bang **k** was in value and not only just a factor for the first time. It is **k** that is in the process of development as singularity produce space in decline of time by extending the line **k**. The development of the Sun is **k** and the influence, which the Sun holds on all the planets are in relation to **k**. The fact that we have life on Earth and on no other planet I put down to the **k** factor the Earth has with **k**. I explain this just mentioned with proof in my book The **Seven days Of Creation ISBN**.**0-9584410-4-9**.

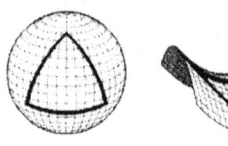 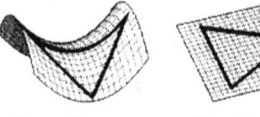

The line can only be if the line fills the space it is not filling when the line is duplicating the filling of the space the line does not hold by applying motion to the filling of the space of the line. Then there will be either or but never both as we now wish to see the Universe. By the duplication it therefore insists on relevancy because without relevancy there can be no motion and no motion means no space. The strongest proof there is about this is the manner in which the Coanda principle applies the reproducing of space taking shape from a round object and involving motion to produce such duplication.

Closed Geometry   Open Geometry   Flat Geometry

In the general relativity the geometry very closely connected to the distribution of matter. In the space of two dimensions, (as if the universe is like a flat rubber tyre) Euclidian geometry applies so that the sum of the eternal angles of a triangle is 180^0. If a massive object is placed on the rubber canvass the sheet will distort and the path of the objects moving on the sheet will become curved. This is in essence what happens in general relativity. Well missed by science is that the water drop proves the point the best.

The only fact that the line is there is because the line is not there. By duplicating the line not being there the line establish the fact that it is there but by duplicating the line the line at the same instant becomes half of what it is at the time it is duplicating what it was not. This is coming straight from the horse's mouth. The Cosmos told Kepler that $a^3 = (T^2k)$. That is it. That is gravity. That is how the cosmos unveiled the cosmos. The cosmos said that from singularity k^0 came space a^3 with the motion establishing time T^2k. If it was this simple why make it so terribly complicated just to please the playing of the games by the mathematicians. The moment space realised a^3 such realising presented itself as motion. Translating Kepler's statement from the mathematical equation of $a^3 = (T^2k)$ to verbally spoken English can only be translated as "space moves" and "space is motion" and "space becomes motion". It is space filling as motion or motion will complete space.

It is just this very issue about where lines start that the Newtonians understand least. They frightfully declare on the web in plane sight for every one to see that they somehow managed to observe a galactica for the first time ever at the edge of the Universe. That point must be where the Universe began. It then must be the point where the Big Bang occurred. If the point is on the edge we must be in the centre and we are not on the centre so the point must be at the centre. Such a point holds the centre because space formed from that centre. When reading a declaration such as Hubble's deepest view of the Universe reveals earliest galaxies, one do realise that Newtonians view that there was one point of origin. Then that point is further form us so when did we cross such a point. We are not where the Universe started because those galactica are where the Universe started. The Newtonians admit we can see past the centre point to where the galactica now are and where the lot started. Or else how did we get so far away and the galactica did not? From this argument they admit that there has to be a specific point where the Big bang started shifting material in directions in relation to other materials that went in numerous other directions. That is not the only admitting they make. From where they stand they as Newtonians are able to see in all directions the edge of the Universe. That puts them as Newtonians of course in the centre of the Universe where they as Newtonians belong. Being in the centre would help them to draw a map of the Universe just as Magellan did when he voyaged around the world. However old Magellan had to cruise around the world and shift his position daily to draw the map while they are so secured in the centre of the Universe they need not move. They can see what there is to see from where they are. If they say they can see what ever is on the edge of the Universe then either we are on some edge where it all began or if we are not on the edge where the Universe began and the Bang occurred at that point must be where we see the edge where the rest of the stars fell off. But there are so many edges in all directions so what must I believe being the incoherent amateur and that's officially according to the Unisa (University Of South Africa) document in my position. The only way the centre would not be crossed is where "we" and "they" went in the same direction after the exploding of the Universe during the Big Bang explosion, but "they" went further than "us" because "they" went faster than us. It is ridiculous, but then they are the professionals and I am an amateur.

The Curvature of space-time is as common as space-time

The Titius Bode principle   Half the Roche principal   The Lagrangian and Titius bode principles

The Sphere.   **The Black Hole.**   **The developing galactica**

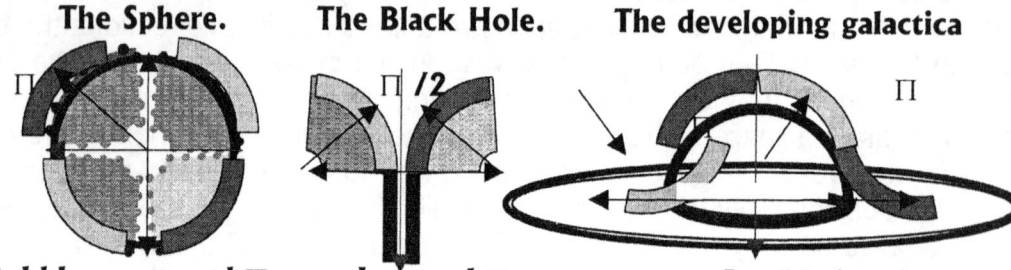

Bubble or normal Π Inverted Π Double inverted Π

The reasons for this applying I explain explicitly in detail in other books. With singularity, being where it is and in a common place as it is the curvature of space-time also becomes a common factor but not yet commonly distinguished because singularity changes the profile to suit space-time.

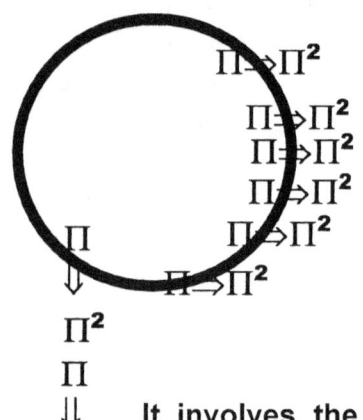

The Coanda principle which in fact should be seen as a law because it is that strong is the principle of gravity duplicating with the motion that provides such duplication a relation of the particles having the seven factor and such a factor of seven produces through motion another three dimension. This total that material fill while in motion is ten and when ten crosses the line of singularity to duplicate the seven in the other side of the Universe the crossing cuts singularity in two as much as it puts singularity in the square.

It involves the motion of concentrating space to be or hold fluids around solids that may or may not move. In this must be a solid, a round basis Π, fluids concentrated in space and motion applying to one or all of the factors.

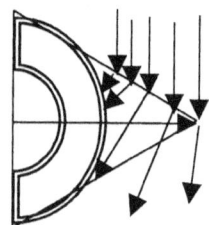

From what this says, it is validating Kepler's formula again. By bringing about Newton's claim distorts facts, because if some distortion must become evident, it should progressively be evident. The Big Bang did not happen yesterday and the progress from then to now must indicate that the Friedmann Universe or the Walker Universe is taking place.

Because of the Coanda principle the motion that light provides will bring a bigger gravity affect than what normally apply to space-time. The higher the motion is the more does the centre gravity contribute to the repelling of space –time by creating space in the confining of a^3. Gravity cannot bend light because light at best is a fluid. The Coanda affect places through the extreme motion that light provides an even bigger restraining of the flow of space-time on the light passing by. The relation forming the duplication of singularity is a duplication but applies as a dimensional forming of Π and placing 7 in relation to 10 forming Π^2

From the space we find the space extends by the layer the liquid increases the space holding the solid as well as the liquid connected to the space. That is when applying Kepler's formula putting everything in the right context of solid controlling liquid $k = a^3 / T^2$ as the liquid flowing controls the boundary space has.

From the liquids position we find quite another and opposing perspective as the liquid attaches to the space by the layer the liquid confirms to the space holding onto the solid while providing the gravity k through the providing of the motion as the liquid connects to the space. That is when applying Kepler's formula putting everything in the right context of solid controlling liquid $k^{-1} = T^2 / a^3$ as the liquid flowing controls the boundary space has.

With the roundness in which Π is formed and by securing Π a centre is created. This is evident in the Coanda effect where gravity comes about in such a manner. From the centre there then is control coming, which resembles the same principle in the spinning top. Much more important worth noticing is that that is what Kepler said. A centre forms when k^0 establish Π From k^0 a centre forms holding a^3 valid by allowing motion T^2 to secure independent space-time which is a bowl normally holding a fluid we know as pure heat. Then there is the motion of the liquid $\Pi^2\Pi$ securing the space by producing the motion and thereby establishing a centre k^0. In the example we see where the Coanda effect establish just this principle by having water flowing around a round bowl the motion of the water forms the liquid part $\Pi^2\Pi$ which generate the space Π^3 or a^3 which validates singularity Π^0 or k^0.

The motion of the water becomes the liquid part $\Pi^2\Pi$, which generates the space Π^3 or a^3, which validates singularity Π^0 or k^0. Looking at what happens in the Coanda effect we find a centre that is activated by motion of liquid since the centre takes charge of the space within the bowl. That is k^0 that produces a^3. Then on the fringe of the space motion the space-time becomes part of the space holding the centre which can be seen in a formula coming from Kepler's formula which is $k^0 = a^3 / T^2 k$ and when using relative cosmic values read as $\Pi^0 = \Pi^3 / \Pi^2\Pi$

The extension of Π is well received as a dimensional implication to matter holding seven positions from singularity and space having four quarters through out the rotation of singularity forming the centre to the five dimensions (one side lost to the cube's six sides connecting to the five remaining sides) making the total sides facing space from the point holding singularity at any given instant at a value of twenty (4 X 5 = 20). Then adding the singularity cross of Π being (1+1) = 2 the relation becomes 22/7. This is crude because in more precise calculations it becomes 20 = 0.91 + 1 = 21.91/7 = Π

The relation forming the duplication of singularity is a duplication but applies as a dimensional forming of Π and placing 7 in relation to 10 forming Π^2

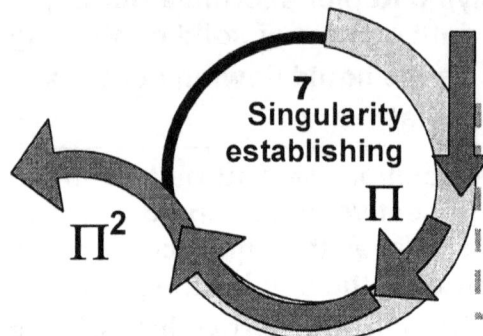

The liquid applying motion forms the 10 disciplines. No motion leaves no Coanda as well as no gravity because gravity is motion that duplicates singularity.

Because of the Coanda principle the motion that light provides will bring a bigger gravity affect than what normally apply to space-time. The higher the motion is the more does the centre gravity contribute to the repelling of space –time by creating space in the confining of a^3. Gravity cannot bend light because light at best is a fluid. The Coanda affect places through the extreme motion that light provides an even bigger restraining of the flow of space-time on the light passing by. The relation forming the duplication of singularity is a duplication but applies as a dimensional forming of Π and placing 7 in relation to 10 forming Π^2

Conditions that prescribes the enactment of the Coanda effect is that the one surface has to duplicate singularity by establishing Π as a form. The round surface Π will bring about the shape of singularity Π that becomes enticed by the action of the motion of the liquid or of the solid or the motion of both around Π, which then establish and confirm singularity by form. The next factor is the presence of liquid. Air or atmosphere is liquid and water is liquid. Heat is liquid. The third factor being just as important is the motion establishing Π^2 by duplicating singularity as singularity becomes relevant through the applied motion that produces gravity from the singularity spot that provides the form.

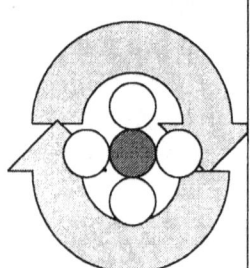

When the very first motion came about as Π^0 extended by overheating to form Π by creating Π^2, where that action produced the Coanda effect as the heat distributed space by cooling a centre. The Coanda effect was the very first principle that combined the other principles, which produced the initial gravity. The cooling centre contracted as cooling always does while the heat produced a flow by expanding. The circle flow established a contracting cold centre.

That too forms the answer about the question concerning the Titius Bode gravity implicating of cosmology. The seven sides are linked by rotation nothing changes because there is a steady linking to the inside centre of the sphere. But it is to the outside that this rotation brings about dimensional complications. There are five T_1 points moving to five T_2 making contact with five moving points. The moving non fixed points is the point before reducing by five to the point after reducing by five that bring along the ten points in stead of the five to one point as it is the case with the Lagrangian system. Two points relating to seven points coming from by continuing in the same direction it is going to remaining seven points as going. That means in matter there are five times two points relating to seven in a moving constant and seven fixed rotating points.

Gravity is just the retracting of cold and the expanding of heat. Heat finds new locations in order to distribute the overload while cold is the point that found relief and contracts because of cold. Gravity is the interaction behind games heat and cold play.

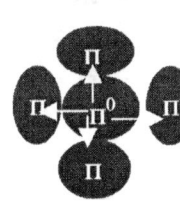

By moving from 1^0 to 1^1 and from $1^0\Pi^0$ to $1^1\Pi$ requires space. Yet such moving did not leave the realm or the domain of singularity. The motion was still within singularity because moving involved forming a relevancy between heat and cold between infinity and eternity, between space and time and most of all producing what will in the far future develop into a Universe that can even be a host for life albeit on a very small spot for a very short while in relation to the vastness space has and the duration cosmic time has. However, the distribution of the heat overload required a simulation of motion, which was interpreted as liquid and serves as such since then.

Relevancies came about when the dot moved away from the spot but had no space to move into. All that was possible was to charge singularity by relevance to comply in being activated into complying. Charging came in the way of distributing the heat overload to four set points forming time. That formed the basis of one solid controlling four liquids relevancies. Space-time is motion and movement are all the same things only separated by dimensions and dimensions are formed space, where the dimensions become space being in motion and the space is motion by contraction or by expansion but because time is almost eternal at k^0 our perception of the universe we are in is a stable and steady eternal structure. Gravity is motion and motion creates space to the third by the third in the third that interacts with one but establishes ten.

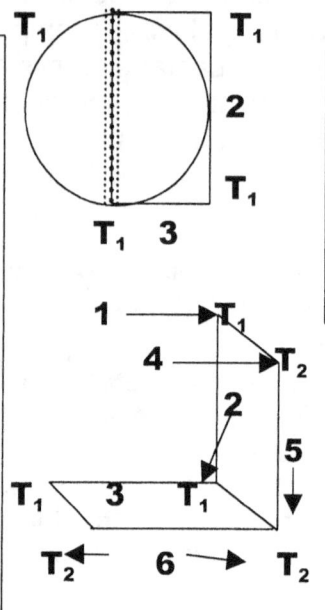

Time or spin or rotation, call it what you like but it is the moving from T_1 to T_2 using time that provides the dimension of depth to the dimension of distance between dimensions.

In the motionless Universe there will be on point in time and that point will represent k more than anything else. Every point being T_1 will only show the extending of k from singularity to that specific point. The fact that T_1 indicates no motion brings the universe to a stand still and to a flat Universe.

It is that which give **k** the coming from the first dimension and by only extending from singularity it the forms two more positions becoming a^3. $k = a^3 / T^2$ but remove T_1 to T_2 from he equation and only k remains $a^3 / T^2 = k$. By the effort of spin or motion the universe becomes the three-dimensional object it all seems to be.

Motion is what **gives space stability** and **security** in supporting **six sides** where **three sides** are **opposing three sides**. This Kepler shows so very clearly in his statement $a^3 = T^2 k$. If there is anyone out there that cannot see this and miss the interpretation such a person cannot read mathematics.

Duplicating material in space in measure per time in motion.
The less time per movement required to duplicate the less space can be filled in the time of duplication

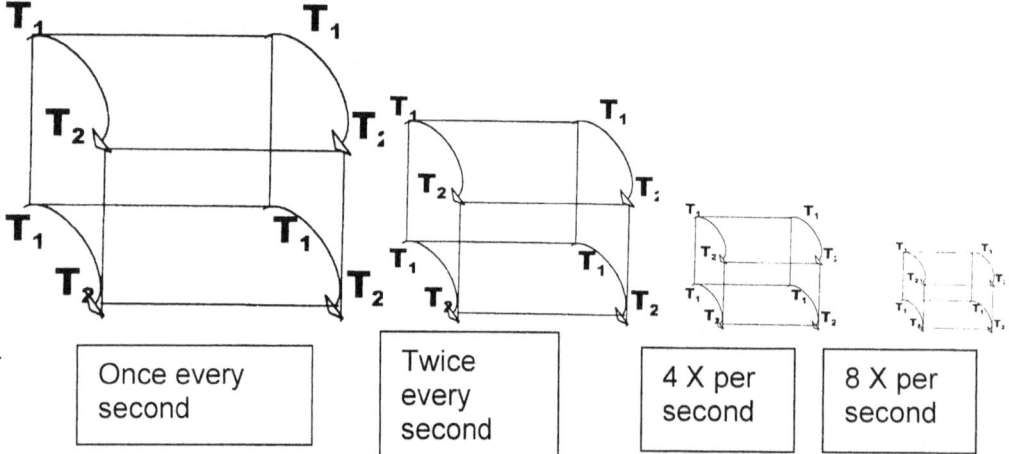

| Once every second | Twice every second | 4 X per second | 8 X per second |

This is the measure that duplication is processed in the Universe. The numbers seems meaningless at first glance but as soon as one unlock the coded values one find meaning in the formula as to tell the story of happening about the Universe at the time singularity stood in this ratio. It is where duplication became a sphere with a specific in and outside having the future roundness we now are accustomed to. It indicate that it is where space filled moved about in space not filled defining the space that filled with material.

This is a tractor, which is a man made device where man found the way to extend the manipulating influence of life into the tractor. The tractor can move at mans will. However that takes more than the thought would allow. For this tractor to move from one point it holds to another point it wishes to hold in the future, the tractor has to break up every atom it claims to have and every sub particle in the group where the sub particle forms a part of the entire space that claims the space occupied by the tractor on behalf of the group that then forms the tractor. The atoms must break down subatomic particle by subatomic particle fragmenting their composite that they are made of to the smallest detail and relocate that composite to the newly acquired location that the moving will establish in the following time duration. If I whish to use less time to do the breaking and reassembling then there must be less to break down and reassemble or else the harmony of moving is no longer in frequency with time. However if man not wishes to shift the tractor from one location to another then the tractor still has to do all that which I just mentioned and that is gravity. In a specific time it must un-develop and redevelop a certain structure holding a certain number of particles grouped in a certain form. The tractor is manufactured from atoms which is a combination of heat densely packed and stored in a container that forms a larger part of a group of containers that each combine to form the overall container which we named to be the smallest thing possible but is not the smallest thing possible. However the fact that life can manage to move is a part of life and not something connecting to the Universe. In the second book Xepted Astronomical Mistakes the book shows what life is about in the art of physics and what influences physics has on human life in particular. When does life end being life and where does death enter the equation.

Because motion is part of life and we are life we with life take motion for granted. We think of motion a simply not an issue but for all else in the Universe that motion is an impossible issue. Those atheists that puts life everywhere and so common all around must be sporting enough firstly to define life. What is life? I put life down as motion independent from the motion the cosmos produce. When something has no motion that is independent of the motion of the Universe that thing has no life. From that we may conclude that life is about manipulating space-time to a degree higher than the cosmos is manipulating space-time. Life has the ability to fragment, that which is in the moving ability of life, and relocate that in harmony and in frequency to that duplication trend that the cosmos set. That makes Newton's falling apple not a cosmic affair but an instigated manipulated motion that was harmonised with life's ability to manipulate space-time. If that goes array life meets its demise in an explosion of heat. Motion is more complicated than life and nothing on Earth or the Universe is more complicated than life especially the moodswings of a woman.

There were liquids and there were solids and the liquid defined the solid within a containing dome. Our ability to understand is exclusively vested in understanding the formation of space-time development prior to the Big Bang. We have to understand the phases of development where every phase brought about a new line of development. We have to see where space and time parted and at what point did space and time part. We must see what happened when space and time parted. For instance the fact that space- time and matter parted was when space had another motion to time, which had to move different to provision for the defining material, and thus compromised the position time had. Incidentally this took place at $10 / 7(\Pi^2/ 2)(\Pi^2+\Pi^2) = 139.15$ Time formed separate from the material that formed $7(\Pi^2+\Pi^2) = 138.1744$ in relation to the heat forming a liquid $7 / 10(2)(\Pi^2/ 2)(\Pi^2+\Pi^2) = 136.37$

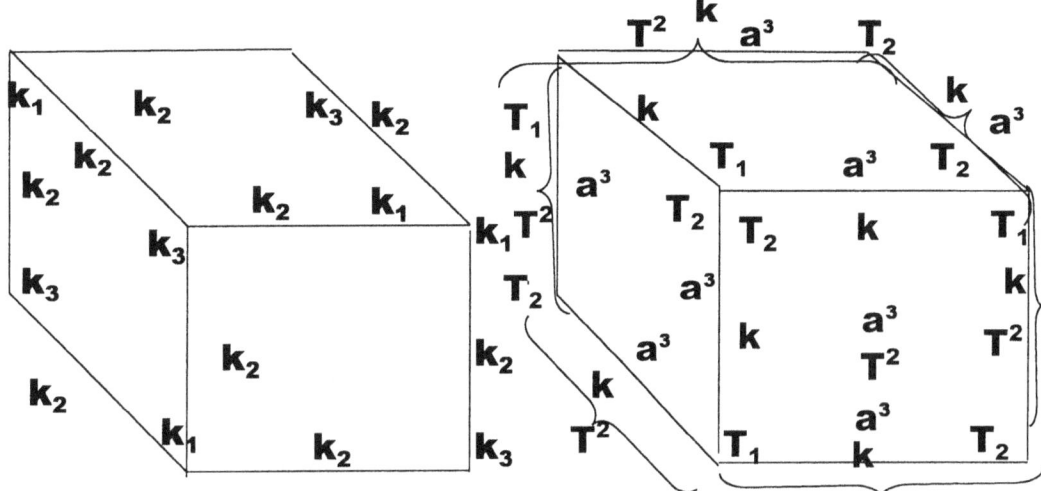

$a^3 = k \cdot T^2$ As I indicated previously it is all a dimensional differentiation. The Universe we know shows space having six sides and even that Kepler proved as he proved space in spin $T^2 = T^{3-1}$. It is the same, therefore it is dimensions repeating to form $a^6 = (a^3 \times a^{2+1})$. This becomes a factor of space forming six sides when Π forms Π^6 in relation to the six sides of space ($\Pi^6 / 6$) in the presence of gravity supplying motion of 7 / 10. That is outer space $7/10\Pi^6/ 6$ forming the element space-timer limit of **112** in which we find all the known elements.

$T^2 = a^3/k^1$

$a^2 = a^{3-1}$

$T^2 = k^{3-1}$

$k^2 = k^{3-1}$

Einstein said gravity is the strongest where space disappears and even he, the Master, misinterpreted his mathematics. The space mathematics show is not the Universe we see but space located at a point where space departs from singularity applying the value of k^0. It is a point within all atoms secured in a dimension smaller than the spin of the proton. It is where $k^0 = a^3 / k T^2$ It is at that very point that the one side is where space disappears and on the very other side motion gives space dimension stability holding size sides in form. It is beyond the micro cosmos at a point not mentioned yet or named yet.

$a^3 = a^{2+1}$

Our six-sided Universe is about space-time. It is space in motion and the motion is placing relevancies that are applying the time. Time even to human standards at present is the positioning of objects after some other arrangement of positioning took place and repeating by duplication brought about by transmitting space as the transmitting or the moving and the pace of the moving of the space is repeatedly duplicated to the previous order it was.

If it was not possible for space to use time in providing space for space to move too in duplicating the Coanda effect was not possible. But the Coanda effect is only applying gravity in a way we are not use to because we see our motion as being in a straight line following the curvature of the Earth. In our way of thinking about gravity is that we go down wards by not going downwards. This then according to our misconception comes about a means of pulling is very incorrect because mass is the result of the lack of space we must duplicate to still remain in the cosmos. By cosmos law we must move to fill the space we are moving towards and by our not applying we create the mass we are in. As we are part of the Earth space we are duplicating in time with the Earth because we have to comply too cosmic law. Instead we double the space by standing still of space as motion insist of replacing the space we have. By not being able to duplicate the space as we move onto the next space our motion creates we establish mass as a means to cheat gravity out of the space we should be duplicating. Mass is the effort we have to bring about since we cannot duplicate our space with motion. The duplication we are using to double our space stems from the protons up to the stars where singularity is providing such duplication.

The Universe we know shows space having six sides and even that Kepler proved as he proved space in spin $T^2 = T^{3-1}$. It is the same therefore it is dimensions repeating to form $a^6 = (a^3 \times a^{2+1})$ This becomes a factor of space forming six sides when Π forms Π^6 in relation to the six sides of space ($\Pi^6 / 6$) in the presence of gravity supplying motion of 7 / 10. That is outer space $7/10\Pi^6/ 6$ forming the element space-timer limit of **112** in which we find all the known elements.

One part of space is time, which is the pace of the duplication of the space duplicated or the spin of heat and the other part of time is the dismissing that is establishing the flow of space towards a dominating centre. If there were no space moving or space to move within or available to move towards, then the next time period will find no space to fill or find itself unable to fill the vacant space that is not available to fill. If there were no space to fill the space with space being the filler, the Universe will collapse back to singularity where all is without space being without motion thereof. It is what Kepler said it is, being $a^3 = T^2k$

The reason why man can never fully create the complete 3D is because of mans' inability to recreate motion that we find in time. The duration it takes any one point T_1 to move to any other point T_1 will have at the time of the arriving of T_2 produces the 3D Universe we are in. By such means does the Titius Bode changing five relating to seven to ten relating to seven bring about gravity or time or moving singularity. Use the name you like but it is all the same being Π^2

One part of space is time, which is the pace of the duplication of the space duplicated or the spin of heat and the other part of time is the dismissing that is establishing the flow of space towards a dominating centre. If there were no space moving or space to move within or available to move towards, then the next time period will find no space to fill or find itself unable to fill the vacant space and if there were no space to fill the space with space being the filler the

Universe will collapse back to singularity where all is without space being without motion thereof. It is what Kepler said it is being $a^3 = T^2k$

Space/ time in motion and movement are all the same things only separated by dimensions and dimensions are formed space where the dimensions become space being in motion and the space is motion by contraction or by expansion but because time is almost eternal at k^0 our perception of the Universe we are in is a stable and steady eternal structure. Gravity is motion and motion creates space to the third by the third in the third that interacts with one but establishes ten.

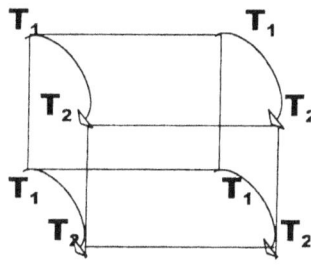

The motion coming about supports the structure material forms. The fact that the rear end will fill the vacant front end when the front end vacate the place where the rear end will move to while the front end drags the rear end into the position the front end vacate is the pillar that unites the structure. While the one behind fills the other in front empty but the other empty to allow the one to fill.

In this very manner is gravity the very same as speed where gravity is the moving

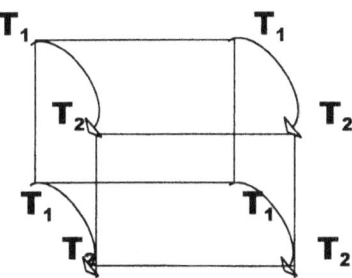

The reason why man can never fully create the complete 3D is because of mans' inability to recreate motion that we find in time. The duration it takes any one point T_1 to move to any other point T_1 will have at the time of the arriving of T_2 produces the 3D Universe we are in. By such means does the Titius Bode changing five relating to seven to ten relating to seven bring about gravity or time or moving singularity. Use the name you like but it is all the same being Π^2 **Space is created from one position to another position and the duration it takes to complete the distance is time.**

The **k** that science sees that positions an object in relation to the centre of the Sun is not the only **k** applying. The line is **k** and there are many lines forming. There are the lines forming **k** by motion. The only way **k** can extend is by revaluing T^2 to produce an individual gravity T^2 in the space a^3 that the occupier occupies. By increasing the heat that supplies the cosmic structure individuality, the structure then receives a new gravity T^2 because $k_1 < k_2$ in producing $k_1 = a^3 / T^2$ compared to $k_2 = a^3 / T^2$. It is not surprising that T^2 takes on such a different value because the surrounding space of a^3 that influences the ratio does change considerably. By changing the value of **k** brings a new cosmos about with a new gravity T^2 and a new value to space a^3. By motion increasing or decreasing gravity changes values. This is very evident in the

Coanda effect. It is also very apparent in the sound barrier as well as the re-entry of objects entering the atmosphere. An object entering brings about that the **k** of outer space changes to the **k** of the Earth. The Earth **k** secures a new time component T^2 but such a time component also bring about a revaluing of the space a^3 occupied by the structure. If the structure does not adhere to specific rules the Earth will reduce the material to heat as we can see with all objects coming through the atmospheric barrier. It is an interlinking formula $k = a^3 / T^2$ where any changes to one aspect brings about changes to all aspects.

By recalling the space, it is also reducing the space because it is counter acting the time expansion provides. That then is clarifying the reason why gravity will always on the limit be stronger than light. At a point it slows the time component down to such an extent, that the space reduces faster in that time than what light can produce motion. Gravity and antigravity must be seen in the speed relations that come about from the process. Gravity is speed or velocity applying. It is space in volume in relation to time in motion. It is $k = a^3 / T^2$

the Earth

New positioning k

> When the object is stationary on the Earth a certain value applies **k** in relation between the object and the Earth to the object. The space a^3 surrounding the craft influences the craft to maintain the time T^2 that the Earth applies

By departing to outer space the relevant **k,** that is influencing the space a^3 in the time T^2 changes as the time T^2. This is the result coming from the changing of **k** changes and since **k** changes all relevant factors change.

Atmosphere

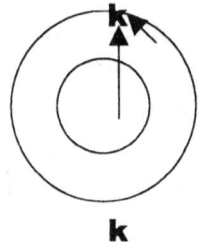

k

the Earth

As the craft extends the **k** influencing the space a^3 the craft requires the extending of k brings about a bigger time T^2 component. There is a slight change in the space occupied by material that composes a lot of fluid such as the human body but the time T^2 is the big factor that changes. T^2 is the gravity applying and it is all clear that the Earth gravity dynamics do not apply. The object then secures a gravity depending on the **k** it receives from the sun. In that manner the six sidedness of space apply and not the Earth's sphere having a seventh point that produces a new **k** and separate gravity

There is a time component present to activate one singularity to the next singularity as singularity is generated or exited by motion of the space moving it hold by the motion, which generates the singularity to control the space. Considering when a body enters the atmosphere there is a tremendous re-adjusting going on about in the governing singularity that the group or unit elected and that spirals down to individual atomic singularity and therefore all the atoms forming part of the craft must undergo space-time re-adjudging to match the singularity that the Earth has in time related space occupying.

There is not enough material in that area of the space we call the atmosphere limit to bring about the friction required to unleash the heat we see develop from such an entry into the Earth's atmosphere. The flames surrounding the structure is the area a^3 that reduces because of the changes brought on by the new time component and the new **k** factor changing the relevancies between the factors. Space travel is limited and the distance to time endured is gravity. For us to be able to leave the confinement of the solar system will require anti gravity equal to the gravity in the centre of the Sun. It is the same requirement there is when we wish to leave the Earth centre.

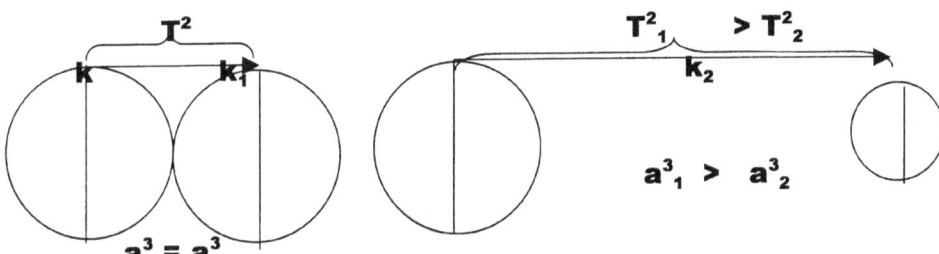

This also comes about and influence travelling in space. There is no unlimited speed that can be achieved once an object is in space. By increasing the time T^2 or decreasing the movement **k** such changes will bring changes on all factors and will demand revaluing of all aspects. This we find by using Kepler's formula **k** = a^3 / T^2. This is what happens to objects entering the atmosphere. There is no particles touching which introduces friction and such thoughts are senseless, however there are friction brought about by motion restriction.

By looking carefully at Kepler one can see that space comes about from matter moving by which space is created through the motion causing time to become an integral part of space. There is no chance of wiping out time or turning time in reverse, because by fiddling with time motion moves space back to space closer to the realms within singularity. Changing motion is taking gravity back to singularity and that reducing of **k** will lead eventually to conditions that prevailed before the Big Bang event.

Gravity starts where gravity started in the very first instant. The very place where **k** left singularity and stepped into the 3D **k** extended ever so slightly but never more influential. The point is where **k** formed a line outside singularity and went from k^0 to form **k**. The length of the extending of k^0 to **k** is so small it is beyond any manner of human conception or mathematical comprehension but so it is vital the Universe it is the result that comes in the form of space we call the Universe and the Universe come from that action. With that action of extending **k** by the very utmost, utmost slightest of margins the Universe we know came into being what it is. The factor **k** went from k^0 to **k.** By the extending of **k** space a^3 came into place. But that space only came about by T^2 producing motion. By **k** extending by motion, space came into place. Space came into place only through the motion of gravity and antigravity.

The motion T^2 produced brought **k** the independence which a^3 secured. The factor is documented as $k = a^3/T^2$. Seen from the point singularity holds every aspect there is in the cosmos including other singularity dots is space-time. It is space generated by motion attached to the centre or governing singularity. From where **k** starts expanding **k** produces space a^3 through time or motion T^2. It is $k = a^3/T^2$. To one point holding singularity that point of singularity holds everything there is and there is only the space-time extending by **k** producing $k = a^3/T^2$. There cannot be space without motion. There cannot be motion without gravity and there cannot be space without gravity. Gravity is motion producing time forming space capturing gravity. It is a confined unit belonging to every singularity extending.

Accepting gravity to be motion is half the story and that is very much literally half the story.

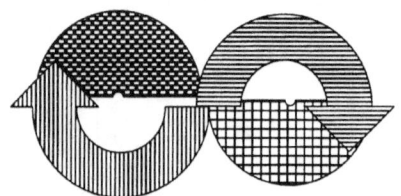

The heat brings about expanding singularity from a one sided affair to filling a volumetric Universe. All of it is a relevancy where ten positions will sacrifice individuality and compromise singularity in order to secure two positions in singularity that holds a relevance of one.

That which I refer too at this point being ten to seven is material in unoccupied space. There always is a relevancy of singularity being relevant to singularity and one takes charge of seven while the relevant partner is taking charge of the other singularity (7) in the unit plus the three in motion between the two in relevance. There is another relevancy in place that only takes space occupied by material into account and in that confirms the sphere as forming space occupied. In this second relevancy the same principle duplicates once more because of the direct attachment to singularity and then there is the space of ten duplicating twenty in a dimension above. The first dynamic involving the Titius Bode principle was relevant to space filled with material in relation to space providing the motion and thereby filling the space.

By using the Titius Bode principle of seven on the one hand relating to ten and all ten coming about to sacrifice their position in order to save one vantage point material as well as space has to cross the border of singularity and fall in the other side of the Universe. The centre point holding singularity still forms the divide but all other points have a task to perform. Securing the centre singularity and maintaining the centre singularity brings about securing all singularity and maintaining all singularity. It is six relating to six where three is on the one side and three is on the other side. There is space formed and three is motion in space formed.

That which I refer to at this point being ten to seven is material in unoccupied space. There is always a relevancy of singularity being relevant to singularity and one takes charge of seven while the relevant partner is taking charge of the other singularity (7) in the unit plus the three in motion between the two in relevance. There is another relevancy in place that only takes space occupied by material into account and in that confirms the sphere as forming space occupied. In this second relevancy the same principle duplicates once more because of the direct attachment to singularity and then there is the space of ten duplicating twenty in a dimension above. The first dynamic involving the

Titius Bode principle was relevant to space filled with material in relation to space providing the motion and thereby filling the space.

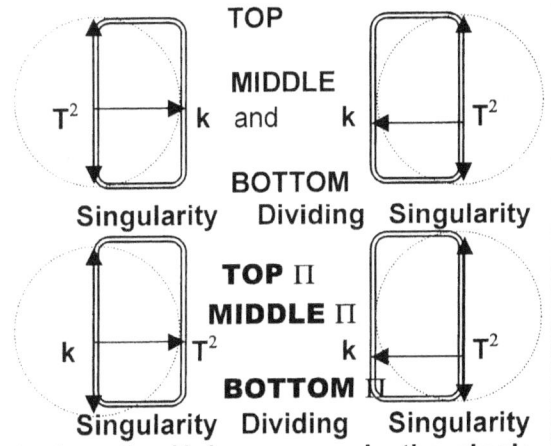

The names I use in TOP, MIDDLE and BOTTOM must not be viewed as sides but merely as terminology using names to implicate divisions. Direction depends on positions and positions form a value only when the observer forms part of the cosmos and not part of the observing.

The universe divides into two separate issues because of singularity. Nothing can be in two places at the same time where as all the rest in the Universe has to confine to the law applied by singularity.

But when the Universe was in the single dimension, all values were Π, therefore every value related to $\Pi\Pi\Pi\Pi$ forming three of the same that was very different because it was where Universes met and formed relations. Every spot formed an individual dot or Universe and every dot was another new Universe. Objects can only be in one side of the universe holding three parts or in the other side of the universe holding three parts. From the totality three will be a double with six sides to show, but that forms 3D. From singularity it is flat with three sides forming on either side of singularity. During the Big Bang two things happened. Particles all overheated. Heat established motion wherever overheating applied. By overheating, material enables the securing or claiming of more space. Only by overheating or increasing heat can material claim more space. In order to supply **k** with any reason to grow into where **k** then becomes the fibre of material, there had to be a way fitting natural processes to do such extending. The precondition of material to grow and claim space is to accumulate heat, and that must have happened because we can trace the excessive heat there was even today. If **k** grew, the temperature had to rise. All temperatures were the same in singularity. Singularity is homogeny in all areas including heat. Singularity can generate heat on one condition and that is by producing more spin. The Coanda effect is vivid proof of my statement. To generate heat it must spin at a higher rate and by spinning singularity then produces more heat. That is in motion and all proof still exists that singularity had and has no movement. Spinning the top will bring the top to become more exited and then being more excited the top will attempt to elevate the motion of the top to separate from the Earth gravity. In singularity there was no motion yet and there was no heat yet. Still, to get **k** to extend, the temperature had to rise. Let us have a good look at gravity.

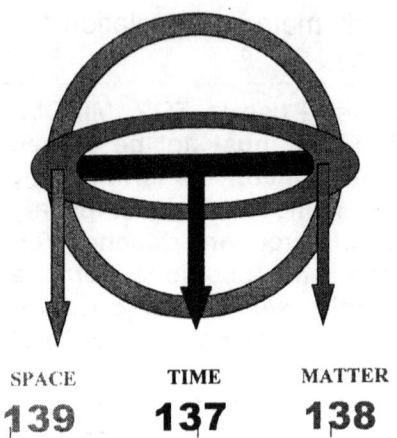

SPACE   TIME   MATTER
139   **137**   **138**

The period I refer in the space above was when space parted with time. It was when Π^0 moved to Π or the spot became the dot. It must no be confused when matter entered the equation. In a while I shall explain what happened at a proton displacement value of 139, 138 and 137 when space filled with the heat that we now think of as material. What I refer to above was in process where cold for the first time parted from heart before any time delays became a reality in space-time.

On the one side of the Universe ten positions form a web with one and a range of dots smaller than one (0.9991) to form the protecting of singularity. On the other side there is the same number and that number form a unit 10 + 1 + .0991 + 10 = 21 that forms the space effected by the seven forming material and forms the motion of the Titius Bode principal. The factor of space that is surrounding material (always in the sphere to protect singularity by applying gravity) brings singularity back to the original value of Π by maintaining Π^2 in the motion and space required to maintain Π. This produces the dome covering the sphere forming the circle mimicking singularity. The atmosphere is as much part of the sphere as the sphere is part of gravity forming an extension of singularity being the eternal Π.

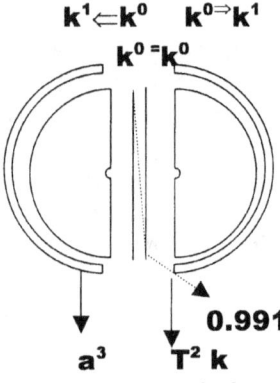

$k^1 \Leftarrow k^0 \quad k^0 \Rightarrow k^1$
$k^0 = k^0$

0.991 Singularity eternally running to the inside

$a^3 \quad T^2 k$

From the relevancy of the overheating which is bringing about space by creating motion there are positions taken in space occupied and controlled by singularity as well as positions influenced by singularity. There are those on the one side forming ten and then the eternal divide and there are ten on the other side also involving the one and the infinite.

On the one side there are $k^0 = a^3 / T^2k$ and when that adds it is seven points. Then there is another three either being space or being motion creating space a^3 or $T^2 k$. It is holding either but not both and will establish duplicating once more.

It is seven relating to five, which is duplicating five in either sides of the Universe by involving the Roche limit. The second one is duplicating ten that relates to the line representing singularity as one and the line reducing towards singularity. The second dynamic involves the curvature of space-time by producing through duplication the singularity dynamic of Π.

The atom once was as round as any sphere but with the enormous development of space-time and the massive favouring of space in relation to material the dome form grew flatter. This does not change the singularity forming Π to Π^2 in any way. We can see this very same tendency in almost all galactica of substance. The centre still holds a dome or a sphere while the edges grows flat with gravity growth reconstructing the original form. It is important to realise that because motion establish space as well as control space where the ten factors is representing the square of space.

The square will always grow much more than the seven factors do in the relevancy. We can even follow how the proton relevancy reapplies in the shape that the galactica form or as the galactica takes on a "normal" neutron form. The control of motion producing time is essentially locked in singularity where time starts the motion as it is establishing space. From our position outside singularity the motion creating space is the point, which is where the slowest time can be when not being eternal because there time is periodic and not eternal which, singularity is. Singularity cannot shift because singularity is the first dimension and as such immovable. Therefore anything not eternal is temporarily created and beyond being permanent although I agree that we are seeing the Universe as structurally solid.

The Universe we see as solid is solid because of the motion being the flicker or stroke relevancies and as time duration slows down as space expands, space relatively remains longer in one position of duplication than in the other position of resembling positional correlating. By it going out of frequency we see the space as space only because it is relative longer remaining in space and that is what confirms the 3D qualities of the space we enjoy.

$$2T^2 = 2a^3/k$$

On the one side of the Universe ten positions form a web with one and a range of dots smaller than one **(0.9991)** to form the protecting of singularity. On the other side there is the same number and that number forms a unit **10 + 1 + .0991 + 10 = 21** that forms the space effected by the seven forming material and forms the motion of the Titius Bode principal. The factor of space that is surrounding material (always in the sphere to protect singularity by applying gravity) brings singularity back to the original value of Π by maintaining Π^2 in the motion and space required to maintain Π. This produces the dome covering the sphere forming the circle mimicking singularity. The atmosphere is as much part of the sphere as the sphere is part of gravity forming an extension of singularity being the eternal Π.

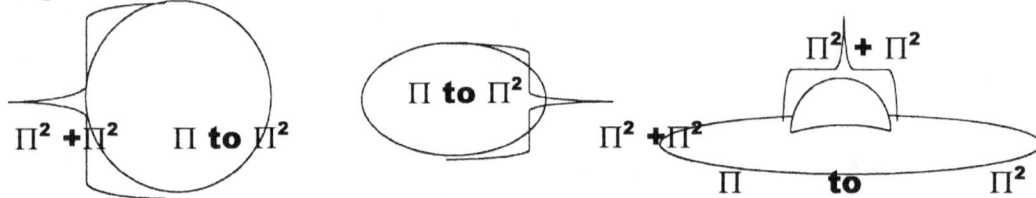

The atom once was as round as any sphere but with the enormous development of space-time and the massive favouring of space in relation to material the dome form grew flatter. This does not change the singularity forming Π to Π^2 in any way. We can see this very same tendency in almost all galactica of substance. The centre still holds a dome or a sphere while the edges grow flat with gravity growth reconstructing the original form. It is important to realise that because motion establishes space as well as control space where the ten factors is representing the square of space. The square will always grow much more than the seven factors do in the relevancy.

That only comes about as our perception. In stead singularity create space and time and by allowing space to reduce from singularity the reducing is placing the motion establishing space to reinforce as well as secure the immovability of singularity. It is providing singularity with permanency while singularity is removing the permanency from the creation sector to the power of three to the power of two. Where the time excels beyond our measure is within the atom structure. That which we think of as a mass difference in the atom, is a reducing of space in stages and a reducing of space is an increase in the duration of time by motion. The proton reduces space 1836 times to the electron and that we see as mass. In fact that is space-time differentiation because all mass is just space-time differentiation. Things go even more reduced as the proton sheds all space to enter singularity.

As the one proton moves into singularity and allow space to disappear the motion brings about the other proton to move into the place of the proton that disappeared. At that point time is next too eternal because the very next spot time becomes eternal as it enters the immovable singularity, which is eternal. This is like watching your nails grow. After time one can see the nails showed growth but the growth is so time consuming we find the growth of the nails to be next to eternal. By the motion of the proton, the proton is doubling the next proton that disappeared as it connected to singularity. This motion at that point is 1836 times to the power of three to the power of two more time consuming than is the speed of light because the electron is representing the increase of space-time displacing to the speed of light and at the point where the proton becomes singularity the motion is so much time consuming the time created to indicate the motion presents us with a permanently secured structured. It seems as if it might be $((1836)^3)^2$ times the speed of light but explaining this is far too time consuming at this point.

This period of space relative to motion being $((1836)^3)^2$ is the motion securing space in the Universe and is from our position eternally fixed. We see the proton as vibrating because the proton is securing the permanency of singularity. Any light shining permanent is actually flickering and the flickering that is so fast gives us the impression that it is permanent. Anything less permanent will seem broken into fragments when holding comparison to what we find to be permanent and at the same time anything flickering $((1836)^3)^2$ faster than light becomes solid.

At the present moment science is looking at the speed of light as being three hundred thousand kilometres of space being displaced every second of time on Earth. This again is a speed. The speed of light is a form of gravity that has gone anti-gravity. It places distance in relation to time but according to science they place the emphasis on space where the emphasis should rather be on the time factor. That the space factor what we see when we are looking at the larger cosmos in outer space we can't see any of the space which we think we see. We can see in the time that light brings and misinterpret that information as space. It is however the light coming across time, which we interpret as space that we vision as space but we stand to be corrected because it is in fact time we see.

It is a light holding no space **a³** because it is pure motion **T²k** that is coming across the vastness of space. When saying space it is just a linguistic cultural expression because we know by now that space is time and motion of space. What we see the light representing in the image it brings across, is the space but the space is since long gone as it disappeared from the Universe. In the meantime all that remains from the time of the space is the flatness of time delivering the light that represent not the space of back then but the time now forming an image of the time then as it was time that was crossed and not space that was crossed in a flat dimension of time. The light is about space. Yes but the space is history that can never repeat and such history is not time because time is what it took to carry the light and not the space from there set in when to here set in now.

The picture of space we about space as well as time because the light is the image of space where the motion of space only left the image of a vision of light representing what the space was while light by duplicating an image but it is carrying the image of space gone by. The space in motion is in a completely different location. History is once again repeating as it is duplicated through the effort of light. The light is duplicating an image through time (it is not the space duplicating but the light coming from time of the space that is duplicating) and that is not the same thing. In the Sun on the very inside, the displacing tempo might take three hundred thousand years to displace one kilometre of space or said in another way, it takes the photon three hundred thousand years to travel one kilometre. The space creates time and if the space is dense the time that creates the space must be slow moving. Einstein declared that time moves slower in larger gravity. In this it takes light an effort of about ten million yeas to escape from the inner core of the Sun to the outer ridge. When space reduces time expands and when space expands time reduces. $k^0 = a^3 / T^2 k$ therefore the bigger **k** will bring about a bigger T^2 since it reduces a^3 in the process. On the other hand will a bigger a^3 bring about a longer **k** and thereby shortening T^2.

From what we read into Kepler it says that time contracts space at a specific value and that value remains a unit that brings about a rule. The factor **k** moves one specific distance creating a very specific space a^3 that takes a specific time T^2 to duplicate. This ratio applies through out the Universe but the relevancies of the factors as the factors stands is in sequence with singularity extending. That is why the atomic conditions applying inside a Black hole is almost unrelated to atomic structures in the Sun or on Earth.

The applying conditions demand different requirements on space-time brought about as the space-time is influenced by singularity at that point. The atom changes space-time to comply to standards the singularity allows and with what standards are prevailing within the borders that singularity establishes. It is because of this reason exclusively that stars vary so much in the way they present and react to space-time. If it is motion that contributes to space-time differentiating in the form we call gravity and gravity is exclusively coming from within the atom it then is highly advisable to find the properties and influence of gravity within the atomic relevancy. I elaborate this aspect much more in another book of mine called **A Cosmic Birth...Dismissing Nothing I.S.B.N.0- 620- 31609**

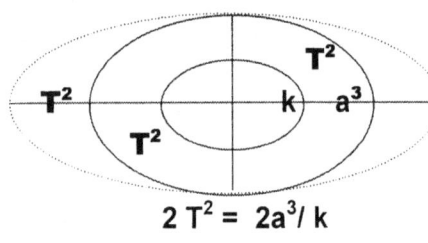

That is the relation there is At the point of cosmic birth the cosmos is Π^3 as Π extends from Π^0 becoming 3D through spin. Where spin ends space ends and gravity ends. It is a unit undividable. Linking Π^3 to Π^0 is the motion Π^2 brought about by spin or rotary motion.

The proton forms the first line of contact from singularity into space-time. That is the manner in which the atom sets space-time as well as rules about space-time.

There are four time sectors in the Universe coming about from singularity and singularity fills every one. In the one sector there is a proton Π^2 connecting to singularity Π^3, which is connecting singularity in the form of Π to singularity in the form of Π^3. In that there is the Kepler formula $\Pi^3 (a^3) = \Pi^2 \Pi^0 (k\ T^2)$. Then there is the other proton filling the second opposing quarter of time implementing the same procedure with the same result coming about. In the third quarter is the first neutron connection following an identical path but linking to a proton forming the space-time. In the forth quarter the motion divided even further by splitting the motion as it conforms space-time from **3 to Π,** which in the end is the motion of **Π².** In all instances space **a³** is confirming singularity **k⁰** in time **T²** except in the one that produced the Big Bang.

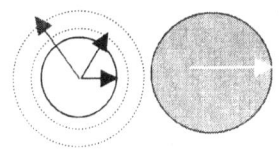

The Universe is filled with points each holding lesser singularity. We must see how the atom filled with heat. Heat is a liquid and a liquid will build by filling without leaving any traces of differentiating. With this in mind we then must place the filled atom in relation to a built up location by using points holding singularity.

The circle to the left would come about from a straight line r growing influencing the appreciation of Π, but to influence Π would lead to a breakdown in r as Π and r are different entities. The circles to the right shows a continuous growth by extending Π every time and since Π is the same part as the previous Π, only extending that billionth of a millimetre each time, the circle will be truly continuous without any signs of a break

The **k⁰** confirms **T₁** to **T₂**. Singularity is and remains Π therefore Kepler takes on Π but the dimensional impact remains the same. This connecting happens on both sides of the Universe Time has four parts to fill space and maintain singularity by criss-crossing fills singularity by connecting directly space-time through the proton to proton (Π² + Π²) or the proton to neutron (Π²) or the link of the neutron to electron link (ΠX3). The proton removes space by doubling space to a point of the oblivious. The space the proton duplicates is so little and the duration so long that eternity devours space into the oblivious. By the proton circle reducing as it constantly duplicates, the duplication becomes so fast that the period it takes to duplicate the space becomes eternal in the centre circle. The atom takes space where space reduces into the electron as a factor of 3, where the neutron further reduces the space and extending the time by the factor of $\pi^2\pi$ and in the end the proton diminish the space-time to $(\pi^2\pi^2)$. Adding this total confirms the mass difference between the proton and the electron at 1836 times. The proton has a mass of 1,673 X 10^{-27} which is 1836,12 times greater than the electron's mass of 9,109X 10^{-31}. (Π² + Π²) X (Π²) X (ΠX3) = **1836**.

With singularity placed in infinity within the centre of every rotating object every atom and its relation to its surroundings including other atoms form space-time diverting from the point holding singularity as far as rotation goes because every object holds three relative positions in as far as where it was, where it is and where it will be in relation to singularity providing time. I elaborate on this else where.

Any point will be opposing itself within the **rotating of 180°** where it **then change every aspect** of its **previous flowing** characteristics it had or **will once again have** in **360⁰** from there.

By reducing the space-time the lesser singularity is claiming singularity independence by offering reduced space, which will result in promoted time with heat being the net result. That heat is filling the space, which should then be entered by the independent space in motion if the motion of duplicating is brought closer in a relation to what the matching tempo requires.

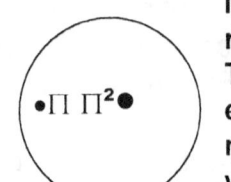

Space generates the mass where the space has to reduce the size by becoming more intense and concentrates space-time to the time of 1836 time more when entering the point of the proton being on the verge of singularity. In single dimension seen from one aspect, with single dimension contacting the edges forming the sphere it will still keep the seven positions because the sphere remains a unified structure though apart because of singularity. In the core of the sphere the proton connects in alliances as $\Pi^2 + \Pi^2$ with the solidity of the neutron holding Π^2 as a second forming a π value. That brings the atom unit of π to a number of seven.

It is clear that the density of material in motion is $\Pi^2 + \Pi^2$ but since that is k, which extended we know that that extending cannot sustain the initial speed. Since the speed is reduced the space in motion will value less. Taking these atomic relevancies into account, we can detect what relevancies brought about the atomic Universe of $(\pi^2 + \pi^2)(\pi^2\pi)3 = 1836$. The first substance that formed from singularity was solid and if that were the case the contra substance would then be a fluid with less motion filling more space taking shorter time duration in duplicating. The fluid substance that then formed was one less than the proton in motion which makes it slightly more in mass since it duplicates more with less space that then has to form $\Pi^2\Pi$, which has one Π less from $(\pi^2+\pi^2)$ which is resolved becoming a fluid like substance relevant to the first solid substance which is the proton. The loss of the one Π then became the factor claiming more space that is holding less substance. In this fluid state the neutron has more duplicating of the substance than is required of the proton. That what we find in space we also must find in the atom because the cosmos is not keen on inventing but is passionate on duplicating. This fact will also apply to space-time in many forms. That means investigation must prove the same results and what we find in the atom then also has to present in the cosmos at large.

While in rotation from the view point of a bystander it all may seem static and never changing but to the object in spin every next instant in time will be diverting from every aspect it had every second passing, and the direction it held in relation to the direction it held the previous mille, mille second will totally be incompatible with the direction it holds the very next mille, mille second of rotation. This is why we can use degrees measuring the circle by (6^2) (forming the square relating to matter through singularity) X 10 (square if space) = 360^0 however it is always in motion. That proves no point can be static or constant, though it may seem that way to outsiders.

Although matter is matter, matter can also be anti-matter and moreover form its own anti-matter at the same time. This degeneration of structure is very likely to occur with overheating. Revaluing Π to Π² will bring about a new contact point where Π meets **r** forming another relation in Π² **Time is the changes in relation** where Π **contacts a different r** not withstanding the many r points there may form because **every r constitutes a different value** to the universe through other ratios and relevancies brought about **by heat and light. Time is the duration it takes Π to rotate between any two given points of r** and therefore must always amount to **a square (T^2)** moving from point to point through the **cube of space (a^3)** in that **duration of time (k)**. With that it proves **Kepler's a^3 (space) $=T^2$ k (time in the instant of motion)** but motion must continue through a specific value in space where the space-time is maintaining relevant equilibriums throughout singularity connecting.

It is the mass that space generates where the space has to reduce the size by becoming more intense and concentrates space-time to the time of 1836 time more when entering the point of the proton being on the verge of singularity. In single dimension seen from one aspect, with single dimension contacting the edges forming the sphere it will still keep the seven positions because the sphere remains a unified structure though apart because of singularity. In the core of the sphere connects the proton alliances Π² +Π² with the solidity of the neutron holding Π² as a second forming a π value. That brings the atom unit of π to a number of seven.

From the centre to the outside there is a connecting of **k**, running the length of space-time in relation to the **T^2** that brings about the liquid or neutron form. The fact of the liquid neutron is to produce duplication by repeating **k** as **k = a^3T^2** whereas the role of the proton is about relinquishing **k** by **$k^{-1} = T^2/a^3$**. In the centre of all the atoms space is relinquishing a position through the dissolving of heat by means of maintaining singularity. But through it all, another singularity forms in the very centre of the structure, claiming the position where space is the least available that bind the singularity of all the atoms sharing space as a unit. With that evidence I realised there are a connecting of singularity and that connection is electricity.

Later on I prove that electricity in its most intense form is gravity, which we will find in the very, very centre of the Earth. In the cosmos all objects form a sphere. Some solids do not seem to be a sphere and space is no sphere, but the truth is hiding in the way of connecting. At the centre connects **T^2** forming the base of the solid. At any one specific given point forming the surface of the sphere is another marker holding the connecting relevancy of **k**. When there is no sufficient heat to form space that will part **k** from the other holding of **k**, the two will combine in a solid joining connecting as 2 **k** that translates to **T^2**.

In the way space and the sphere connects the sphere will have 7 points holding a relation to 3 points not within the sphere forming the 10 that creation started with. This will mean there is a division forever, and such a division may run smaller as it runs everlasting. With fluids connecting it is simple to recognise the sphere, as **k** will indicate the point-to-point location of **k** as the form of the sphere. By gas forming the connection there are the three points of space being apart and not forming **k**, but still holds a relevancy to T^2 through the value of **k**. The gravity applies as much to material as it applies to form installing form. In this way stars are spheres while the stars are just one more cosmic atom and all rules apply as much to stars being just cosmic atoms as the rules apply to atoms being just individual stars. Remember that big and small, hot and cold, tall and short or any other boundaries we humans can see so clearly is no valid boundaries to the cosmos. In the cosmos a star fill with atoms and with the compliment of atoms working in conjunction all the atoms eventually unite and combine as a unit. In the very end of their developing of the star, the selected and developing singularity will again, as it was before the beginning of time, combine all singularity in one structure fitting and holding all singularity of every atom within the star unit. The Universe will end with a united singularity and the Universe started with a united singularity where from there is no size but there is just space-time.

The rules in the cosmos are the same applying to all in the same manner. The relevancy of the atom being $(\Pi^2 + \Pi^2) \times (\Pi^2) \times (\Pi \times 3)$ = **1836** remains but the structure of the atom relevancy demise as space demise by prolonging time in the growing star. It might be beneficial to compare the atom and the Earth in composition. At the spot where $k^0 = \Pi^0$ becomes $k = \Pi$ space comes about in the form of $a^3 = \Pi^3$ by the measure of $T^2 = \Pi^2$. That puts space a^3 or Π^3 in relation to three sides and it will forever be three sides coming about from one point. $a^3 = k T^2$ where $k = T$ and therefore $k^2 + k = k^3$ or $\Pi^3 = \Pi^2 \Pi$

The sphere was the first to come about and only after the sphere could not produce the gravity required to suppress the overheating forming the antigravity did the Universe try other options. The double proton formed but by that the rising heat excelled. The proton deformed where one became a neutron and the other became space in heat. With the demise of many protons favouring the forming of other particles and substances such formations united. Then the atom came in to form…did the atom clusters form…did the material clusters form…did the antigravity apply enough to allow overheating bring about expanding where the relevancy will produce one softer and one firmer structure…from which the antigravity expanded into space.

When the heat turned to apply antigravity that eventually produced space, as we now know space to be the Big Bang was happening. But before the Big Bang there were some mighty jolts, cracks and jerks announcing the Big Bang to come. It was the proton $\Pi^2+\Pi^2$, which afterwards connect to other protons which became independent Π^2 connecting yet again to other Π^2 particle that with help establishing further antigravity which then overheated and expanded to Π^3. One can see there were many stages to produce what we now have. However most important is the fact that if the laws that once implemented the stages of developing all periods in the Universe then those laws must still apply and must still be present in our Universe in the same way as what it was when Creation started.

How many dots was there is a question no person can answer because everything was un-dividable solid and yet it did group together to form every sub atom particle located in the 3D. From what is now we should be able to trace what was back then when Creation was still fresh. Let us dissect how our Earthly atmosphere brings about duplication or dismissing in relation to singularity boundaries applying on the ground and in the atmosphere.

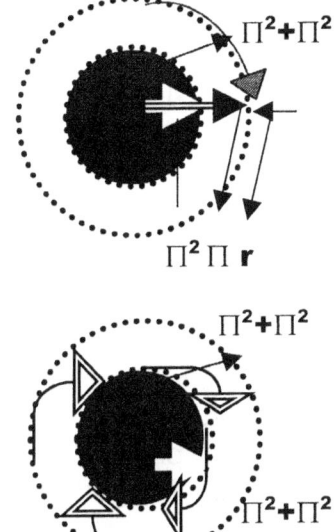

In the circle $\Pi^2\Pi$ which consists of the atmosphere the space surrounding the rotating object will also extend by Π as the concentration of the spinning motion draw or drag on past Π^2 extending the influence of Π^2 by the value of Π. Through very clear evidence about this one can see as the Coanda effect. This extending of Π^2 to accommodate Π we refer to as the atmosphere, but physics apply to this extending in the normal fashion. The soil of the structure represents the solid proton being $\Pi^2+\Pi^2$.

From the spinning motion Π^2 does not stop at the end of the solid structure but the influence of Π extends and this then becomes the atmosphere. The influence of Π^2 or gravity does not stop at the end of the solid structure but the influence of Π is extending and eventually plays a most dominant role in the local cosmos, although not yet recognised and that factor is most crucial to a better understanding of the implications of laws governing the cosmos.

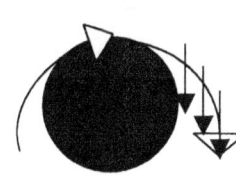

With the circle being $\Pi^2\ \Pi$ the Π^2 will reflect the circle in the square with Π forming the extending of Π^2. This is an extending of the six Π forming in alliance with the centre Π. This produces that any extension of 6 forming material one further extending goes into space and relates to a seventh dimension.

The extending of Π will not end immediately but will carry to the surrounding space the circle influence through rotation. The influence immediately above the circle will have the biggest influence and reduce gradually as the value of Π reduces in the leverage that the space has on Π and a gradual but definite change from Π to r will affect the extending of Π progressively more. The decline of Π will follow the same contour of the circle at 7^0. Every one of the dimensions indicates an individual significance as I shall show later and the increase into space runs by 7^0.

Individual singularity and governing singularity and group singularity enhancing the gravity every time singularity find an accumulation of influences. With our ability to look at Kepler's in a mathematical sense it is clear that from singularity comes space by three duplicating space in time by three. **$k^0 = a^3$ /**

(T^2k). Very clearly the dimensions produced space and produced more space by applying time and gravity as motion. The space comes about as three and the time coming from singularity that is standing still to the motion, which we call gravity, three duplicates three. The Universe came into position by deploying dots supporting other dots and some dots remained dots while other dots went on to become dots of hybrids as it was supporting dots through claiming dots of lesser density and pass that on to dots with larger density.

We must also see that space within particles stand related to singularity, which through material also implicate space, which does not hold material as dense as we recognise material to be. It is in this that three comes into contact with three that is in contact with three plus one. That gives a total of ten and that brings about that singularity contacts the Universe in formations of ten. It is one of the most crucial properties we have to recognise to understand the cosmic relevancies. I do not see that the Creation started with one particular or another specific but relevancies brought about that all particles grouped and as relevancies dictated different circumstances so the dynamics changed to accommodate the Creation requests. But laws were in place to guide the development according to precise lines to follow. It is however quite clear when looking at stars advancing the development that there is the tendency to reverse the sequence followed during the Big Bang.

Looking at stars and in particular Neutron stars we see sufficient contraction of space-time where such reducing of space-time brings about a full scale demising of the Neutrons in the space-time the Neutron star claims. This is a final phase of the entire star's development where there finally is no space-time left to diminish through fusion. Then as the star progresses to the final development, we see the dismissing the proton by ejecting the double proton out into outer space. Following the trend we see as the Black hole does it, it does seem that a method of producing came into place. Remember the star is the Big Bang in reverse.

The fact that there are six defining positions seems to bring a structure to order and disallow the five looser connected Lagrangian point relation. It is not necessarily means u Unit but it seems to indicate where six is in the line, there is an indication of a group forming a unit. This idea allows for motion to take place but motion is the most complex issue there may be in the Universe.

It would be useful to think of space as a line and time, which is a line as two circles. Thus may sound incorrect but please allow me to explain. Singularity does become valid and established only by motion. It is quite true that the sphere holds singularity in a natural position but to have singularity become a controlling factor singularity must be charged by motion. We have to consider what we see in our Universe and then use that which we see in our surroundings to try and further what we should see in space that we couldn't see because the space is invisible and the line is infinite.

When we consider Pythagoras it seems most likely that space is in a 90^0 angle with time. This I say because it seems that every time space crosses time to be space through the delay of time. Space in material is charged in the same manner as one would charge electricity. We know that material is compiled of minutely small invisible but immensely densely packed pieces (if I may use such calling it as such because the terminology falls acutely short with terms to use in order to say what I wish to say and would be pardoned for doing so) The effort to move that which by no effort are able to move has to be created in as much as being excited and the exiting will render the notion that motion is taking place.

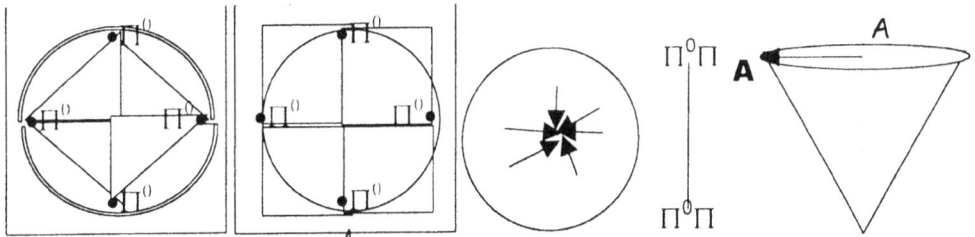

To generate any form of the above singularity from the singularity naturally found in the sphere a charging of singularity by motion has to take place.

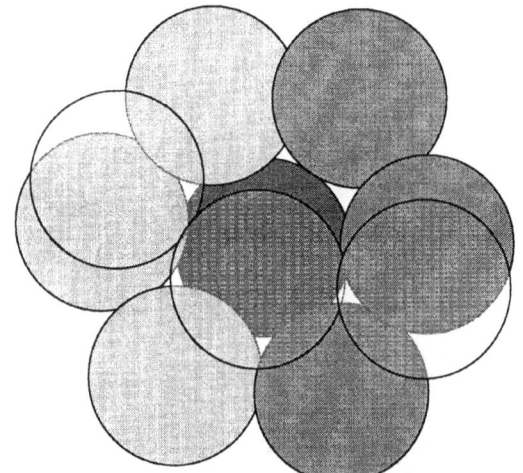

On the one side of the Universe in relevance to all the dots that came before, three dots landed forming one side while three dots formed the second side and three dots formed the third side, all relating to a centre dot which in turn related to the original centre dot from which all the dots came and developed. One part holds relevancy to material, which is three parts in contact with the one of singularity.

The intention to produce motion, which eventually became motion from singularity, brings about another three dimensions through motion producing seven points to material. Another three holds space and from singularity The Universe has a value of ten always. That gives the Titius Bode law of seven to ten from singularity

During the Big Bang two things happened. Particles not in direct singularity control using the Coanda principle all overheated. Heat established uncontrolled motion wherever overheating apply. By overheating material enable the securing or claiming of more space. Only by overheating or increasing heat can material claim more space. In order to supply **k** with any reason to grow into where **k** then become the fibre of material there had to be a way fitting natural processes to do such extending. The precondition of material to grow and claim space is to accumulate heat, and that must have happened because we can trace the excessive heat there was even today. If **k** grew, the temperature had to rise. But all temperature was the same in singularity. Singularity is homogeny in all areas including heat. Singularity can generate heat on one condition and that is that is by producing more spin. The Coanda effect is vivid proof of my statement. However to control and manage the heat, the heat must be contained by the Coanda process as it must spin at a higher rate and by spinning singularity then produce more heat. That is in motion and all proof still exist that singularity had and has no movement. Spinning the top will bring the top to become more excited and then attempt to elevate the motion of the top to separate from the Earth gravity. In singularity there was no motion yet and there was no heat yet. Still to get **k** to extend the temperature had to rise. Let us have a good look at gravity.

There then formed material ΠΠΠ = Π³ and there formed time (matter in relation to heat surrounding material) ΠΠ² and then there formed the space to fit this lot into ΠΠΠ = 3, which with the singularity including Π makes the total value space-time in motion has being 10Π relating to the Π³ bringing about gravity or contracting Π². The cosmos formed in equal part forming from the same substance, which might in some sectors be more or be less concentrated but remains equal. The entire cosmos resulted from one basic form of material (π^3) that formed motion ($\pi^2\pi$) as well as space $\pi\pi\pi$ to fit the lot but initially it was all the same ingredient. The substance forming the entire Universe came about from a natural occurrence: heat in solidity contracting heat in solidity that caused friction. It formed not antimatter but a natural occurrence: heat and it did not vanish from sight it became a natural substance: space-time. The results are the gravity and the anti-gravity we know.

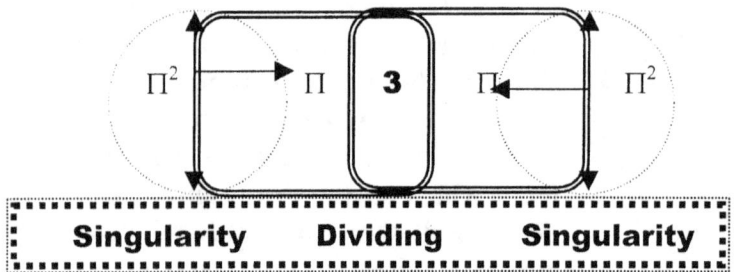

Why would a water drop floating in a space capsule in space, in micro gravity always form a sphere when left to capture free form? We all accept that the true cosmic form would be and most probably will be the sphere...but why would the sphere form as the original form when matter is not pre-cast to have any specific form and therefore take on by cosmic pre-cast the sphere as form? In the past, all intellectuals looked passed this question but in this concept we find the origins of the Universe.

As the relevancy between the particles promote overheating or applying antigravity (overheating) to the responding cooling or applying of gravity, the one repels material into space-time while the other is collecting material into space-time. The motion to extend comes from overheating, which brings cooling and the cooling produces an urge to retract the extending. The one cannot be without the other. The one loses material and ensures a model of preventing overheating while the other gains material and sustain a model of overheating is prevented The one principle we named the Hubble constant where overheating produces space and the other one we called gravity where gravity is demolishing space, but both phenomenon is at present dominating the flow of time in the Universe and will do so until eternal relating again comes about.

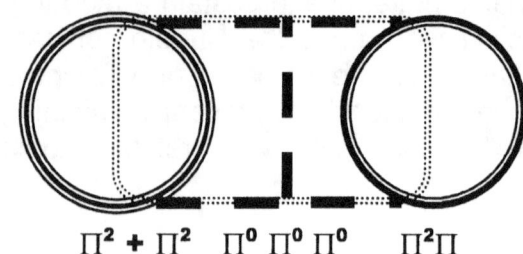

$\Pi^2 + \Pi^2 \quad \Pi^0\ \Pi^0\ \Pi^0 \quad \Pi^2\Pi$

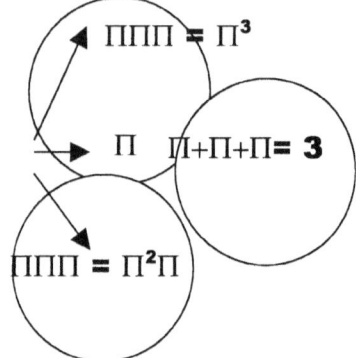

It is all about relevancies applying the relations gained and lost through relations. If one place $\Pi^2 + \Pi^2$ on one side then $\Pi^2\Pi$ is the related form, where $\Pi^2 + \Pi^2$ is in the other side of the Universe being on the other side of the relevancy. Then $\Pi^0\ \Pi^0\ \Pi^0$ will again relate to the other two factors forming the "outside" of the other two being the "inside".

Space formed as motion came about through singularity overheating. Singularity k^0 produced motion at the point where k^0 became k and a^3 became T^2 by motion duplicating space. According to Kepler, a^3 is equal to k relating to T^2.

Using that, I have already shown how three is a duplication because $a^3 = T^2 k$ and this shows that every aspect of creation used three Π to create space in time through motion and space Π^3 secured by time $\Pi^2\Pi$ applying motion and space using time to form space $\Pi^2\Pi$ **X3**. This formula is the atom, which is the universe because every secluded atom forms a Universe apart from other Universes.

$(\Pi^2+\Pi^2)\ \Pi^2\Pi$ **X3**

Matter formed where matter had to have $\quad \Pi\Pi\Pi = a^3$

space to occupy since matter was to use in some space $\Pi\Pi\Pi = T^2 k$

in the space outside the atom $\Pi^0+\Pi^0+\Pi^0 = 3$

therefore $\Pi\Pi\Pi$ met with $\Pi\Pi\Pi$ to form the proton in $\Pi^2 + \Pi^2$ because the matter is within the space it holds and another Π^2 employs Π as a representative of singularity. This then placed the seven positions of singularity as the ending of matter and the three squares ($\Pi^2 +\Pi^2$ **and** Π^2) of singularity as the limit of material. The last $\Pi\Pi\Pi$ became k^0, k^0, k^0 and that became the space producing heat without occupying matter in order to allow heat to be restrained inside the dome singularity provides. When I refer to an atom it includes all unified cosmic structures holding an excluding formation such as an atom or a star or a galactica does. This is where k defines many but as a whole also one a^3 / T^2 determining space within time holding space within time to the value of unifying the lot in one cosmos container. There is little difference in the cosmos in comparison with for instance a lead atom or a giant star or a large galactica. It is all space confined to time extending singularity because in the cosmos there is no big or small. From that the effect of gravity as a restraining on the exploding of space came into effect.

Five was space because space was one removed from time. Space is the distortion of time and one outside time would bring a time delay or a time distortion of four plus one which is five, hence the principle behind the Lagrangian system.

Five was space because space formed only as a value within a value of time where was one removed from time. Space is the distortion of time and one outside time would bring a time delay or a time distortion of four plus one which is five, hence the principle behind the Lagrangian system.
Because material was the square of space material was a crossing of three plus three forming six.

Space-time is the four of time, plus the three in singularity around which the four of time turns, therefore space-time is seven.

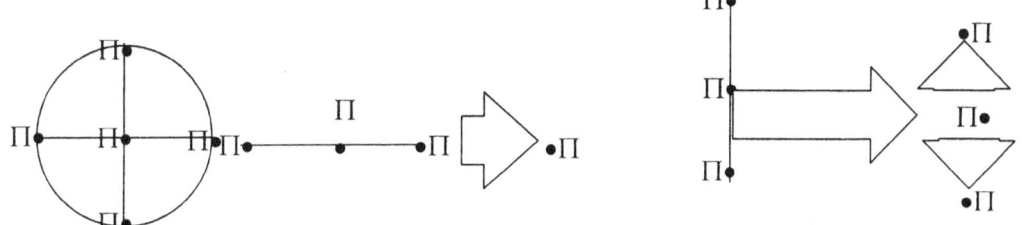

In the circle using $r^2\Pi$ the r has to have distinctive qualities placing it as a factor apart from Π. Where the growth shows no separate distinction but a continuous flow from the precise centre to the precise edge the flow would become in relation with Π depicting the circle and Π replacing r as reference to any point on the circle.

By using r, distinction in the circle is possible but by using, Π there is no distinction possible. Therefore, in the beginning when time formed space there was only relevance coming about from $\bullet\Pi^0$ and the dot was $\bullet\Pi^1$ with no mention of any possible r. The fact of r representing a radius represents space and what we refer to be long before the Big Bang introduced space or mathematics using space.

Before the Big bang the lot was form without dimensions playing any part. It was the point lining up and forming positions that was spinning faster than the speed of light can ever achieve. However every point today still serve the role it took on at that stage and serves in the position that it had during the time it had no space with eternal time.

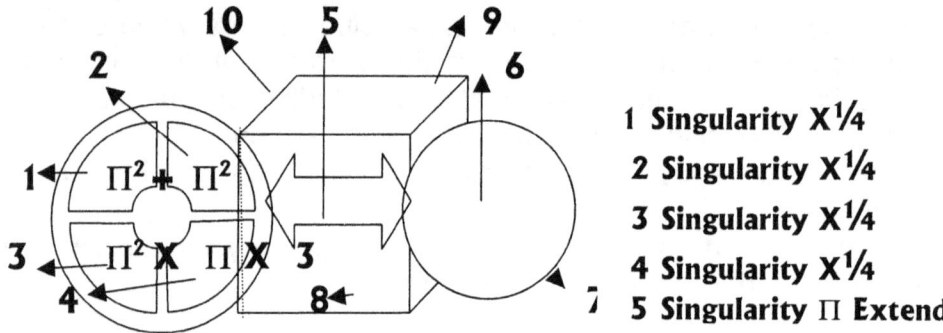

1 Singularity X¼
2 Singularity X¼
3 Singularity X¼
4 Singularity X¼
5 Singularity Π Extend

1, 2, 3) Singularity and supporting 1
(4) Time
(5) Space -time
(6) Matter
(7) Formation
(8, 9, 10) Dimension

From the sketch above one may calculate the value singularity has to apply.

These relevancies developed as part of a Universe we shall never understand. The Universe had no sides and a line was equal to a triangle, which was equal to a half circle.

Singularity holds the double space-time position of five times two (matter and space duplicating singularity) which then is ten.

How did the Universe liberate material and heat space from singularity because with singularity comes an unchangeable eternal condition that is non-changing-everlasting in all conditions and aspects that is remaining in absolute equilibrium. This equilibrium maintains because all development extends form precisely in a detailed equal equilibrium throughout. Think about what brought the cosmos out from the eternal rest in which it was. The eternal rest still maintains and is therefore our detection. What inspired the eternal rest the cosmos was in and inspired change to the state of eternal rest? What evoked change? That is the question the Atheist will never be able to answer but that too is the most basic and ever-lasting fundamentals of the Universe. What changed in this split second start before the official start?

I do not wish to ponder on this matter in the book I am writing at this minute, as there are other books where I delve into this matter. From the deep freeze of creation came the Hot Big Bang and the 3D Universal displacement came about with the relatives being $10 \div 7(4(\pi^2 + \pi^2)) = 112.795$ and then a second one established 3D by introducing the six to seven sides Universe at a density point of $7/10 \, \pi^6 / 6 = 112.162$. There is of course a lot more information about this establishing of the Universe than what I mention at this point. The question is what made the Universe freeze, to form the Universe in space and through time. It had to start with a specific reason applying, which brought about space-time. Once the process started there was no stop to it, but there is no chance that the initiation of the start was spontaneous by nature. With everything being in one spot, all within that spot was in a state of eternal rest. While all remains the same and nothing changing, what brought on the sudden change of everything, shocking

At first when material presented one side of the Universe matter had three sides to show. Matter had to have space to keep matter somewhere in some part of some Universe and that made up three positions. Between the two Universes **k** and **T²** placed a value but since only singularity applied any values the value therefore was **k T²** where **T²** = Π^2 indicated time coming from 7/10 in relation to 10/7 and $\Pi^2/2$ (proof of that is somewhere in the book) and **k** = Π valued by singularity. When space-time developed 3D the dimensions falling outside the sphere becoming space-heat formed as $\Pi^0 = 1$.

The electron holds a relevancy of 3 relating to the Neutron being $\Pi^2\Pi$ and the three keeps the electrons in different Universes relating to separate or individual singularity. Every aspect of cosmology is influenced by singularity and singularity carries the value of Π in many disciplines. The ratio started at the start and that ratio still applies.

The Universe divides into two separate issues because of singularity. Nothing can be in two places at the same time where as all the rest in the Universe has to confine to the law applied by singularity. Objects can only be in one side of the Universe holding three parts or in the other side of the Universe holding three parts. From the totality three will be a double with six sides too shows, but that forms 3D. From singularity it is flat with three sides forming on either side of singularity.

The alliances in form formed which was the prelude of events that made the Big bang event unavoidable. A set of singularity grouped loosely to become time without control of time in control.

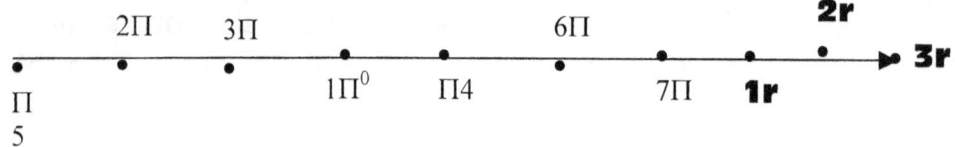

1•Π^0 Singularity governing

2Π• Singularity in relevance

3Π• Singularity in relevance

•4Π Singularity forming motion to become time

•5Π Singularity forming time distortion becoming space.

•6Π Singularity forming time distortion becoming material in space

•7Π Singularity forming time distortion becoming material ending space

• **1r** • **2r** •**3r** Time going into uncontrolled retarding of complete overheating

The motion will produce more duplication of space and the duplication of space is the gravity we experience as a contracting direction of motion where as heating is the expanding of space through motion. But it had to have started with a space less motionless dimension-less Universe wrapped in singularity. The differentiation coming from motion is a dimensional barrier that changes many aspects in cosmology. The dimensions came about as the Universe came about and each had its individual introduction period.

Space and time parted at $(\Pi^3)^2=961$. At that point there was no material because, time containing material as space formed $\Pi \times \Pi^2 \times \Pi^3/5=192$ and $\Pi^2 \times \Pi^2 \times \Pi^2/5=192$. Then time developed as a line and a containing space until material entered the calculation at $7(\pi^2+\pi^2)=138$ where spaces either had material or had heat without material and space separated from heat and matter at space holding $10/7\pi^2/2(\pi^2+\pi^2)=139$ material $7(\pi^2+\pi^2)=138$ and space having liquid within $7/10\ \pi^2/2(\pi^2+\pi^2)=136$

This is suggesting that these are meaning this was the first time liquid became part of the cosmos while all were still part of the same unit as the Roche principle would suggest $(\pi^2/2)$ as well as the (**7/10**) and the (**10/7**).

Second Over All Overview 427

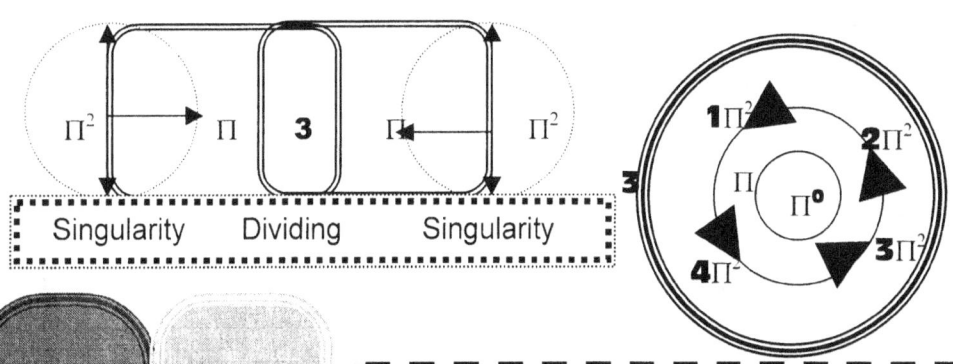

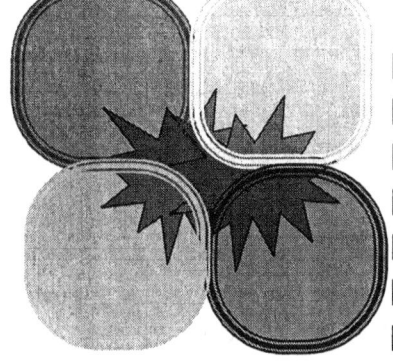

There then came a tendency to remain solid or become softer and more flexible. That is evident by the simple deducing of logic. If there was no space available and everything was solid

Kepler show that motion and space is the same therefore all space involves time and all space evolved through time. By repeated duplicating space grows using time or motion of space. The Big Bang is the proof. But through time motion places discrepancies and that also brings synchronised harmony. That could not have been in place when the very first motion brought about the very first space. With all space coming into birth from singularity all space was filled as some space was spinning less, which had to bring about more space developing by heat expanding and other space contracted by reducing the space through material motion.

something had to give in. We have a solid state (freezing at $?^0C$) and we have a liquid (boiling at $?^0C$) therefore liquid is the addition to a solid where the solid state forms a liquid after a certain relevancy has been bridged. It is not the solid element of whatever element that becomes liquid because all elements are natural solids. After conditions have been bridged and heat surges the liquid engulfs the solid to protect the solid from expanding beyond borders. Some had to apply a more fore giving form to allow shape or form to come about. We still find evidence in this since some elements hold heat relating more rigged while other comply much more willing to adapt and comply with heat deforming the structures. In this there are matter that grinded the other antimatter into heat and destroyed most of the antimatter that destroyed form to space not connecting to form demanding singularity any longer.

By the same token dismissing brought on reducing of space-time where no space was available in any event. With no space available and a demand for space as big as the Universe self in size some had to go soft to allow others to remain solid and that set the course for the next stage to last the one eternity to come. There were to groups then formed and the two groups are still today present. The one group went on to form two groups and that we call the Big Bang event where space and heat separated for the third time. It was all about securing space by demand of a requirement for space. The proton and the neutron came about and one softer applying motion by duplicating part came in relation to another dismissing space to remain solid part. But further collaboration was required since the demand on space was far from fulfilling to satisfy the need that was developing. Proton /neutron clusters grouped as the successful cluster drove the unsuccessful particles to form more heat in the separating of space and the heat still went softer. As the dismissing proton clusters demanded space-time to stop allowing further overheating and the duplicating draw heat around the sphere the neutron formed as protection from the overheating. The heat formed space, which the neutron duplicated and passed on to the proton to dismiss as its part in the sustaining of singularity. The proton cluster $\Pi^2+\Pi^2$ entertained the space producing neutron $\Pi^2\Pi$ to engulf the sphere which now is called the atom with heat that relates as three from where the neutron taps heat to duplicate as well as send some of it on to be dismissed by the protons. The accumulating heat in the third relevancy then turned to uncontrolled space producing from that the Big Bang. In the 10^{-42} to time zero that is accounted for a development took place that took many eternities to take place. Today we find secluded space that we call atoms covered by space e call space and the secluded atoms group to form secluded space we call stars in between space we call space.

There then formed material $\Pi\Pi\Pi = \Pi^3$ and there formed time (matter in relation to heat surrounding material) $\Pi\Pi^2$ and then there formed the space to fit this lot into $\Pi\Pi\Pi = 3$, which with the singularity including Π makes the total value space-time in motion has being 10Π relating to the Π^3 bringing about gravity or contracting Π^2. The cosmos formed in equal part forming from the same substance, which might in some sectors be more or be less concentrated but remains equal. The entire cosmos resulted from one basic form of material (π^3) that formed motion ($\pi^2\pi$) as well as space $\pi\pi\pi$ to fit the lot but initially it was all the same ingredient. The substance forming the entire Universe came about from a natural occurrence: heat in solidity contracting heat in solidity that caused friction. It formed not antimatter but a natural occurrence: heat and it did not vanish from sight it became a natural substance: space-time. The results are the gravity and the anti-gravity we know.

By freezing **k** such a value came about and set the Universe into a concept other than being the one unified lump of not being anything of sorts. When creation came about and when the spot moved pi from the spot π^0 to the dot π heat came about, but being heat it froze solid by motion. As everything is heat, motion is what freezes the heat solid. The freezing action brought on **k** and **k** brought on the freezing of space through spin. By the extending off **k** did the Universe obtain space-time $\mathbf{a^3 / T^2}$. However, it had to produce **k** by freezing **k** into existing from where **k** produced $\mathbf{a^3 / T^2}$. This we have to understand about fusion. Fusion is freezing liquid to solid and space into eternity. Fusion is about creating a freeze in the deepest of heat and where all is engulfing in heat surrounding all, from that freezing must come about what is to apply the freezing we call fusion. It is this first action ever that has to repeat to establish fusion. It is in gravity attempting to secure the most heat under the prevailing conditions where gravity eliminates the most space to establish a freezing centre in creating fusion. But that does not solve the indicating action.

How did the Universe liberate material and heat/ space from singularity because with singularity comes an unchangeable eternal condition that is non-changing-everlasting in all conditions and aspects that is remaining in absolute equilibrium. This equilibrium maintains because all development extends form precisely a detailed equal equilibrium through out. Think about what brought the cosmos out from the eternal rest it was in. The eternal rest still maintains and is therefore our detection. What inspired the eternal rest the cosmos was in and inspired change to the state of eternal of eternal rest. What evoked change? That is the question the Atheist will never be able to answer but that too is the most basic and ever-lasting fundamentals of the Universe. What changed in this split second start before the official start? I do not wish to ponder on this matter in the book I am writing at this minute, as there are other books where I delve into this matter. From the deep freeze of creation came the Hot Big Bang and the 3D Universal displacement came about with the relatives being $10 \div 7(4(\pi^2 + \pi^2)) = 112.795$ and then a second one established 3D by introducing the six to seven sides Universe at a density point of $7 / 10$ $\pi^6 / 6 = 112.162$. There is of course a lot more information about this establishing of the Universe than what I mention at this point. The question is what made the Universe freeze to form the Universe in space and through time. It had to start with a specific reason applying, which brought about space-time. Once the process started there was no stop to it, but there is no chance that the initiation of the start was spontaneous by nature. With everything being in one spot, all within that spot was in a state of eternal rest. While all remains the same and nothing changing, what brought on the sudden change of everything, shocking everything from and out of the eternal rest? With $k^0 = a^0 / T^0$ all stood still in singularity and that factor is still with us controlling and generating the cosmos. It is there for all not to see and for every one to establish.

The effect coming from $k^0 = a^0 / T^0$ and about $k^0 = a^0 / T^0$ is beyond denial. The part of proof being beyond denial is because it is still there for all not to be able to see it. That centre spot in the rotating of objects is up to this point of cosmic development as incapable of having space-time and can only secure space-time as it was capable of at the start and as it will be capable of up to the very end. In only $k^0 = a^0 / T^0$ being present it was as stable as it still proves its stability. There were something external and from the outside of the **Universe** that was outside controlling the $k^0 = a^0 / T^0$ Universe that set the lot into space being motion. That control coming from the outside of the Universe is still a provable fact; which I am about to prove in this very book. That is the one aspect natural physics will never answer. What forced the cosmos out of the stability of singularity as $k^0 = a^0 / T^0$. There are two controlling cosmic principles it control of all. The one is gravity and the other is the light performing as the example of the epitome of antigravity.

There has never been any explanation offered about gravity being able to bend light except mathematical calculations producing the factor prevailing and the principle produced. Einstein declared that large objects producing massive gravity could bend light. Has any one got the answer to the question about light bending? If light can bend what is light then made of? If light is what science portrait about being a simple line running along a very flat piece of paper one might just find correctness in such an arrangement but that is if light is that simple to explain. Taking more facts into account, there is a lot more to light than such simplicity can ever cover.

Light can bend if light was a solid with the flexibility of a taint of liquid mixed in between. But if you think of the broad spectrum of information that one stream of photons release when coming through my eye about a Universe being so wide in its entirety it will take me several lifetime periods just to inspect the Universe close up where the Universe is hiding information about the most complex parts there is out in the Universe. It once more shows how simple mathematicians make concepts to portray the most complex issues in the most meagre sense. Human abilities are meaningless when brought in line with the totality of the cosmos therefore what we perceive as massive and what we perceive as little is in cosmic standard amounting to about the same. It is our norm we create bringing on our true Human incompetence in realising limits prevailing in cosmic standards. If large gravity can bend light all gravity can bend light by affecting the flow of light. Gravity does not bend light because light is not a solid that can bend. Yet again it is relevancies in space-time in differentiating velocity that change because gravity is velocity differentiation between two cosmic principles filling space-time.

By dismissing the space through which light travels such dismissing must lead to rerouting or redirecting the path light will displace space-time. Also by duplicating space and duplicating photons in the photons travelling across space the photons are duplicating as they travel by changing frequencies of singularity. The path they then follow, will affect the harmony and therefore the flow direction of light. The closer the light travel or pass the centre core of any object the closer it comes to the main gravity within the structure because the less space there is available the stronger the pressure of gravity is and that alone will bring on a greater effect as the more reducing of the space conforms to density in heat. Gravity is bringing space in relation with time through applied motion.

It is about space becoming reduced or redundant because of motion either by space moving or by matter moving in space-time through which it is moving. The duration of the time in relation to the space affected bring about the gravity applied. Gravity is space in relation to the centre while at the same time it is the centre coming in new relevancies to space surrounding the material structure. There are always two components to gravity in space being $a^3 = k\,T^2$. It is k in the way the centre stands in the space changing in relation to a centre a^3 but at the same time it is a centre T^2 changing relation of that centre to another centre k applying control. It is a constant re-matching prevailing centre influences on space occupied by material and space not occupied directly by material.

It is the time the space takes to bring about new positions in space occupied by motion and through motion that takes certain duration in time to move from point to point. Gravity is motion of space towards and in relation with a centre and the time is the period such motion takes while that centre is attempting to produce space by doubling space through motion leading space away from a specific controlling centre to another specific controlling centre. The time it takes to complete such attempt provides space the opportunity to double its status. Moreover, the time stands affected by this motion material creates to duplicate space-time by generating singularity and activating different locations holding singularity. Gravity is speed and speed in space in motion through time duration. Gravity is motion combating heat expansion and supplying space with space producing through motion providing the space the opportunity to expand while remaining in relevance.

At the start the gravity part invested heavy in material. However in space light came about since there is no gravity in space. Light is the attempt to establish motion not controlled by any centre and the time it takes to establish space between such a centre of space control and the light finding an ability to dispense the space by reactive motion. Gravity produced space by allowing as much as producing overheating with a feeble attempt to combat the overheating. But what is the meaning of heat if there is no cold to set the standard for the heat to become a value, which is then related to the other end, where such another end must be the limit in cold. The moment singularity produced space-time heat distanced from the cold factor. Space produced a cold base to have heat within. How did singularity part the shared principle of being the unification of heat and cold as a unit? Singularity froze in applying gravity bringing about particle separation within singularity. Heat and cold parted to produce frozen heat by atoms forming and captured heat by unleashing uncontrolled material that ended as space in time.

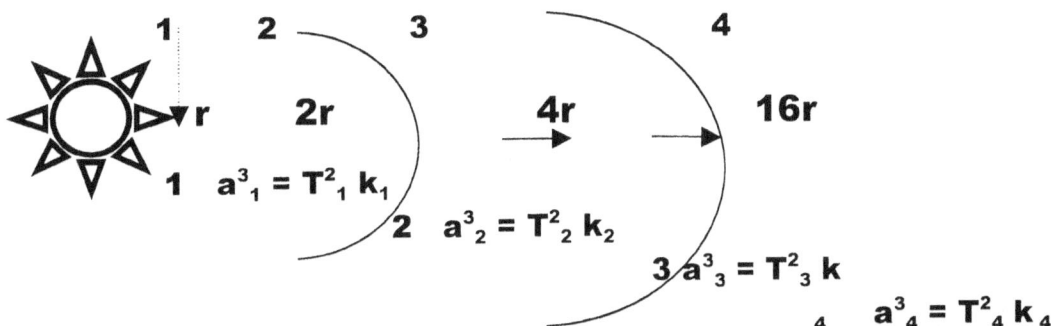

As a^3 increases, so does T^2 as well as **k** increase and with that the influence of gravity per space unit increases with the concentration demise of a^3. Why would that be and what are we missing? Light shows there is an influence out there in outer space, that redirects light's route through space when passing large gravity fields. It is about the relevancy of **k** influencing the a^3 to allow the T^2 of light to divert in route because of influences established by **k** on a^3 and slowing down or increasing the line diverting. In this measure one may also find the Roche limit applying, but to truly understand how the Roche limit comes in place and how the Roche limit works, one have to replace Kepler's factors with singularity and singularity extending being Π^3 $\Pi^2\Pi$ **and 3**. The relevancy brings about the reducing of space to a smaller factor each time.

In the investigation of light and gravity as well as objects and gravity, the mathematical rule of the invert square law must apply without question. However, according to the observation of Roche factor that is not the case. From what one gather through the Roche limit implicating two orbiting structure the opposite is applying. One must accept that although **k** proves to be as an indicator it is also much more when complying the thin influences brought about by singularity in the values carried on by singularity.

Even by reducing space as a result of gravity, space expands because **k** can never retreat to a previous position. Space will expand until time fazes out as being so short it has no influence any more.

The fact that k^0 moved to **k** produced a new value to **k**. By **k** having a new value T^2 will remain eternal and space will be in singularity. If **k** =1 then a^3 =1 bringing eternity to time with time being $T^2 = 1$. This shows that **k** have to increase in order to relive the Universe from eternity by arranging a time component. The expanding will forever bring about motion and the motion will kill the space.

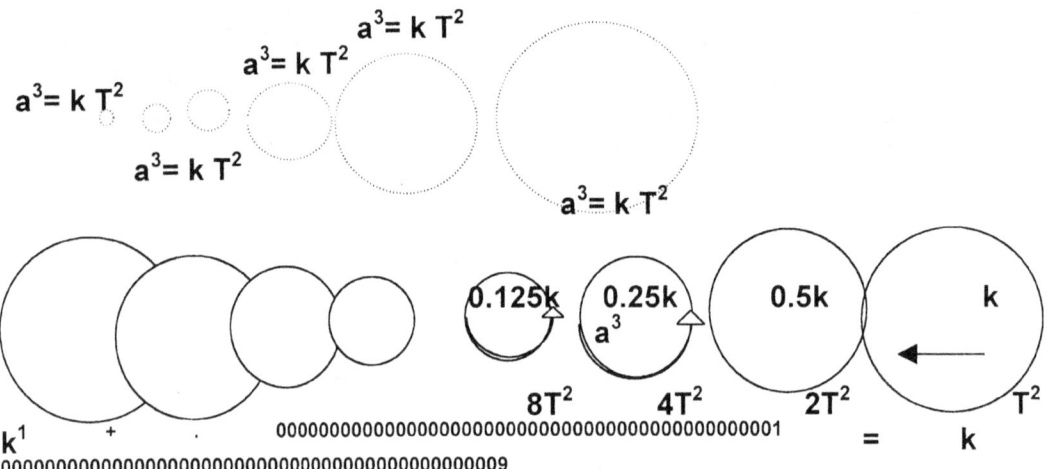

k^1 + . 00000000000000000000000000000000000000000001 = k -
0000000000000000000000000000000000000000009

$k = a^3 / T^2$ when space $a^3 = 10^-$ 00000000000000000000000000000000000000000001 then time being a square factor of the third dimensional space T^2 =.0000000000000000000000000000000000000000009

The space that the dominating singularity captures is also beneficial to the minor singularity because of the motion and therefore the committing of increased flow in space-time. I call this negative space-time displacement. In the space-time arrangement Kepler introduced the factor **k** increase by doubling k in the relation of $\mathbf{k = a^3 / T^2}$ which increases space by quadrupling a^3. Motion by T^2 reduces space by half and that brings **k** back to one and a half times what it was previously. From the fact that T^2 advances (multiplying by X), to a new location T^2 indicates a growth a^3 **X k** in space. In there is the Hubble growth but the growth is in the space claimed by material as much as the space not claimed by material.

Heat forces singularity to expand, which thereby is producing space. The expanding of singularity causes motion of space by duplicating space through time. The motion of space realises that space has to be in any position lesser than the time would demand if space were less motional. That produces space being in a shorter time in one place. But with time remaining motion the motion then becomes the standard and that reduces space because the motion is actually the duplication of the space and with such space dividing between periods forming differentiation of the time, the time differentiation brings about space reducing. By occupying two spots of space in the same time split into two parts the space has two divide in size even if it still grows by overheating.

The shrinking of motion creates heat. The heat brings about space. The space creates motion and the motion reduces space but the reducing is never to the same state it was previously because of material cannibalism. This arises from the materials of compromised heat and secure growth of matter that provides cooling. In this manner the moving of space will capture some of the space it is occupying through motion. The space will enlarge just because the space captures more space than which it previously had. By motion it reduces space and by motion it captures space it moves into. The motion will be there even if the motion is capturing space through expanding by means of overheating or whether the motion comes about by space duplicating by using time reducing thereof. Motion of space is as critical to space as space being space and space is motion that is forming space

Space a^3 is reducing through the motion of space by duplicating space in times motion and the reducing of T^2 by dividing time by half will have k as the constant. But afterward the scenario changes back, as space then is the constant with time reducing k. It comes about as the Earth spins around the Earth axis. I call this positive space-time displacement

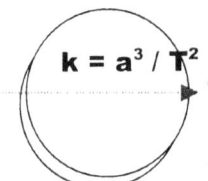

$k = a^3 / T^2$

Space a^3 is reducing by the motion of space with the implementing of T^2 having k as the constant. It comes about as the earth spins around the Earth axis. I call this positive space-time displacement

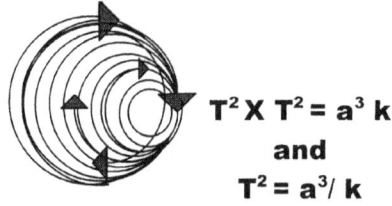

$T^2 \times T^2 = a^3 k$

and

$T^2 = a^3 / k$

Space a^3 is reducing by applying motion with the repositioning of k, having T^2 as the constant. In the reducing as well as forming duplication and in the reducing singularity captures as much as conforms some heat. This comes about in the form of securing singularity since growth is a natural dynamic part of gravity as Kepler's formula indicate $k = a^3 / T^2$.

Proven by Kepler's formula, we see that space will grow by the single dimension because space is material that is captured in terms of one dimension of space when each cycle is completed outside the space occupied by material in space. That process involves the phenominon, which we call gravity. It comes about as the Earth spins around the Sun axis and the space between the Sun and the Earth captures by the Sun as well as the orbiting Earth. After completing every cycle, the atom secured one dimension of heat by reducing one dimension of space. The fact that one dimension of heat removes one dimension of space and conforms that to singularity, it can only be the benefit of singularity as it secures the growth by the single dimension. $k = a^3/T^2$ shows a growth by the single dimension which can only contribute to securing singularity because singularity is the single dimension.

The second one also fits in the singularity influence on the Universe.

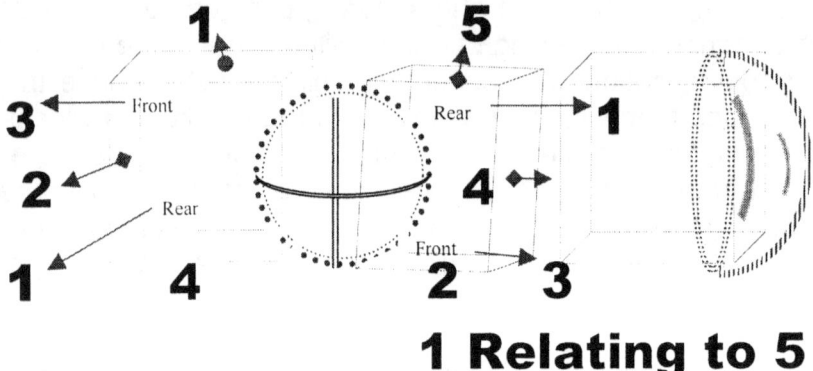

1 Relating to 5

The space that the dominating singularity captures is also beneficial to the minor singularity because of the motion and therefore the committing of increased flow in space-time. I call this negative space-time displacement. In the space-time arrangement Kepler introduced the factor **k** increase by doubling **k** in the relation of **k = a³ / T²** which increases space by quadrupling **a³**. Motion by **T²** reduces space by half and that brings **k** back to one and a half times what it was previously. From the fact that **T²** advances (X) to a new location **T²** indicates a growth **a³ X k** in space. In there is the Hubble growth but the growth is in the space claimed by material as much as the space not claimed by material. The factor **k** could never move back to the original because singularity shifted into that slot. If **k** moved back to the starting value **T²** would stop and the one half of space **a³** that comes about through motion will vanish taking space **a³** altogether with the vanishing.

The relevancy falls on motion **k** and spin **T²** bringing about space **k = a³ / T²** by reducing and expanding space. Overheating brings about that **k** doubles. With k doubling space becomes four times larger. Then time applies motion and divides space in half. That brings about that every instant motion applying space becomes four times larger and then reduces by half. But such motion **T²** is at the prime of space **a³** and the eternal beginning **k⁰** of time where time just moved from eternity **T** one step to **T²**. It is precisely where Einstein placed the strongest gravity. It is next to singularity. It is where space disappears and the motion that is providing us **a³ = T² k** time begins. It is where $\Pi^0 \rightarrow \Pi$. There cannot be light without gravity and there cannot be gravity without light but that is on condition that the light we perceive to be light is only the symptom of what the Universe use as light confirmed. Light is not what we think light is but explaining that will be too extended to explain in this book. Light is heat concentrated to a form almost material and space is heat dismissed. Therefore light, heat and space is the very same thing and all form the extending of the antigravity that is applying. However, gravity is more the redeeming and the re cooping of light where light then is antigravity.

Light is space concentrated in motion at speed and gravity is reforming space by motion forming speed. Gravity is speed and light is space at speed. In drawing a most basic picture of light passing the gravity lines extending from any structure, I felt it was most insightful that the Brainpower in cosmology was not able to see why light does not bend in the presence of increasing gravity.

More surprising was that I found the mathematicians had to call on Einstein for advise on a most ordinary problem. Light does not bend when passing large objects. It is Kepler's formula applying, and the evidence is clearly in front of the searcher for truth. But one has to go back to Kepler to re-apply the principle what Kepler formulised and change the significant from Newton's significance.

The Roche limit 5/2 becoming = ($\Pi/2$ X $\Pi/2$) = 2.4674 as singularity interferes.

Creation started with the Roche limit and in conjunction with everything else came about. In the Roche limit the space factor provides space to a solid structure and therefore the value of r is replaced by the value of Π bringing about a square in half of Π. The cube holding 5 to either side removes allowing the extending of Π to indicate position to space. The space between the spheres divide in half, but because of the extending of Π and not applying r as ordinary mathematics will suggest where Π replaces r the singularity extending from Π^0 will be half of Π in the square of $\Pi = (\Pi/2)^2 = $ **2.4674.** In this lies the dynamics why planets have a positional (be it rather a dimensional) relation of 7/10. Half of the five of the Lagrangian points is the Roche in conjunction with singularity. With singularity coming involved singularity will enforce the value change to fit Π.

When a sphere establishes as form it took on the laws of Pythagoras. The Laws of Pythagoras to provide a centre principle all lines running through the centre will be effectively related by groups forming 180^0 and 90^0.

We take a line running between two points as being 180^0 and the rest of the explaining is saved in the accepting part of mathematics. Any one of the two points where the line starts or ends at, is a point in infinity. The start and the end depend on the viewer putting the relevance to favour the side of choice. That puts the point of end or beginning in the spectrum of choice and not fact. Any direction is as equal as all other directions.

The TITIUS BODE Principle Inside the sphere

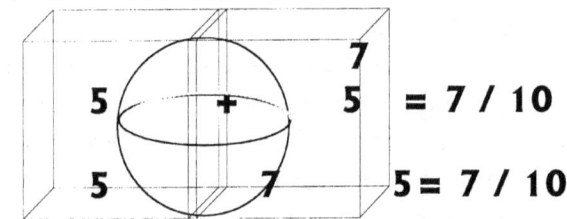

Space-time is a four dimensional position of the universe where the position of an object is specified by three coordinates in space and one position in time

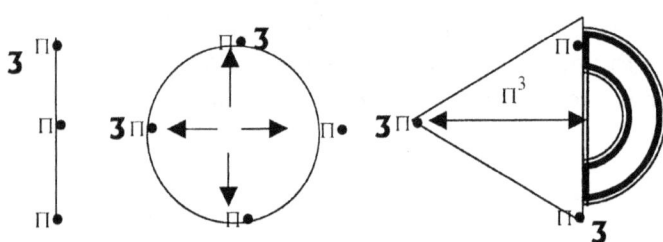

Because of the Coanda principle that is assembling the three dimensions in space in a single line dimension of relevancy to the square of motion, it is the double square as well as the square in relevance that keep space time in a relation with both 180^0 as wall as 90^0.

The eternal relevancy of three that is committing singularity to a space les motionless line The motion is committing a time, which are four to a relevancy of three spinning to one dismissing bringing the three into line with three acting in motion and committing time to duplicating

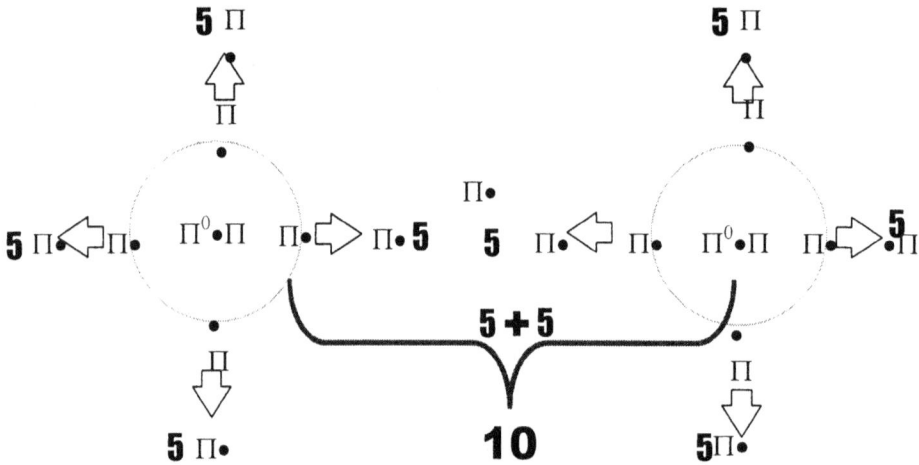

If it were four points serving the centre it would form a unit by means of being a solid combining unit, as does the Coanda unit form with motion concentrating a centre. It would then become the Coanda principle with the three points activating the space factor. By employing a fifth position outside the four of time the rules change and the gravity becomes much more forgiving allowing five loosely connected objects still performing as liquid by having motion, but not connecting to the solid centre. By using five positions not activating the three in the centre it becomes a loosely connected unit with less stringent gravity, whereas if it would form six positions with the seventh being the centre the rules would insist on forming another sphere. But having four plus one the unit remains a disc shape and only connecting as a time delay and not as a full time compliment as it would have done when it kept to the Coanda four or forming six with one centre spot.

When the initial and ultimate break came about where singularity moved from the spot π^0 to the dot π singularity established space in motion to the value of π and formed Π as Π parted from the rest. When looking at it from the other side there are ten points in reference to singularity. There are the four factors of time and by moving one point away the dot released space from time. In one dimension after time in the four came singularity by producing the motion of space in five, but the

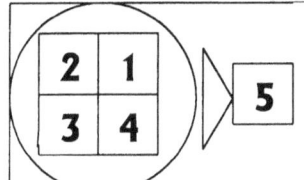

Time holds four positions and the first position in the dimension of space outside time is five.

Universe is about duplicating and repeating what was again and again afterwards. The second dimension the duplication involved five (see the Lagrangian explanation) above and below which is the half of ten and as ten rotor around an axis of one on the one side and a relevant .9991 on the other side bringing a representation to singularity which is always being present and when dividing this space not including the singularity by the four time factor an individual five is established. By adding the lot the total forming 21.9991 and as this is rotating in the relevance of the previous dimension where the sphere holds seven point one see that Π once again form. But this time Π is always rotating from one side to the other and therefore again the Roche limit proved to apply as the one side had ($\Pi / 2$) that had to duplicate on the other side ($\Pi / 2$) X ($\Pi / 2$) = ($\Pi / 2$)^2. My discovery of this let me to believe that all cosmic laws played a part as they repeated time after time with the developing of the cosmos through out all eternities that might come about. That meant the square of space (5+5) = 10 in relation to singularity on both sides of the divide forming 1and adding that to the eternal 1 minus one in relation to the square of space being 10, which is on the inside and on the other side of the divide as singularity are forever about reducing which means it is .9991 and that forms once more 21.9991. When placing that in relation to the sphere having seven markers the square of space becomes in contact with material forming ($\Pi / 2$)^2.

In my book " Starstuffin" I explain the principle, which I called the Lagrangian atom structure and show with proof that atoms indeed follow the connection of proton clusters using the manner, as does the Lagrangian material compiling do when forming cosmic star units. It has nothing to do with planets though! Again I have to caution those in search of the wonders of planets that there is no such a natural phenominon as planets in the sense as we know planets.

There is older stars contracting heat with micro stars in the Lagrangian formation benefiting from the contracting, but it is micro stars such as our five "gas planets" that is far from planets but developing stars. As of the rocky solid structures such as we have to our benefit of use and on which we live, that there is not as a natural phenominon anywhere.

In <u>*"The Seven days Of Creation ISBN 0-9584410-4-9"*</u> I prove how it came to be that such unnatural phenominon as our solar system is, did come into the Universe. However, having them and those as a natural condition is a bit Newtonian and very far-fetched. Should there be any person insisting on planets being, those persons then should indicate what rules did apply to form planets and what purpose planets has other than fitting the Newtonian imagination with giving life another sporting chance to be elsewhere. There are cosmic rules serving specific purposes and adhering to the rules serves that purpose. Putting life in a pivotal role in the cosmos is very "Einsteinium" and super Newtonian and later on I explain why it is so typical of those blessed with the "Einsteinium" Mathematical genius to fall into such a trap. However more of this a little later...

5/2
Five sides divided by two spheres.

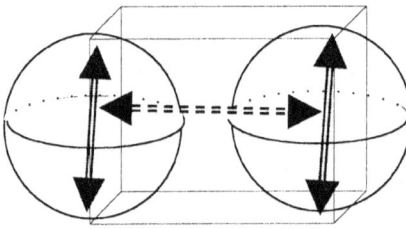

Where Π extends to lock onto the next sphere's extending indicator, Π has to connect to Π forming the square of space and translating that to the half of Π being (Π/2)².

The Roche limit 5/2 becoming = (Π/2 X Π/2) = 2.4674 as singularity interferes

Time holds four positions and the first position in the dimension of space outside time is five.

But this time Π is always rotating from one side to the other and therefore again the Roche limit proved to apply as the one side had (Π / 2) that had to duplicate on the other side (Π / 2) X (Π / 2) = (Π / 2)². My discovery of this lead me to believe that all cosmic laws played a part as they repeated time after time with the developing of the cosmos through out all eternities that might come about. That meant the square of space is double five by two (5+5) = 10. This ten is in relation to singularity being on both sides of the divide. The divide also has two implications forming 1 and adding that to the eternal 1 minus one in relation to the square of space being 10, which is on the inside and on the other side of the divide as singularity are forever about reducing which means it is .9991 and that forms once more 21.9991. When placing that in relation to the sphere having seven markers the square of space is exposed to material forming

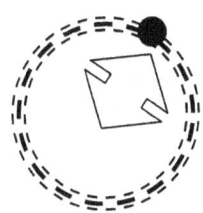 This gave birth to the relevancy between contraction or recouping space-time that is vested in the four times square of the proton and the fifth position the electron later employed. Today the entire Universe is in principle bonded by the layout the Lagrangian system of a fifth point outside time used.

Independent singularity = .99991

Controlling singularity = 1

In all units there is a singularity seeking independence in relation to singularity elected seeking dominance. From one side and any singularity there will be present in the unit one factor of singularity carrying the value of 1 but also there will be one divided by space square singularity absent because only one singularity can apply to the unit in dimension. Therefore one tenth of the space is absent where singularity is one and holding .9991 valid as part of the other side of the Universe it is attached to but not connected to. With the unit being connected to motion the motion will stand related to seven from the one side as well as the full ten from the other side on both sides of the Universe. That relates as ten in space-time on both sides of the Universe (10 + 10 = 20) plus one factor of singularity present and one factor in the tenth not present (1+.9+.09+.009+.0009+.00009-.0001=1.99991)

The Lagrangian ratio is normal dimensional particle layout brought with development through time as a result of the Roche principle. One can see the Lagrangian system by looking at the elements developing clusters of five forming such characteristics of similarity in groups. Looking at how singularity applied connection in 3D there will always be five relating to a centre where that centre carries half of the value. This forms the principle behind the Lagrangian system. The will be singularity taking a centre with one point to the top of the centre and one point to the side of the centre. The centre then becomes the one side in corresponding to the side the centre takes. From the centre another connecting will be to the bottom the front and the back. Every point will form a position where it will support the centre in displacing space by providing heat in an attempt to secure the prevention of overheating. The function of the linking of five to a centre as a group effort is with the group work individually to secure the survival of the individuals forming the group and being the group but as a group as well as the securing the group as a unit by a mutual concentrating of space.

By halving this effort in the Roche proves the motive behind the forming of the five in the Lagrangian points because of the destruction or development that space less ness then in Roche bring about. The five to one forming the centre of the governing gravity by establishing a principle singularity such a centre improves the gravity effort by all taking part to dismiss heat and create a concentrating heat flow to the centre whereby the group as individuals and as a structure will survive through mutual gravity. This is the effort of the whole cluster in underwriting the centre spot control. It is plugging five spots same as electrons do and in that accelerate the space flow by reducing the space flow as a motorcar carburettor does. In producing mutual support it underlines the individual support required by all participants to prevent future overheating or antigravity. The extending of space collecting and the extending of space reducing prevents overheating by establishing matter in six by connecting matter in six to a centre one. That centre one has the factor of less than one forming the Alfa singularity to the cluster forming the gravity.

Where the Roche factor is the singularity influenced half of the Lagrangian system, the Titius Bode is the dimensional duplication of the doubling of the Lagrangian system in space occupied and space not occupies by material.

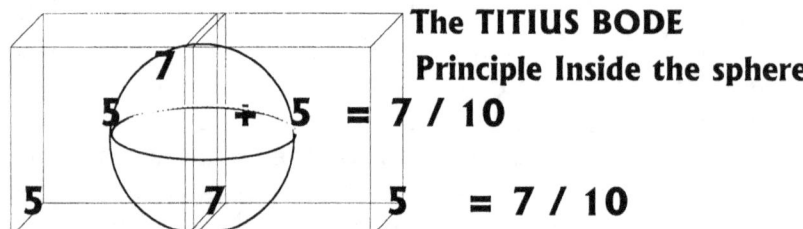

Space-time is a four dimensional position of the universe where the position of an object is specified by three coordinates in space and one position in time

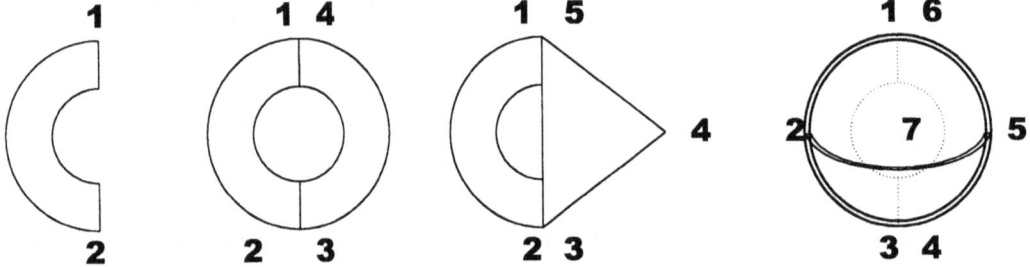

The whole principle of formation is vested in form and relevancies where one dot turns the whole concept around. That is why the Pythagoras principle still forms the basis for all mathematics. It is because before mathematics there was only form

When concerning a sphere by establish the Laws of Pythagoras to provide a centre principle all lines running through the centre will be effectively related by groups forming $180°$ and $90°$.

We take a line running between two points as being $180°$ and the rest of the explaining is saved in the accepting part of mathematics. Any one of the two points where the line start or ends at is a point in infinity, The start and the end depends on the viewer putting the relevance to favour the side of choice. That puts the point of end or beginning in the spectrum of choice and not fact. Any direction is as equal as all other directions. In view of that one must remember the fact that singularity cannot move. In order to relocate singularity requires a shaft in relevancy where the line providing the space that moved the distance to allocate the space it represents has to recline in the status it has with singularity. Then singularity has to re-establish a new alliance in rotation $T^2 \times T^2$ in order to provide the new forming of **k** as well as provide singularity with a new concept of space-time that shifts from and to singularity as the new space that singularity awards in $k \times a^3 = T^2 \times T^2$. If the rotation of a^3 is not quite the same in value as the rest of a^3 space, an unbalance will occur that will come about as a distortion. If there is one spot in a^3 that is bigger than the rest such rotating of that spot in synchrony with the rest of a^3 will demand a larger k in place of that position and that will bring about in that a bigger space ($k \times a^3$) must come about when time $T^2 \times T^2$ allocates a new position relating to that specific point. This breaking down and rebuilding of space-time happens as often as change come about but the effect of the demand on a new k depends on the ratio changing that will come about when a^3, which represents the overall space that is all equal is allowing a new space a^3 position to the heavier space a^3 that comes in unbalance. But that very same principle will be in place when all material is in perfect balance and motion is harmonised.

The Lagrangian Principle

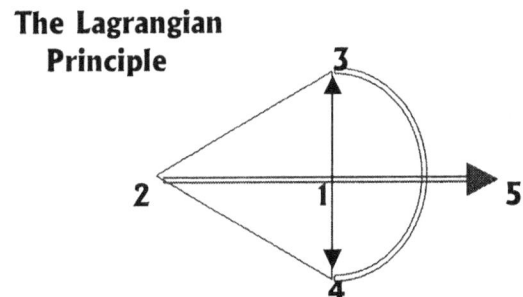

The Lagrangian Principle combines the triangle, the straight line and the half circle like a glove. It is just the reason why it is used as building method through out the cosmos.

That proves no point can be static or constant, though it may seem that way to outsiders. Although matter is material, material can also be opposing material and moreover form its own opposing matter at the same time. Which means that if material is being anti-material such a statement does not necessarily means that expressing something as being anti, anti must be expressing a situation where Anti material is devouring the material it is anti too. Being and being anti also can mean the two is in support of each other as much as opposing each other since the one that is not anti cannot be if it is not there to oppose the other which then is the anti to be. By opposing in being anti it is in supporting the one the not anti. The degeneration of structures are more likely to occur with overheating than with opposing because the opposing of spin in material causes friction to come in place, but it is the friction part that destructs both material in opposing or anti spin. Where one may be more vulnerable to friction such vulnerability will lead to destruction but in such a case it is not the opposing or the anti that brings on the destruction. However the friction part is taking π to π^2. Revaluing Π to Π^2 will bring about a new contact point where Π meets r forming another relation in Π^2. Every time material swap sides it also qualifies as anti matter to matter because if it goes out of orbiting rotation frequency, it has the ability to collide with the same matter it forms union with but is located on the other part of the spin.

It then becomes in a situation where Π **revalue to r.** I refer to r as the distance which Kepler referred to as **k** which is just another line running from the one edge to the other edge through the designated centre. This proves that my method that I used to locate singularity the first time was the correct way since that is the way nature goes about improvising for time. **Time is the changes in relation** where Π **contacts a different r** not withstanding the many r points there may form because **every r constitutes a different value** to the Universe through other ratios and relevancies brought about **by heat and light. Time is the duration it takes Π to rotate between any two given points of r** and therefore must always amount to **a square (T^2)** moving from point to point through the **cube of space (a^3)** in that **duration of time (k)**. With that it proves **Kepler's a^3 (space) =T^2 k (time in the instant of motion)** but motion must continue through a specific value in space where the space-time is maintaining relevant equilibriums throughout singularity connecting. That provides the motive to have another look at how cosmos produce a straight line from time long before mathematics came into place.

A straight line, triangle and half a circle will always have equality in dimensional capacity providing equilibrium being 180^0 because each one shares a common denominator in singularity to the value of Π. As the straight line averts a zero it holds another straight line in place to set about such an averting where the two lines will always carry a relevancy in elation to progress (the triangle) and a common denominator in the start from singularity. This concept we apply as the graph or the vector. By going back to a line, any line and all lines, the line is a connection of dots in infinity with every dot and all dots not excluding one possible dot being left out, running from one specific to another specific and avoiding zero in that manner. At every point in infinity it dips into infinity coming out on the other side of infinity by choice of direction and the direction is unforced and change presents any angle including the straight line, which incidentally is just another angle. We have to see the Universe as optimists. The Universe is not filled with nothing but filled with possibilities.

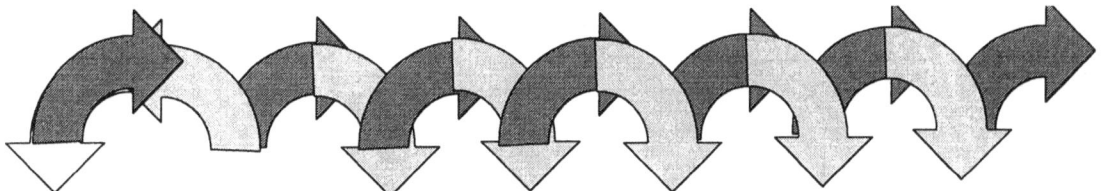

Following the flow of any line such a line is an extension of the previous dot in infinity to the next dot in infinity without any ability to skip or bypass any of the other dots in the connecting line. Any direction change including the remaining of travelling in the same direction is in relation to a line travelling all being the very same. Change does not affect the line.

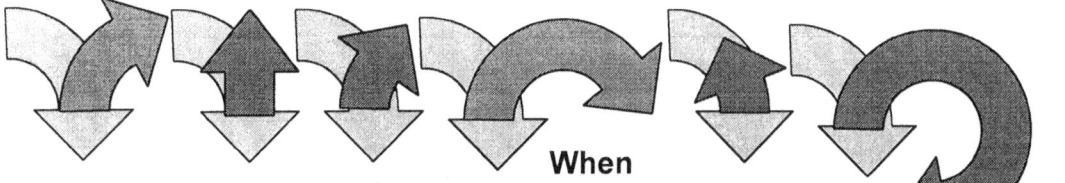

When connecting to the dot representing infinity the flow can be in any and all possible directions, including in the same direction. We all live in a graph as the universe with all in it is nothing less tha*n* a three-dimensional graph flowing according to time. That means in the case of Pythagoras the mere fact that the line shows changes in direction does not implicate or affect the line as a tool of mathematics. Whether the line changes into a half circle meeting at the other end again or meeting in a triangle in forming a half square by joining the point where it began, the result still indicate a line flowing between points.

It is because of this there is no true strait line. There is only singularity in the infinite connecting and any point connecting is a redirecting of what connects. It is our senses of dimension that will have us give it a line value but that is because of the dimensions we are in search of.

In the Roche singularity apply all three components

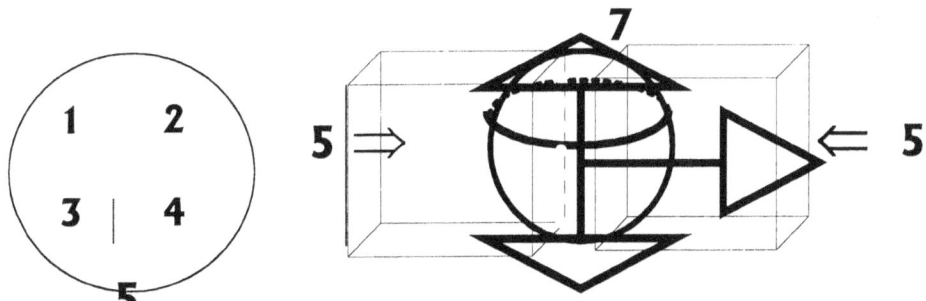

In the Roche limit the **straight line** forms part (1) and the **half circle** is part (2) and the **triangle** forms part (3) to singularity (4) Holding 5 points outside singularity

The influence of singularity as the extending of Π into space links Π² to r and forms 2(5)+2(5) =10+10=20

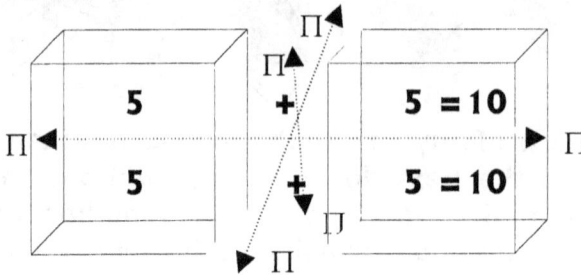

From the position of singularity there are different values in Π where each indicate a position. The value it represents being ΠΠΠ, Π³, Π², Π and Π⁰

From there it influences singularity in the triangle flowing through to the half circle. It is an interaction between circular and linear motion as the value of Π continuous past Π² (at the end of the solid) and every cosmic structure holds an individual and specific singularity. The field where Π extends we call the atmosphere having a value of 21.991 / 7, which is Π.

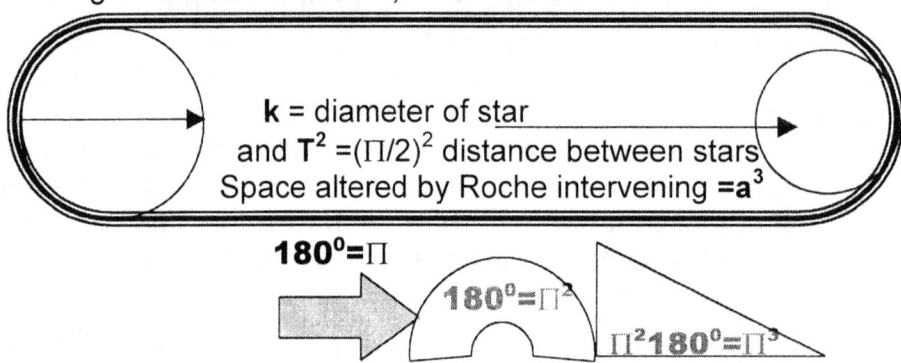

The triangle, the half circle and the straight –line has two things in common, they share 180⁰ as a mutual value and they are part of singularity. This value came about even before mathematics came about. It constitutes the drive engine of the universe.

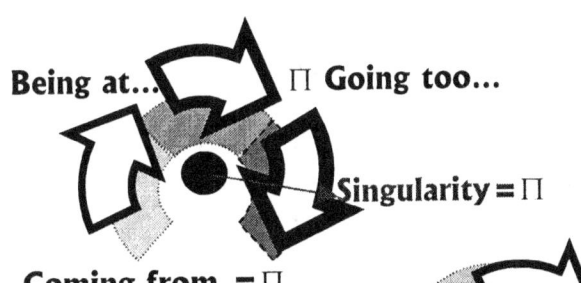

Being at... Π **Going too...**

Singularity = Π

Coming from = Π

Singularity: a mathematical point at which certain physical quantities reach infinite values for example, according to the general relativity the curvature of space-time becomes infinite in a black hole.

With no line starting from zero because there is no zero as a mathematical fact, then all particles hold the point of infinity and not merely the Black Hole of nothing,

From that argument one may conclude that all stars will become Black holes depending on the gravity increase they may generate.

Where singularity holds position in the centre of any and all rotating objects as a value of Π merely applying movement (in the form of atoms) qualifies all matter to be space-time. It does not only fit the description of space within Black Holes but it fits all stars where singularity becomes part of all the stars from the minute to the largest cluster of matter.

Through rotation encircling the point of singularity and matter is (1) coming from, (2) being at, (3) as it is going too in one movement in relation to the specifics of the centre point being singularity, all matter then qualifies to form space-time.

That **confirms our vision that expresses Kepler's formula a³ = k T². a³** holds threeΠ³ point in time **T²** from point to point Π² relating to **k¹** Π.

Firstly singularity expands from Π **to the seven positions it holds in material.**

It is quite significant to note that there is never free energy available any place in the Universe. The fuel driving the cosmos is overheated singularity that now forms space, which back then could not maintain their structure by avoiding friction and by freely applying gravity. In this sense energy is one part of the cosmos that was structurally destroyed as it now moves with motion bringing about gravity to the other part of the Created Universe that can secure the heat from a gas (space) to a liquid (fluid atmospheric heat) to a solid we find present in maintaining the elements within the visible Universe. Material is just frozen heat and space is gas heat leaving fluids as liquid heat. Transforming heat from space as the gas through fluid to frozen material is the energy the Universe run on. By moving heat from one sector to the next a drive engine comes about that drives the Universe from eternity through eternity to eternity. The Sun for instance liquefies (producing light) gas to photons and in that produces energy by applying motion as space creating and gravity.

Using the concept that gravity applies Π as the circle factor Π as well as Π² replacing r², which the replacing by Π that brings two values as Π and Π². That

I found is the case with gravity and will be apparent when explaining the sound barrier as well as the Four Cosmic Pillars. In order to create a distinction I remained using r as the indicator of the cube or non-circle that has vacant space and by vacant space I refer to non-solid structures. In the solid structure I use Π as a value for reasons that will become apparent in due time.

Firstly singularity expands from Π to the seven positions it holds in material.

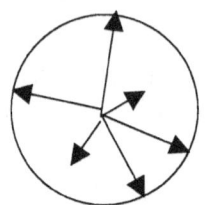

Singularity explodes into space having a value of a^3 as Kepler introduced or Π^3 if the singularity value is used. But that does not introduce the six-sided Universe with the dimensions we came to know. In the process of making contact the Π^3 becomes another Π^3 by abandoning the one side of the Universe and because of the moving of singularity Π^0 through Π^0 a situation develops where the seven points representing Π^0 receives another three pints in motion as the points cross over to the other side of the Universe. Remember the Universe in singularity is a flat Universe with no sides at all and the slightest motion of any point provides the point another Universe to move to. It is not the 180^0 we see as a straight line and at the same time it is the whole issue of 180^0 in all three forms of Pythagoras that forms the issue. The slightest motion becomes a most deliberate crossing of borders since there were no borders to cross and the movement provide the borders to cross. Then by motion of seven dividing into ten and on the same subject the ten dividing into seven as the crossing is implicating singularity $\Pi/2$ the ten and the seven form an alliance that brings about the value of Π^2 in relation to singularity as Π^0

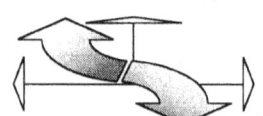

The Roche singularity applies all three components of singularity

The motion discrepancy between the points holding singularity controlling the Earth as it does throughout the Universe in relation to the singularity in control of what we named mass will establish the Roche limit although in another value.

The universe started off as a spot, which was a solid dot without shape, size, form or sides. The universe was wrapped in the single dimension. It was where no mathematics and not even a thought could reach because what the Universe represented at the time still forms our most inner basics we are unable to reach because deep in there it is the I that makes up the me. Later I explain in the second part what is and what are not we the "I" forming "me" and how we have to distinguish between what I am and am not.

The dot came into the age of dimensions. At that point singularity broke free but the Universe kept in tact, secluded from the future in a single dimension where on the fringes of the singularity Π formed. There was then material, which was wrapped in singularity and there was the rest hanging on the fringes of singularity as singularity. There was a relevancy between form and singularity where singularity was the only aspect with form being in singularity. Only Π indicated form outside singularity but from one point having Π the centre of the Universe came about establishing such a centre and the centre use the rest that was established as being space-time indicating a centre.

In this it is clear why the Titius Bode ([10 + 10 + 1 + .991] / 7) and the Lagrangian 5 \\ 1 systems part their ways when applying the different processes they hold. With all the differentiating, the observer must also consider the dual massage that light uses in travelling through the vastness of universal space. The thought of nothing is just what it is, a thought of nothing and although it is in the human mind common nature to present nothing as a value in the recalling of something, nothing is a presentation of the figment in the human mind.

There can be no number such as nothing and that was (possibly) Newton's biggest error. Nothing represent non-existing and that is just what nothing is, it is non-existing. The Titius Bode influence in a manner that on the one side holds the matter-to-matter relation of 7+7/10 whilst on the other side during the same time holds the space-to-matter relation of 10/7 forming equal and opposing values. From this the orbits of cosmic structures are always oval favouring the singularity dynamics of the one structure at one point and switching the favouring to the other structure on the opposing side. Because the structures can never be equal in size (singularity will not permit that where the Roche principle will intervene) the shape is always "off centre" as well. This influences coming about as the Titius Bode principal manifest in other ways proving Kepler's time relation with space through distance from singularity controlling the factors.

A child's toy holds the mystery to what the greatest minds in the world misses. It shows how singularity is charged into "life" from "nothing" by the dynamics of motion.

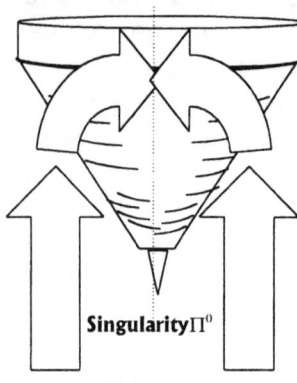

By rotational motion the top creates a line confirming singularity running down the line and by generating the line the line charges gravity. The gravity is what drives the top as the top and as long as the top spins. There is an influence generated by the spin of the top that keeps the top upright while the top is spinning.

The line is generated but the line is far from magic. The line is where the centre of the Universe is which the Universe is then that what the top fill by particles from the line to the edge of the sphere. The particles in motion generate motion by electing a centre from the centre of every particle in the spinning top. Such an elected centre becomes the centre of the Universe as far as the top relates to a Universe because all the atoms in motion elect the centre of the Universe.

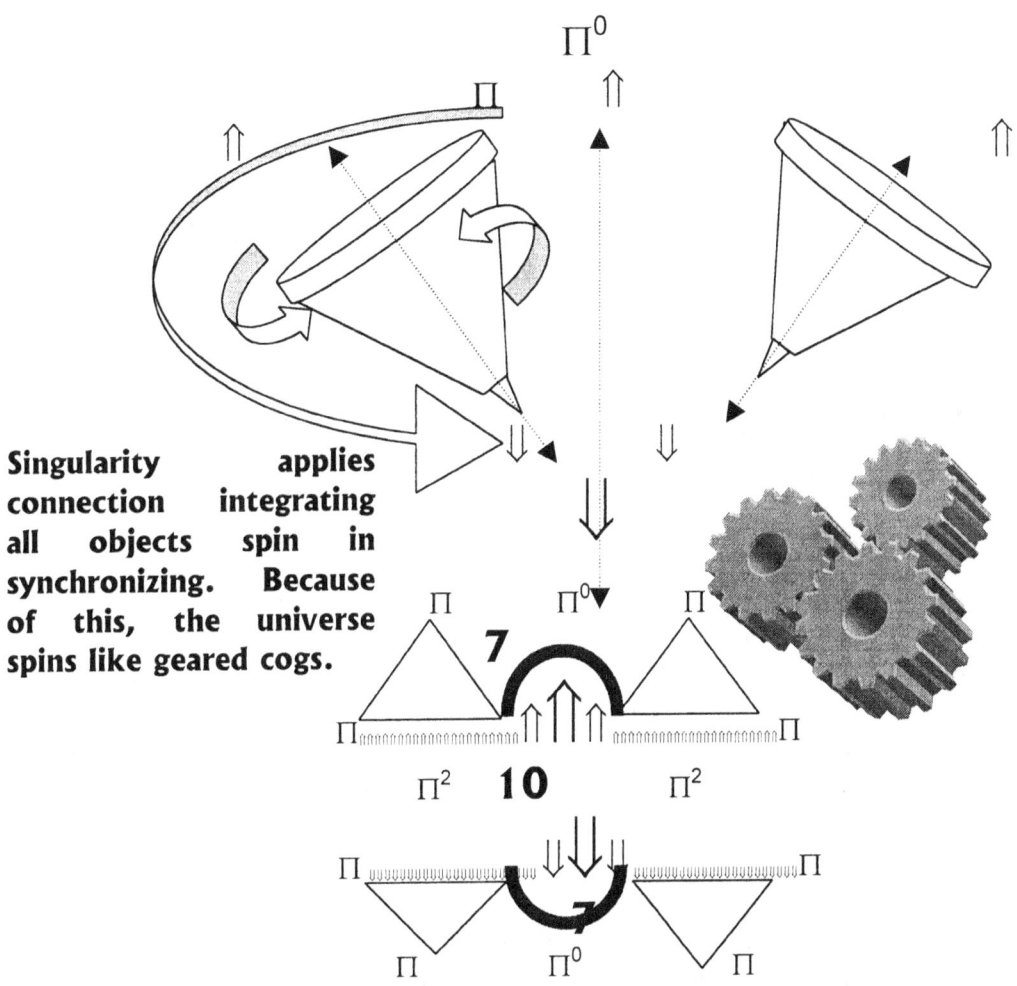

Singularity applies connection integrating all objects spin in synchronizing. Because of this, the universe spins like geared cogs.

The dynamics can be traced directly to the atom and the atom therefore works on the principle that the Coanda principal prescribes.

$(\Pi^2 + \Pi^2)(\Pi^2\Pi). = 7$

$(^2 + ^2)(^2^1) = 7$

$(\Pi^2 + \Pi^2)(\Pi^2\Pi). = 7 + 3 = 10$

7+ 7 + 7 = 21 + singularity in one dimension below +9991 = 21.9991 and the next sphere is relevant to the previous sphere

The way the top spins and the way the atom forming the unit which is the top establishes a six sided Universe of its own initiating by spinning proves two points:

1. The motion establishes a unit of motion in bringing the Coanda effect into reality.
2. The Coanda effect coming about, establishes an elected centre serving singularity to produce an independent six-sided Universe in the established six-sided Universe.

The value of space becomes **twice times ten** being a combination of space on both sides of the border plus **singularity applying** to form an inclusive unit with **three factors** and **a presence** combine as **a unit** of **twenty one point nine** that relates to the **seven of matter** and that reproduce **singularity to the**

value of Π as well as **forming** Π² It is singularity that is replacing space by replica of motion.

The concept in own merit is rather simple to understand and I also, find is very difficult to explain. By the motion of particles within the confinement of any centre of a sphere, such particle motion increases the heat levels by reducing the space factor and that increases other types of reaction and more gravity in the centre of the sphere.

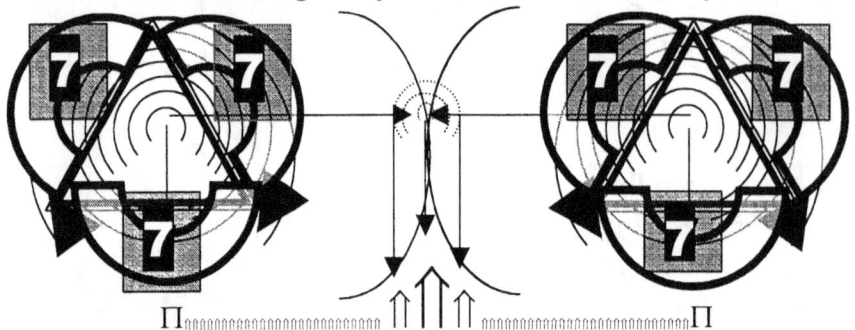

The relevance at first were not space as we now see space in the 3D, it was relevancies of motion duplicating the seven. It is three by motion of where it was, where it is and where it will be because without one not one is valid. By duplication, the atomic relevance came about that was the forerunner of the atom. Every part of individuality in singularity is by form of π a sphere.

Newton might have seen the rotating cycle as accomplishing "nothing" but modern times showed that the very opposite is true. Every possible driveline that transmits energy must come about through such a line forming a rotation about a very specific and included centre of rotation. No driveline can come about is a restating process drive around a precise axis in a straight line. By rotating the driving shaft that sends the power down a straight line is the only manner in which a drive can be established. This fact comes from the principle Newton could name as gravity but failed to see because Newton was unable to recognise the motion correlating with the straight line, which forms the basis for gravity. I say this because he denounced the existing of the importance of rotation principle by announcing the value after completing one cycle delivering an accomplishment by the principle of achieving zero. He might have foreseen something, which he named but the principle Newton killed the meaning of the principle as he failed to recognise the manner in which gravity works. By completing a cycle in rotation the cycle establishes singularity that comes about through the motion. The motion claims a centre and space using the process that carries the name of the Coanda effect breaks down space-time as the motion elect a centre that brings focus to such an elected centre. The motion re-establishes a governing singularity by re-installing a new elected centre that controls the space coming from the space-time rotation. It is important to recognise that by motion a new Universe is charged by motion still within the confining of a domineering singularity. The breaking down and rebuilding of space-time does not only apply to the driving of electricity but stems from singularity no being able to move but still forced to apply rotation on gravitational demand. A simple throw of a ball or the spin of a top or riding a bicycle is motion enough to generate a new singularity in charge of a newly established Universe. Driving or transfer of energy can only come about through the rotation that Newton gave the principle value of zero or "nothing". Blame me for "anything". We can see this by once again witness the spinning top applying motion in accordance with the establishing of the Coanda effect. It is necessary to remember that any motion and all motion including the driving motion is reforming gravity, countering gravity by forming gravity that is linking a generated singularity to form a newly established gravity and is therefore a fact that all machine-motion is gravity manifesting in some form. Just as the spinning top shows independence. So does the space place the Universe in a double sided with two halves that at the same time is also in a balance.

Once again the following proves that mass is a result of gravity and gravity does not come about through mass, because by using a new a^3 it can establish a new k, which will convert that gravity T^2 to apply to the new a^3. In this manner does motion release the spinning object from the Earths containing gravity. **However** the motion must bring about differentiation between that which act as solids and those taking the place of the liquid and in that forms relevancies. It is of much importance to recognise the **fluids** that stands apart from the **solids** when differentiating the differences is some times just a concept. Water is sometimes a solid but can perform as a liquid**, which smoke and dense heat also are. On that and other grounds I maintain that the Sun on the inside is liquid. Gravity and the establishing thereof is not a God given rite of birth bestowed on all the heaviest to create.**

The gravity it develops is a "cosmic life" not to be confused with carbon life we find and have and are on Earth, but that which makes the Universe alive. That establishing or creating or exciting of singularity by applying motion which separates new heat in space by separating distance by spin is a new cosmos entity standing apart from the rest of the Cosmos. Every singularity is a Universe that can apply new values and rules by changing any of the Kepler formula factors. By generating heat singularity comes "alive" or energetic. The response to that is that motion comes about to rescue singularity and bring about gravity. Gravity is motion through space of space in space using time as a measure to secure the motion. The motion creating time must never be interpreted as a secluded event. Motion is always relation applying between various singularity points. Gravity can only come about when motion changes a relevancy between two or more structures and such change will increase the coming together or the moving apart of the structures in question. But without applying heat by increasing the levels of heat motion cannot come about and "cosmic life" or energising singularity cannot produce such motion, as gravity will bring about. There has to be a relevant solid rotating about an axis in a relevant liquid that permit motion. There will always be seven to the one side as gravity reduce the space bringing about a "cooler side" and on the other side of the relation the space will be "hotter" and motion will come into place.

By applying heat to a spot the spot has more heat than the rest of the spots surrounding the spot applied with heat. By receiving more heat the heat will turn to space and the growth in space will bring about motion since the space has to go somewhere, as it is growing bigger than what the space was. The applying of heat will increase the effect of singularity on **k**. The increase of **k** will turn the increase of heat into an increase in space. The space becomes more therefore it has to go somewhere larger than what it was before. This we call expanding and the more ferocious motion we call exploding.

The Lagrangian system provides a means for particles to move about in an independent Universe by attaching to a domineering Universe. It is the forth dimension that provides the space for the independent object to function in the security of the forth but not being part of the sixth dimension. There is time forming the forth and one outside of time is space being independent of time.

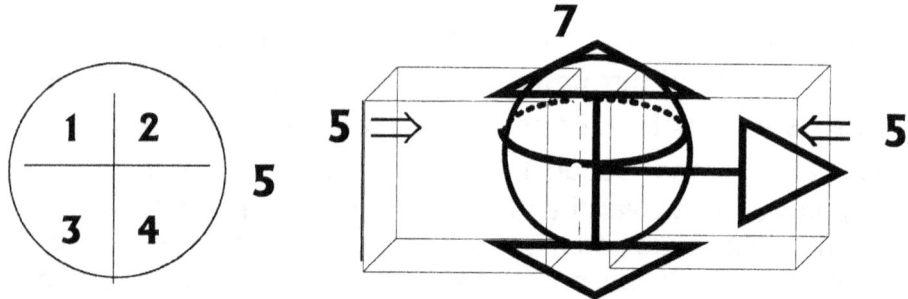

In the Roche limit, the **straight line** forms part (1) and the **half circle** is part (2) and the **triangle** forms part (3) to singularity (4) Holding 5 points outside singularity

The formation of the sixth dimension

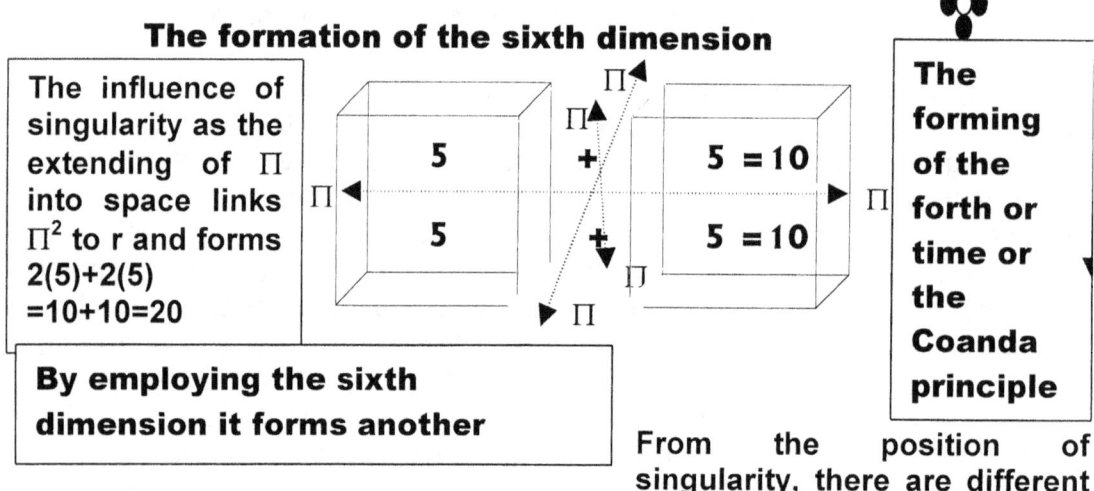

The influence of singularity as the extending of Π into space links Π^2 to r and forms
2(5)+2(5)
=10+10=20

By employing the sixth dimension it forms another

The forming of the forth or time or the Coanda principle

From the position of singularity, there are different values in Π where each indicates a position. The value it represents being $\Pi\Pi\Pi$, Π^3, Π^2, Π and Π^0

From there, it influences singularity in the triangle flowing through to the half circle. It is an interaction between circular and linear motion as the value of Π continuous past Π^2 (at the end of the solid) and every cosmic structure holds an individual and specific singularity. The field where Π extends we call the atmosphere having a value of 21.991 / 7, which is Π.

By altering any of the three Kepler factors space-time can establish a new significant gravity in the midst of gravity applying. By creating a new spin in the presence of the Earth gravity, the spin creates a gravity that will encourage the release of the major gravity in order to find independence in example the spinning top to try (it can never happen but it is trying all the same) through a newly charged singularity to develop a gravity that will produce such vigorous movement T^2 that will take the top to a position apart from the rotation it normally has with the Earth. When the speed of the rotation exceeds the limitation Earth the Earth allow the spinning top will start wobbling from side to side indicating a maximum effort to create lift and go in a separate spec at a separate distance from the Earth. In the most vigorous attempt the top will fight for release by jumping in the air. When the top slows down the wobble will become present again, as the gravity established through the spin will fight to stay alive and apart from that of the Earth. But it shows that in Kepler's formula new space comes about from establishing a new T^2, which the spinning then forms in the alliance of space created through the manifestation of a new $a^3 = k\ T^2$ in the boundaries of the Earth. Make no mistake about the fact that the spin is new gravity that comes about in the area the top occupies and the **k** is now the rotation coming about from the centre of the new spin. The Earth has replaced the role the Sun had before the spin commenced and the top by spinning resumes the role that the Earth had but involving much less dominance. The wobbling at the bottom and at the peak of the spin effort of the rotation is a gravity struggling for independence either to maintain independence or at the top to establish ultimate independence. The Earth will respond by loosing the battle as long as the Earth receive the asked bounty in the form of that the escaped has to release to accomplish the escaping or if the release in heat is insufficient the Earth will kill the rebellion by destroying the motion and reinstate the resistance we call mass.

The Newtonian view about the top standing upright while spinning is that the top is in balance...but in balance of what? The "what" part Newtonians never get to answer. It is like Darwin's evolution theory where Darwin was of the opinion that one species evolved in another ...and yet in almost two hundred years there was not one example of a bee transforming to a butterfly or a cockroach becoming a beetle. There is no evidence that the one became the other as per sample. There is no evidence to prove that this or that specie died when the transformation was in process and we now can show the fossil as per photo proving the cross over in progress. Yes the building ingredients used correspond but how did a chameleon become a crocodile. Surely with all the digging of fossils one such example must be on record. It is as saying all buildings are precisely the same because all buildings use the same material and was therefore built on the same day or with the same purpose in mind? Surely there has to be more proof with substantiating evidence before any idea is accepted as religiosity. The top spins because of a balance and everyone seems very satisfied with such a brilliant and masterly though through conclusion. That is so Newtonian where a mere suggestion with the correct and with some appropriate backers can be accepted as God given truth. It began with accepting gravity as a fact without substantiated proof and see how far did it go off course into a blur of suggestions and nuances. Never is true detail required but speculation about a notion becomes a fashion and in the correct circles with the correct backers the fashion becomes fact. How does and why does the top stay upright when spinning or fall down when not spinning? If it is a balance then what is in balance. If it is a balance within the top then there has to be a divide within the top...so what is the divide within the top.

From such deductions one may speculate in order to learn. There is the fact that is a result of speed discrepancy between objects. The speed is a result of duplication. When particles are closer than singularity will allow the Universe standing affected by this, then falls back to conditions that presumably were in place when conditions instated the Big Bang scenario. From what we see in the Roche limit we can project back to what applied just before the Big Bang came in place. When two objects are very close to each other, they either have to spin at a harmonious pace in synchronized motion, or friction will come about leaving heat as the consequence there of. Gravity is motion discrepancy and the Roche limit shows that heat comes about as particle friction causes the heat in such a mismatch of spin. Since space was non-existing which brought on the obvious cause of the heat being in over supply, it is safe to make deductions based on such facts.

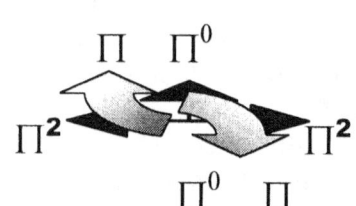

The Lagrangian system allows motion independent from time components of singularity

The motion discrepancy between the points holding singularity controlling the Earth as it does throughout the Universe in relation to the singularity in control of what we named mass will establish the Roche limit although in another value.

The ten dimensions I named the atomic relevancy is also showing the double value of singularity as singularity extends into as well as beyond space. The atomic relevancy is $(\Pi^2+\Pi^2)(\Pi^2 \times \Pi \times 3) = 1836$ that is the mass relation between the electron (3) and the proton. Proton = $(\Pi^2+\Pi^2)$ Neutron = $\Pi^2 \Pi$. Then the electron comes as 3 where π lost form.

The atomic relevancy holds the dynamics of what singularity controls. In the ratio and dimensions we find in the atom, all space-time derives from the atom, which is resulting from whatever the atom is. It is very important to realise that every molecule, every atom, every particle and every cosmic structure holds singularity and as it holds singularity it is a Universe which it has the absolute task to secure and maintain that singularity in the future. Such maintaining is the prevention of further overheating by securing gravity opportunities under cosmos rules and laws. The accumulation of influencing all surrounding and all other singularity forms the unit we think of as the Universe but that is not the Universe. That is an accumulation of all Universes and therefore in the centre of every molecule, every atom, every particle and every cosmic structure that holds singularity is the centre of the Universe. Singularity formed a unit once that formed a Universe. Then every part took with it the Universe it represents and still form the centre of the Universe it departed from.

The influence that singularity has on the surrounding space-time or the involving of such surrounding singularity is part of the basic issue of gravity in the cosmos. The process is that which should be based at the core of our fundamental understanding. The concept in own merit is rather simple to understand and I also find is very difficult to explain. By the motion of particles within the confinement of any centre of a sphere such particle motion increases the heat levels by reducing the space factor and that increases other types of reaction and more gravity in the centre of the sphere. Every part of individuality in singularity is by form of π a sphere. It is motion that produces a new Universe by allowing a fluid to form space that separates the Universe in fragments. Where our part remained solid by gravity forming the basis on which the Coanda effect rests another part had to compromise the solidity to become fluid and in relation with the solid, apply the Coanda effect of liquid forming motion in the presence of solid applying a basis for Kepler's formula to determine the Universe at $\mathbf{a^3=T^2k}$. The neutron in going liquid, sacrificed one part of $\pi^2 + \pi^2$ to form $\pi^2\pi$ and in that lost π heat became the virtue of the atomic enclosure. Such heat in the centre will establish the value of singularity at the point such heat concentrates by reducing the space towards the centre. The atomic motion serves as the cosmos in motion. The cosmos is about motion and the neutron is the atom's motion, therefore the atom can only be when there is a neutron attached. The only motion not resulting from the atomic effort in motion is life. Life has the ability to manipulate and dictate the motion of the atoms serving life. *B*ecause life is not part of the cosmos the cosmos does not recognise the difference between any efforts coming from the intervention of life as an energy standing apart from the cosmos or the cosmos bringing about such heat by applying the laws it created to govern the cosmos. By securing a spot through motion holding singularity ten spots surrounding such a centre becomes valid in a task of generating space-time. In such an event of creating a heated gravity spot, the independence of such a spot starts to try and bring about independence and establish a new dominating singularity where as the task the dominating singularity has is to subdue or "pacify" space-time captured (using the expression the U.S.A generals so fondly use to bomb the living daylight out of those persons whom they invaded, murdered and plundered and then has the tenacity not be satisfied with American domination and American rape of their liberty). Wind devils and hurricanes forms another part of the Coanda effect.

If space were zero or nothing as Mainstream science so affectively teach us then Kepler's principle formula would need the changes Newton brought about. But it is true and stands tested like no other research ever coming either before or after Brahe and Kepler's work.

If space were zero or nothing as Mainstream science so affectively teaches us, then Kepler's principle formula would need the changes Newton brought about. It is true and stands tested like no other research ever coming either before or after Brae and Kepler's work. By reducing the line to infinity and raising the line again back in the direction of space, the line would erupt as a natural sphere having Π as the natural basic value. That is the value Kepler interpreted. However not realising what he saw he chose to use different symbols.

$$\begin{array}{llll}
k = k^{3-2} = k^1 & k = a^3 / T^2 & a^3 = T^2 k & T^2 = a^3 / k \\
a^3 = a^{2+1} = a^3 & k = a^{3-2}(T^2) & a^3 = T^2 k^1 & T^2 = a^3 / k^1 \\
T^2 = T^{3-1=2} & k = a^{3-2} = k^1 & a^3 = T^{2+1}(k^1) & T^2 = a^{3-1} = T^2 \\
& k = k^{3-2} = k^1 & a^3 = a^{2+1} = a^3 & T^2 = T^{3-1=2}
\end{array}$$

is the same as   is the same as   It is all the same

$k = k^{3-2} = k^1$ is in direct relation to $a^3 = a^{2+1}$ is in direct relation to $a^3 = T^2 = T^{3-1=2}$. With this information staring mainstream science in the face and scream pleading at them to recognise the information they turn around and ask why can man not fly off to other galactica at the speed of light.

When the astronaut is departing from space on Earth or filling Earth space it will take the departing astronaut k^2 time to reach k^1 and fill out k^3. At present and in this moment our most impressive astronautic engineers will devise an engine that would cut k^1 by say half. This achievement will come as they increase the power output say for argument sake to double what it is at present. There was no friction of particles destroying the frame of the craft because there are not enough particles in space to do it, the space became too small to allow the time it takes to enter because the distance k decreased faster than the space a^3 could compromise with the time T^2 changing from what is present in outer space comparing that to the time in to atmospheric space. With the information in hand for a period of four hundred years and where the information forms the basis of modern cosmology since the information formulated gravity and not merely produced a name for gravity as our English friend did it is amazing that such accidents can happen and it is more amazing that no one in Mainstream physics has the slightest idea why this is taking place! Our most impressive astronautic engineers are assembling a machine that will scramble the ratio Kepler introduced to a level in outer space where the ratio will be more than what the ratio in the Sun is. Surprisingly they are not in the least surprised that not one object in outer space is using an excessive velocity.

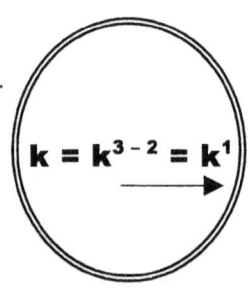

It takes time for space to fill k in the distance. In fact it takes the distance that k, developed since the Big Bang $k = k^{3-2} = k^1$ to fill the distance.

It also takes time $T^2 = T^{3-1=2}$ to produce the distance forming k^2

It takes space $a^3 = a^{2+1} = a^3$ to form k^3 since coming from the Big Bang

The big thing about the evidence that $k = k^3 / k^2$ or $a^3 = a^2 a$ and $T^2 = T^3 / T$ is that the only incorrect attribution Kepler made was in this that he did not foresee the fact that what ever it was that was the indicating value was the same indicating value for all the factors in hand. The rest was as correct as the cosmos gave it down to Kepler. Kepler saw four hundred years ago that an object accelerating becomes larger by becoming smaller. Since the object has to become smaller due to the fact that it duplicate more while the rest of the Universe remain the same size, the object cannot change the size of the Universe so the moving object therefore has to reduce the size it gains in order to allow relevancies to apply.

Kepler's work is phenomenal. Newton put Kepler in a frame where everyone viewed the work of Kepler as some astronomical mistake, which the Newtonian perception about the incident is that thankfully Newton could correct and after the correcting, Newton was able to find a use for the corrected work of Kepler. Newtonian view to this day is one of Kepler making a mistake because it came as part of Kepler's inferior insight when compared to the supremely gifted Newton's super human mathematical skills. Every intellectual sees Kepler's work only connecting to the Newton's rescue effort and through the Master of Newton's greatness Newton could rescue some of the work and make sense of Kepler's mess in order to find some degree of use for the corrected work of Kepler. The greatness of Newton was further underlined and acclimated by applying Kepler as some intellectual extension that Newton could fit into Newton's own work to give substance to the already perfect theory Newton delivered on gravity. Yet the work of Kepler was an accumulation of his lifetime achievement as well as the lifetime achievement of Tyco Brae. What Kepler found is what keeps the solar system in tack and what keeps the solar system in place, which can only be seen as nothing other than gravity. Kepler found what kept the Universe glued in one constructed unit and put order to what would otherwise comprise of total chaos. That can and has to be gravity.

The space became too small to allow the time it takes to enter because the distance k decreased faster than the space a^3 could compromise with the time T^2 changing from what is present in outer space comparing that to the time in to atmospheric space. With this information being in hand for a period of four hundred years, one should think that the wise could derive a conclusion. Where the information forms the basis of modern cosmology since the information formulated gravity and not merely produced a name for gravity as our English friend did, it is amazing that such accidents can happen and it is more amazing that no one in Mainstream physics has the slightest idea why this is taking place! Our most impressive astronautic engineers are assembling a machine that will scramble the ratio Kepler introduced to a level in outer space where the ratio will be more than what the ratio in the sun is. Surprisingly they are not in the least surprised that not one object in outer space is using an excessive velocity.

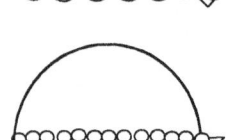

$a^3 = \dfrac{k}{T^2}$

$k = 4$
$a^3 / 2$
$T^2 = 1$

In realistic physics it means double the space will fill in half the time. We know that that is not possible because it can only bring about half the space in double the time or twice the distance in half the time. Space time and distance is a mesh where the lot integrate because Kepler said so. Kepler said the space forming space is the same space forming the distance of the space and that is the same space taking the time to fill the space. If the ratio changes then changes come about the entire ratio.

$k = 4$ and $a^3 / 2$ if T^2 remains the same but that will not happen and that we know from past experiences. If that happens, we have the challenger 2004 disaster repeating once more.

Increasing space-time displacement by six will decrease space by six and the distance the space progresses from a centre by twelve. The heat factor of the craft will rise by twelve times as the space decreases by six times.

Increasing space-time displacement by twelve will decrease space by twelve and the distance the space progresses from a centre by twenty-four. The heat factor of the craft will rise by twenty four times as the space decreases by twelve times.

Motion of anything in any form is about duplicating the existing into following on images of the same thing. That is connecting space to last a certain period in relation to a specific point holding singularity before the next singularity is enticed or charged to maintain the space-time in motion. Every time (and in this case the referring to time proves to be most accurate) is having another singularity building and breaking down the space it represents for that duration of time. The time duration leaves singularity selected in charge of producing the roving space the extent in which it can duplicate the space it has to duplicate. By reducing the period the particular singularity may lay claim to the space, will inadvertently produce smaller space it is able to reproduce in the shorter period of time.

However, to achieve such duplication standard, some heat will have to be sacrificed in order to secure such early time release from the singularity that takes charge of the duplication. Only by paying a price of compromising heat can such duplication be sustained. The compromise of heat in natural conditions will be sacrificed because of the enormity of the gravity that the singularity charged with duplicating space-time can unleash. That means the singularity therefore must have grown to the enormity as to be able to accumulate the heat in order to produce the duplication that will entice such enormous motion.

This effect we see happens when waterfalls onto a hot stove plate. From the plate flows heat that charge the singularity that captured the space-time being water, to break from the unit forming the water drop and seek independence from the unit by enforcing motion of the water drop. The water drop running on the stove finds the plate charging the singularity of the plate to entice motion and the motion of the water is the singularity finding additional space in heat to duplicate the space faster as to create the motion that accelerates.

Unfortunately the compromising involves all aspects and that includes the gravity or time also. The Coanda effect shows that gravity in space can charge and change space and this is because Kepler said space time and distance form the same thing because the three is part of the same thing. Change one aspect in Kepler's formula and the outcome is a mess. I have a book on this aspect called **Inter Galactica Space Travel. ISBN 0-9584410-2-2.** The Coanda effect and the Roche limit results from this implication

Henri Coanda holds esteem in aviation circles but beyond that his work is slightly known. Henri Coanda should be amongst Kepler Copernicus and Newton if his work is correctly categorised. His work is that of a giant. He demonstrated gravity, he produced gravity, he introduced gravity and no one bothered to take notice because he was not a mathematician. Henri Coanda might not really have realised that he stumbled on the oldest principle in all of creation and that this effect he penned and named is the absolute basis of gravity, but his realising thereof the importance did not pass him by.

Coanda realised that by reducing the propeller blades of the aeroplane the rotation speed becomes amplified. Concentrating space to amplify the heat comes from the concentration in space by redirecting the flow of air (space), which also can be water since both are liquids. By confining the space displaced by the propeller shaft the space duplication produces a rise in the heat within the space. What this implies is that the process is basically placing the Big Bag in reverse. The Big Bang is releasing heat to produce space and the Coanda effect is confining heat to produce space.

The role of the impeller is to reduce the space by applying motion that creates a higher concentration of heat in a reduced space during the same time interval. When this is mismatched the impeller burns holes in the blade as air produce heat that removes metal. The rotation of the impeller forms the dual prong of cosmic motion. The propeller has the role of the protons that reduce space by directing space to demolish space through implementing motion. The second part of the double prong is the confining of space that will bring about heat increasing in the confined space. Motion brings about gravity but motion is gravity in the duplicating of material claiming space and material holding space through the duplication thereof. The only condition is that a liquid must be present. In relation to a solid the liquid must relate to a point in singularity or a substitute to such a point in singularity by committing form of singularity to one specific concentrated point mimicking singularity. One has to realise that space is the gas part of heat being the liquid part and space and liquid is far apart from material. Material provides the solid structure where such solid material surround itself with dense liquid heat called plasma by those being smart and educated physicians. It forms between the solid atomic structure occupying the inside and the less dens heat on the outside that is regarded as space and is further away from the atom centre.

When the seven occupied by material rotates it removes the ten in contact from influencing space-time and by rotation introduce a totally new ten points in relation by the act of rotating. But such a process also come about on the other side of the divide because it is the divide that cannot rotate and forces space-time to comply with the frequency of breaking down space-time as space-time releases from one side and if necessary shift or only apply new rotation alliances as the seven points become a part of the new ten.

When the seven occupied by material rotates, it removes the ten in contact from influencing space-time and by rotation introduces a very new ten points in relation by the act of rotating. Such a process also comes about on the other side of the divide because it is the divide that cannot rotate and forces space-time to comply with the frequency of breaking down space-time as space-time releases from one side and if necessary shift or only apply new rotation alliances as the seven points become a part of the new ten.

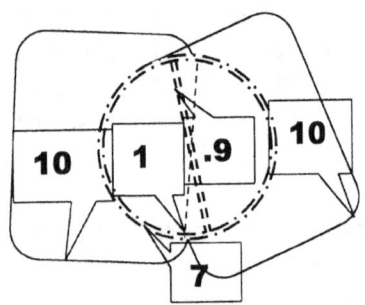

The motion duplicates space (10 + 10) by rolling over singularity (1) from the one side by reducing space (.09991) on the other side. All this motion of space 10 duplicating is directly related to matter 7. The total value established from this motion is the interaction of seven with ten producing Π^2 as the half of singularity $\Pi/2$ on the one side interact with half the singularity $\Pi/2$ on the other side and in this process Π the duplication of Π by means of matter and space produce gravity in the measure of Π^2

The conditions proving singularity is that the circular form will produce a centre point from where gravity will dictate the reproducing of space. The gravity part is the fact that motion must contradict the centre point around which the motion will produce space. The space part is proven by the motion that produces a running line of space created and followed by the liquid producing the space in the motion of the liquid. It shows that relevancy comes from space shared by space and motion separates space shared by space. It shows gravity coming from motion separating shared space.

When further consuming is not possible, when finding any singularity that challenge is the consuming, the Roche limit comes into effect.

Second Over All Overview   461

We must see that we have three dimensions coming from the atom.

In the centre there is the proton ($\pi^2+\pi^2$) that reduces space-time to singularity π^0

Surrounding the proton and installing the Coanda effect space-time is the Neutron ($\pi^2\pi$) conveying gravity to the proton ($\pi^2+\pi^2$)

On the outside of the atom the rest is wasted (3) notwithstanding what such space-time wasted might be.

The centre proton ($\pi^2+\pi^2$) consumes by dismissing space-time.

The neutron ($\pi^2\pi$) duplicates by providing motion through the space-time in which to move as well as serving the consuming proton ($\pi^2+\pi^2$) the heat to grow.

On the outside is what there is to duplicate within the wasted space-time as
Well as to absorb and consume.

When further consuming is not possible, when finding any singularity that challenge is the consuming, the Roche limit comes into effect.

When duplicating of space-time it results in re-connecting of space-time with some space in the centre of the established singularity. The centre dissolves space and also is placing space within a surrounding of space less ness where space then is becoming motionlessness. In the event where space always in a relevancy with time and time being the motion placing relevancies on space such a relevancy places a cup or a limit on what space flight may be able to achieve. With the international media and the international press printing in frenzy about flying sources because they have no better work to do than to go on some flying events that take the visiting aliens through the Universe on some sight seeing tours those visitors find the time to tell us humans tales.

Unfortunately we have amongst us, the mentally impaired that comes out with tales of how those invading men-from-Mars-or-wherever-they-come-from, and travel billions of light years just to have the opportunity to have sex with these mindless storytellers (whenever the halfwits meet extraterrestrial visitors the visitors rape them, take samples of their body tissue and send them home scarred for life. To think that any one would cross one Universe just to have sex with them and then Take their body tissue as samples makes them real special, except on the intellectual front). To top this madness there are scientists with as little mental capacity that echo this insanity by insisting that human crowds will infest far off galactica by swamping the trillions of planets they are yet to find. To them and those I give the free advice: remember that it is useful to take into consideration that gravity also forms relevancies on preventing objects not to come harm-provoking close. Where there is this limitation, there is also one of being too far apart for singularity to allow the escaping.

The criteria which any object has to comply with should such an object wish to leave the domineering of the Sun's gravity displacement, which by the way stretches beyond the Oord cloud and Kuiper belt, that escaping craft has to exceed the motion applying in the centre of the Sun as well as reduce in form and space to fit in a spot such an object will fit into when in the centre of the Sun. The craft is part of the captured space-time the Sun is holding. To find release from that capture the craft has to find release from the governing singularity control within the centre of the Sun. The heat effort the escaping craft has to apply must be stronger than all the atoms within the Sun as an acceleration effort generate what the containing force is that. The escaping craft has to apply more motion $T^2k > a^3$ that the space that the material in the Sun fills. To go beyond the barriers of the governing singularity within the centre of the Sun, which I just mentioned, the craft must produce anti gravity, in the form of heat release that will overshadow the gravity of the Sun. The craft will have to produce more heat than there is in the very centre of the Sun at the point where space disappear because that escaping object has to produce more space through motion than which the Sun can destroy in the very centre where the Sun is destroying the space. From such a centre destroying space-time then is holding the solar system bonded.

The craft whishing to leave the Oord cloud boundary must produce more space through motion that is creating space by duplicating than that which the Sun can destroy because if not it would repeal the motion and bring the escaping object back to the Sun in way comets do. That is the reason why comets have not yet escaped and gone yonder to "nearby" stars and that is why the Oords cloud has been unable to reach the area in the location of the Centauri triple star system. That principle is the law of space forming by motion applying, which Henri Coanda noticed and which rocket propulsion uses to establish motion. The craft must create conditions exceeding the centre of the Sun reduce space to become hotter that the centre of the Sun and go faster than the velocity the Sun can create in the centre of the Sun and even then the object will find a relevancy of say so many millimetres per year to advance past the border of the Sun. This is because fusion does take place within the parameters of the Sun occupying space on a very limited basis. To those aiming to achieve that I bid them good luck for they will need much more than the luck I can offer them. The spinning top I referred to previously and the Coanda effect is the very same principle applying. The Sun is just a large spinning top applying the Coanda effect.

Let us return to the top and find gravity in the behaviour that the top shows to motion. Again I have to stress that one should never forget the fact that the energy attributed to the top is a product of life and although it is in the capabilities of life to manipulate space-time it is also in the ability of life to use and manipulate from cosmic some applying principles that we can milk some behaviour and in doing that, that will tend to benefit life on a micro scale. The spin provided to the top acts like heat activating a singularity centre and from that motion comes about. A more natural result coming about without life's interference will be when the singularity in the centre of the top is all that

Second Over All Overview 463

charged and the heat will entice the singularity as to accelerate the top's independent time to bring much more duplication of the space of the two to come about. However, in the form we see the tops spin; this act by the top is not cosmic produced. It is at best cosmic re-enacting because there is no chance the top will by own sources start spinning and even more so spin violently enough that the top will then manage enough duplication that it can leave the Earth. Gravity is about motion providing space a certain relevancy to duplicate space in relation to the surrounding and that was how space-time came from about the start of creation.

If gravity was in principle about one object hooking another object like fighting boxers pulling at each other as they fight to achieve dominance, well yes, then the space cannot have any influence on the two particles...or can it?

Even if gravity is about particles pulling, which it is not and that I say on lack of evidence brought in support of that statement ever since Newton suggested the idea, some of the gravity "force" must come as a result of the conducting of the force through space being the transmitter of the force. Space is what we find that is between the two opposing structures in complying with gravity. What ever is conducting the gravity is doing so while space is or permitting the force to reach the other side. Allowing some influence to reach another point is what I understand conducting is.

There is a restrain of flow of such conducting when space becomes denser or reduce density. In order to get what ever is being conducted to where it is conducting to, such a force that is conducted with the intent to grip onto or get hold of the other object, must find that the influence has to commit the space between the objects to allow such pulling. There has to be such a medium that can allow the conducting of the massage by the massager to take place but science produced nothing that is in support of proof to confirm this... If the conducting of gravity was about pulling then still the motion of space was the first indicator and without the motion the force will then be nullified. Even the indicating of mass is motion trying to come about but the motion finds moving restrained by a blocking. The sharing space between the objects was the first to be influenced because space has to relay such an influence; the space has to reduce to allow the action of contraction. Unfortunately academics would never previously admit to space deforming through gravity applying. We know that the flow of light is a conducting of light and the speed of light can vary in motion from standing still in certain silicon conducting computer chips, to travelling at C. But where light is anti-gravity, then gravity must be opposing light. Gravity is essentially directly connected to the form of the sphere.

The essence of the sphere is about reducing space from the outside to the reduced inside. The sphere reduces space by applying form and committing space occupied and deforms space unoccupied. This is always the complying of the outer space to relent the form to that of the sphere. In the sphere the form dictates space reducing as space occupation progresses to the inside. From that we have to deduct that if the sphere will apply reducing of space inwards, it will extend the applying to the outside. Present will always be an outer edge and that forms a border that commit space to be in or out. We have only to detect and find proof of such influence being present.

In gravity there are always two relevancies applying in relation to each other. There is always one being part of a holding unit and then there is one other part establishing borders. This is the result of how singularity broke into space. At first there was singularity k^0 holding the Universe in the single dimension. Then singularity k^0 moved out to form singularity k (and this process stands apart from other forms coming out of single dimension at the next step) but in the same action the half circle came into position as well as the triangle forming space. At this where singularity the spot Π^0 moved to singularity the dot Π all other future possible particles were still frozen including mathematics in the rim being to the power of zero. Only Π yet had a value placing a point in the Universe where others still were numberless. The circle was Πr^0. There was only space that Π brought forward as motion and motion was space as well as direction $a^3 = k\, T^2$ where all stood related by relevancy in direction to each other. The size they shared was all the same.

Only the mathematics we now see as trigonometry was in use where the only valid are angles relating in forms and directions. But k was T^2 as much as k was a^3. The motion of expansion came in place but the expansion had nowhere to go than to rotate because the expansion was simulated in one move through out the Universe as one orchestrated motion where everything coming from singularity strived for independence instantaneously. It was motion that was creating space but it was motion bridging singularity, splitting the possibility of singularity into sectors. But part of this was relevancies where the one formed the other sectors space. The one was rein acting the other side of the Universe because all was so small a line was duplicating a half circle because the other half circle belonged to the other side of the Universe and the two applied the space received.

Simultaneously to that action there was the other side of the other side of another part of the same Universe and from that stance the motion applied by the first mentioned singularity action was received in a much different light to those on the other side of the divide. As the second related singularity is on the opposite of the first Universe it must recollect the procedure very much differently than the way that the first sees the space formation. The first one applied motion and the motion placed spin on the second singularity point. The applying space to the one was at that same time applying motion to the next and the confrontation was establishing space in relevance to the motion that came about from the growth as well as the rotation that was established as a retraction by the opposing particle. From such a duel a relation will come into affect putting one point in singularity in a major position as far as security is concerned and one in a minor relation and from that stance the relation will commit to space where each one performs certain roles in relation to the other. The roles has to do with creating space

At the time of Big Bang motion one came about when independence was struck by forming a value of ($\Pi \div 2$). Later on in cosmic development only when one of the two opposing participating object cross the line of ($\Pi \div 2$) from both sides ($\Pi \div 2$)^2 will the opposing particles bridge the first cosmic law and the contraction will over ride the expansion capturing the second relevancy into the control of the first relevancy. The first instance both particles matched since all particles were at least ($\Pi \div 2$) separated. But growth applies as material incorporate space as heat and from rat find material expanding. One of the particles grew more rapidly than the other and secured dominance but not control. By outgrowing the lesser partner it can capture the dominating of the space the lesser partner claims.

Then there is the other duplicating gravity principle where motion that brings about space. That is also putting distance between objects and this we humans understand to be growth of space or by a term much more commonly used as the Hubble constant but only in a balance where the space displaced and the space created match in time taken relating to space. This formula will apply as the second gravity $k / k^0 = a^3 / T^2 = k /= a^3 / T^2$ because the motion advances the distance between objects in relation but will only dominate when the first option brakes the Roche law of $(\Pi \div 2)^2$. Then the relation will change to where the **k** factor holds a negative and total control. This means the space duplicated are not adequate to the space dismissed and a new space to time balance must be established. $k^0/k = T^2/a^3 = k^{-1} = T^2/a^3$ That is what Newton saw and that is what science recognises but that is not the primary gravity found in outer space at large. In outer space the Newtonian gravity only dominates where the bridging of singularity forcefully brought domination and control of a major singularity in capturing a minor singularity. Where that situation complies with cosmic rules, the president set will be that the major contractor reduced the minor expander to heat and then capture the heat. Only heat applying motion brings particle separation as we may witness with the spinning top. Motion brings about space duplication, which revitalises space dismissing.

However, that forms a relevancy.
$k^0/k = T^2/a^3$ and that is the claiming relevancy

$k/k^0 = a^3 / T^2$ that is the expanding relevancy

There are the one relevancy that applies contraction that recaptures space and will answer to the ratio presented as follows
$k^0/k = T^2/a^3 = k^{-1} = T^2/a^3$

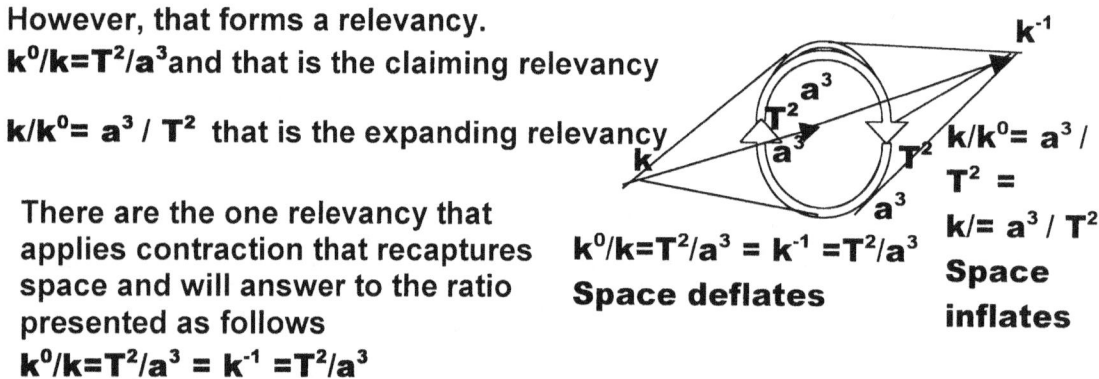

$k^0/k = T^2/a^3 = k^{-1} = T^2/a^3$
Space deflates

$k/k^0 = a^3 / T^2 = k /= a^3 / T^2$
Space inflates

How does one accept and believe in the proof about this, which is said. We find the proof in the Coanda affect where the reducing of the rotating blades and the increase in the rotating speed produces heat as high as heat can come in the form of flames. We find heat burns holes in impeller blades with the impeller blades of the speedboat submerged in water. The blades of the impeller only work by practising the Coanda effect. The blades find the ability to reduce space and concentrate heat in the space by increasing the duplication of the blades in the rotating thereof. In the turbine, the blades provide gravity by reducing the space through increasing motion of the space during the same period in time duration and this gravity establishes the equal antigravity in the form of liquid tongues of flames. If this is happening now it happened very back then...

From the gravity and the opposing antigravity applying came about two interacting translations of material where one formed gravity in securing a position and another by permitting expansion, releasing matter to establish space in relevancy bringing about antimatter in the form of matter performing antigravity. The one was filling space by giving away density and the other was applying density by giving up space. If it was true that matter was drawing matter closer, the securing of space would not stop the contracting and the regrouping of matter would start in the Universe even before expanding could start. The Universe would become one solid structure and remain that way because there would be no reason for it to expand and fight gravity's magical attraction contracting space between particles with mass.

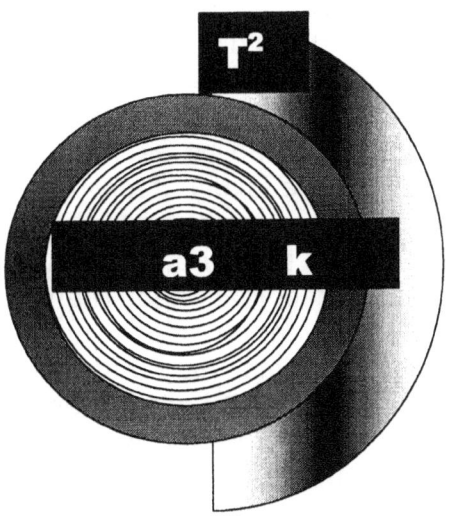

By the motion committing the top to motion, the top forms a secondary alliance with the primary gravity the Earth provides. The motion the top achieves is going beyond as well as including the motion, which the Earth provide. This employs the Coanda effect as the Coanda effect interprets the Pythagoras principle where the square establish a crossing of the singularity line committing singularity to activate the centre line of singularity in forming a division between 180^0 and 90^0. The top secures a centre by lining up in 90^0 with the rotation of the earth that is holding 180^0 position. Every time the top rotated, the top is forming a new graph that is sustaining both the previous and the following graph. Moreover and more important is that motion dictated space boundaries by committing a centre to a specific relevancy.

The fact that motion secures a unit of space-time where one part tends to be without motion while the other part plays the role of a liquid that supply the motion, that became the essence in the pre-Big Bang Universe. Without that information there is no pre-Big Bang realisation.

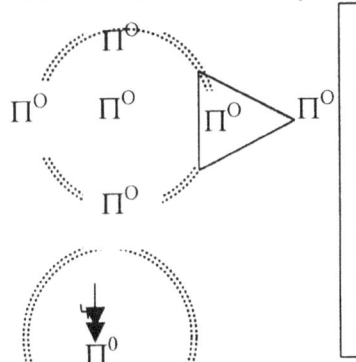

With the object sharing the atmosphere of the Earth, the object shares in the motion that the Earth provides. That is also sharing in the four time factors. By applying independent motion the object can join any of the four points in singularity and even advance to the fifth point putting the motion one point beyond time. As soon as the top starts to spin it follows the range joining motion from one to five.

Within one space unit, we always have two independent positions holding points in singularity that is sharing time but being independent in space at the same time.

The two assert a velocity of $7(3\Pi^2)$ by just moving with the Earth. Since they both move equally in line with the Earth, the two objects hold an independent that is Π^0 in independent speed but the independence may grow to $5X\ \Pi^0$ before the sound barrier becomes an issue of concern.

Some years ago I was reading a book and in the book was mentioned about some discovery Einstein made on gravity. It was about a certain remark that

Einstein once made on a realisation or a conclusion that Einstein came to in his younger days while still being a clerk at the patent office. Apparently the idea Einstein came to was concerning the subject of gravity. This happened while Einstein was still being a patent clerk in his very younger days. Apparently Einstein was looking out a window of the multi story patent office, when Einstein suddenly realised that had he, Einstein fall out of the window from the roof to the ground of the patent office where he was working at the time, then he (Einstein) would feel as if he was weightless during the time of his fall. Not only that but also if all the things in his immediate vicinity also fall with him, then so would all the articles in his office that surrounded him at the time being his office chair, his desk and a pen travel alongside him at the same tempo. Then I conclude that by falling with him those articles would feel equally weightless should they accompany his fall down as being part of the falling process in his imagination. He remarked about him feeling as if...and those other articles was travelling alongside him without the luxury of feelings. As the objects were travelling alongside Einstein down the building to the ground the lot would travel at the same speed from the top to the bottom of the building. That is what Galileo concluded about five hundred years ago. Then I made another inspired conclusion at that moment as I went one step further by supposing the Einstein group's falling was real and no imaginary thoughts were set in the fall, then what was the imaginary factor then? Let's pretend Einstein did fall with his pen, his chair and his desk and Einstein was not imagining his fall. Einstein as a human being can imagine but his falling companions can't. Then during a true fall Einstein may have had an imagination that could tell him about his feeling and in particular about the condition of his weightlessness, but the pen, the chair and the desk had no such imagination and they were travelling at the same speed as he did downwards and therefore had the same weightlessness as he (Einstein) had while they all were being in a downwards fall. If Einstein was imagining his weightlessness, it might be psychological, but in the case of the other travelling companions it was not possible to imagine anything. The falling companions had no such a luxury as having an imagination, however they too had to be weightless as they travelled next to Einstein all the way down to the ground level. There is an immense difference in size between the falling companions and that notwithstanding they travelled the same speed while descending. If they travelled the same speed as Galileo proved and they all hit the Earth the same time, which then indicated that their weight and mass, that which gravity used to drive and what propelled them downwards and that which was causing the drawing of what the mass was instigating to allow the motion of fall to commence, was equal. Size changed nothing to the equality there was in speed. Einstein should only have thought a little further than he did at the time because that would have made him realise what gravity exactly was and what Kepler found gravity to be. Kepler found space a^3 being equal to the motion thereof T^2 in relevancy to a centre point **k**. Kepler found space had to move to form gravity $a^3 = T^2 k$, up down side ways that doesn't matter as long as it was in motion.

When reading this that evening so many years ago, the much younger me at the time came to realise that Einstein could only feel weightless if it was true that he (Einstein) was weightless. He could not feel as if when the as if was part of his imagination because he was truly falling, and in truly falling the falling was then without his imagination doing the pretending. Einstein had to feel his weightlessness as a cosmic fact in the true sense because if he was truly falling, then the part, which was the falling experience, was what he was experiencing in reality by three dimensions with one dimension in time. Then he (Einstein) was feeling weightless through falling and that feeling came as a result of what was happening to him as a cosmic interpretation of reality. He was not pretending to fall whereby he then would feel as if...he was really falling and with that there is no "as ifs". What he then would have experienced came by means of what he was experiencing in reality because of his cosmic state in relation to his relevancy with gravity. If Einstein was experiencing weightless ness, it would be because he was weightless while falling, then Einstein would not imagine the weightless ness because Einstein was truly falling, thus carrying out his cosmic state he was in. His body being in motion **($a^3 = T^2k$)** was at that moment truly weightless while experiencing unrestricted gravitational motion. Einstein, the pen, and the chair had the same weight since they were all weighing the same in falling. This I conclude on the basis that that which propel the falling body acting on behalf of that which instigates gravity being mass administrated the same force on every member of the falling group not withstanding size or preference to volume space occupied or bias to mass. If there were any mass differences there had to be speed differentiation for the force of the one would generate more motion than the force of the other onto the different mass components but since there is not mass discrepancy amongst the falling while the lot was falling, then the lot was at the time of the fall, having the same state of weightless ness, just because they adopt the same speed in the fall. After all it supposedly is the mass that is doing the pulling and more mass does more pulling...except if the mass is not doing the pulling in the first place. With more force applying to different masses there had to be more speed involved and an increase in mass in some participants has to generate more force. All four items including Einstein, would be equally weightless during the falling...that was what Galileo found because objects of different size and different mass travel at an equal pace (distance over time or space moving divided by time flowing while the object changes position in relation to the Earth **($a^3 = T^2k$)**) while descending). From reality we know that the bigger objects do not fall quicker than a smaller objects and that can only be attributed to one fact; it can only be true if the four weighed the same while falling and no one weighed anything while falling. That means the gravity applied while time flow in relation to the space that was applying the motion, which was what gravity is **$k = a^3 / T^2$** according to Kepler. The single line falling is represented by the factor **k** being the relevance of space **a^3** that was relocating its cosmic position while all that was happening in relation to the motion of the Earth **T^2**, which was in relation to the Earth spinning around the Sun and that rotation gives us our time **T^2**. While in motion the four different objects weighed the same since they travelled at equal speed downwards. However, when they stopped moving and came to a standstill, they then weighed different, which then indicated a difference in mass factors amongst them. By standing still the objects had mass differences and when they were in motion they weighed the same. When the motion became frustrated by being blocked by another space, handicapping the gravity-motion applying, and the handicap came from filled space that was also filled with material (being just as solid) and that was holding the spot too where the motion was directed, they then had different weight. The two had different levels of frustration with the larger party being more frustrated in the inability to move. The pushing resulted from the bodies striving to remain independent more so than trying to combat one another in a tussle for a point to move to. It is the independence of the two bodies and the desire the bodies have to remain independent and not to share space that bring about the mass or weight. The bodies rather remain independent and acting as a unit that fight to allow gravity to shift the body into termination the acquired independence of the structural unit. The two objects were in a fight to claim the position each desired, and that was to fill the centre of the Universe. Being **($a^3 = T^2k$)** was being in the centre of the Universe because the centre of the Universe was **$k^0 = a^3/T^2k$.** It may look being a simple mathematical statement but explaining that part to full understanding, requires the reading of ***the entire volumes of* MATTER'S TIME IN SPACE: THE THESIS ISBN 0-9584410-8-1**

From this one can deduct that gravity is motion or the intent to commit motion and mass is when the motion of gravity is frustrated by some solid structure blocking or preventing the continuing of the motion. From that one may also conclude that gravity is motion of space and mass is the restricting of the motion of space. Having mass does not bring about gravity but it does restrict gravity's motion, which is what brings about the mass and weight. Gravity produces mass but mass does not produce gravity or in fact mass produce weight but mass is not responsible for the intended motion. Gravity on the other hand is the intention that the body has to move the very instant the blocking is removed. The intent on moving while being blocked by another object is frustrating the motion of gravity in both cases and the higher the frustration on motion is, the more mass there is coming the way of the bigger object who then has the greater desire to move. The motion is about duplicating and with more to duplicate a more urgent sense in the duplicating effort will translate in a bigger desire on motion. The reason why it has the desire to move and why space is equal to the moving in time of the space in relevance to the centre of the Universe (which at that point might be the Earth or be the Sun) is what interaction there is about heat distribution by motion or expanding and the contracting of cooling by containing space. The cooling comes about by measure of distributing and the distributing brings relief in the form of cooling which results in contracting or preserving. The one action compliments the other and none will be without the other. Mass is the restraining of motion and gravity is material moving about by committing gravity. Mass only comes into the application thereof when two objects filled with space that shows reluctance to compromise form or the unit it formed, moves into a position where both want to claim the very position in space the other occupy. It is the motion and the independence they show to hold onto their individuality as independent cosmic structures that prevent them the sharing of space which in turn prevent further motion that causes mass. Gravity is in essence where mass is present, but instead of performing motion it can only the produce a tendency to commit motion but is then in the frustration of not complying to be in motion and gravity at such a point is the commitment to move once the blocking of space is relinquished. Because the one object that has more "mass" would put in a more assertive effort to move in relation to a smaller object and the effort to move will constitute to a greater resisting effort by the blocking object in a fight not to relinquish its position on the space both object claim that the tendency to move and the tendency to block the movement will bring the effect of greater or smaller mass being present during the effort and in line of resisting the effort. The body with more material in essence has more heat although the spinning of the atoms contains the heat; still it shows a bigger effort to move because it represents more heat in the space it acquires. However while any space filled with material is in motion, the gravity of motion is equal to all and puts everything on an equal basis. Therefore there is no big and small and the big Sun does not pull the small Earth closer. The big Sun allows the small Earth to glide past in a circle year after year without interfering because the two does not claim the space each other has. Mass is when the motion is prevented that a differentiation in motion effort becomes part of the picture.

Do not be fooled by the seemingly innocent explanation that space is the motion thereof which is what gravity produces because of all things the cosmos creates, motion of space through time is the utmost complex manoeuvre and without bringing a restraining of mathematics into science, it is so complex there is no viable explaining in physics about how the cosmos produce the act of motion of space in time. To get every atom to spin as every atom follow the lead of the atom in front and give direction to follow to the atom just behind while giving coherency to the structure the lot of atoms are holding as an individual unit times the units there are going around in the entire Universe is beyond what the human mind can absorb. While the atom in front is vacating space to fill the space of the atom in front is vacating at that instant, the atom behind is filling the space that the atom in front has vacated in order to vacate and relinquish the previous position in favour of the following position to honour the direction gravity is insisting upon. Times that with every atom there is in the Universe and one may grasp the significance of the calculation. Then throw into the complexity the alliances that form by every atom there is changing direction by rotation filling different space by rotating to maintain motion and multiply this lot in addition and see where that gets you. I would love to see the mathematical Brainy Bunch get stuck into that. It might just keep them out of harms way for some time while others with more serious intensions could then get working on cosmic realities. The coordinating of moving one atom from one point to a next point in rotation as well as in a linear route requires the skills that the human mind may never conquer. We may see the moving of object through space being as simple as merely excepting it as a given fact, as science has done in the past, or we may reason about the complexity as civil person's should do, and come to realise that the complexity of motion of matter is beyond scope of human understanding. Removing material from space by filling material into a position of new space sounds simple because the complexity has never been realised. The thought alone is quite revealing about the Newtonian simplicity those in science see the cosmos and the manner that the atheist use such little comprehension they have and show in order to boast about they're being nana- mindedness in thoughts and understanding. The cosmos is created by instant flowing fro instant-to-instant purely on the fact that singularity cannot move and yet it does. That means the motion is generated from some source that is not located within the Universe we find us to be in. Such understanding of reality will reveal what the factors are in understanding the commitment of material to move through time. This was all a result of understanding the dynamics of Einstein's arguing about gravity and mass. The I am not promoting religion and that is the last thing I ever would do because my religion I only share with those I deem worth while, therefore I would never waste it on the blind but I have to say those that are in denial about a greater Being that is being there outside what we see as THE Universe, to then I say there is no silly escaping reality any more. With this information I further realised that gravity is pure and simple motion differentiation between objects. It is the independent motion providing a different speed while sharing a common centre off attracting that allows a discrepancy to establish mass under specific conditions applying between the two in relevancy. While falling the gravity applies as moving of space that is putting time in relation to the distance travelled. That means there is a speed relevancy between particles in motion and synchronised motion would bring about equal orbit around a shared centre.

That is the result of gravity functioning. While the object falls the motion confirm gravity. When motion ends mass sets in and becomes the constraining of the object preventing further motion. The motion is still there but now it is reduced to a tendency to move thus establishing the object mass as the limiting of further motion. Preventing the motion by implementing mass is the resting of objects against each other by resisting the motion to continue, which then is where the mass takes the place of the motion. Where a confronting of objects restricts gravity the action then implements an introducing of the mass as a substituting factor to motion that then replace motion as substitute to the motion that would be and the mass is providing gravity being the motion of space a compromise by substituting the real thing as the tendency of motion. However mass then restricts motion and becomes motion in a tendency to apply motion. While falling gravity applies and motion neutralizes size, mass or weight. Mass counters motion being when the Earth restrains further motion of the falling object and the moving object is stopped from further movement where mass is then preventing or hindering gravity. This is the result of objects claiming an individual and personal claim to space occupied in a dual or in fighting for their individuality and independence of each other while wanting to be in the centre of the Universe.

It has to be a fight for a position being the centre of the Universe because in the centre of the centre of the Universe is where cold meets hot and gravity splits the lot. While falling or moving there is no opposition to the body being independent. The motion proves all singularity equal as $\Pi^0 = 1^0$. When the motion seizes space-time and overheating becomes a predicament but the falling object still retains individuality of singularity as it remains individual and still tends to move while Earth individuality resists further movement of the falling body's movement. Further movement is disallowed as other material fill space that the falling body wants to lay claim to. The only manner to remain independent by the falling object will be to relinquish to motion in the securing of mass as a substitute to motion where it then finally comes to rest. Mass then sets in not causing the motion but substituting the motion and from that motion restriction becomes resistance that becomes mass. While falling the object is experiencing gravity because the object is in gravity but when on the soil the object experience mass which is the restricting of gravity or motion by other space filled with material. It is a fight of objects to secure and retain the position they have of being in the centre of the Universe.

The spinning top places the top in the centre of the Universe because the motion of the top is enforcing singularity to be generated into action in order to maintain the balance between space moving about in equal terms around time in time. The top is establishing individual gravity to which the space-time of the top has to adhere.

We have to realise the fact of relevancies where there always will be a major and a minor factor enjoying the same space-time. Where there are two equal partners in relevancy such as we find in binaries, the two will push development until the ultimate development is reached much quicker than it will be in a single developing star. This is because the two have a near or perfect singularity match with no one able to dominate the other and both developing with one aim and that is to outgrow the other and establish dominance in such a way.

The balance is in the minor particle finding commotion with enough heat that is allowing the particle to seek independence and a major particle removing space by seeking control of the heat that is in the rebelling mode. When the minor particle gathers sufficient heat it will start an attempt to gain independence by placing as much space in relation to motion between the particles whereas the controller will demand the space in which the minor particle is moving by the major partners dismissing space-time. This relation allows infant and toddler stars to be freed from the galactica cocoons and move away from the galactica centre establishing space-time by concentrating heat from the outer space towards the inner core area. It is all about capturing or releasing by establishing independence or surrendering independence.

Since all the objects in motion in the Universe have not surrendered their independence they're in gravity that forms with **k** allowing the maintaining of independence and **k**$^{-1}$ in a lesser but pivotal role securing stability. Only in the dynamics of the Earth that captured all that is within the atmosphere of the Earth will **k**$^{-1}$ be dominant, as all within the Earth becomes part of the Earths motion of **k.** The control of space is about space duplicating and space dismissing. With motion comes space duplicating but the more the object grows the more will the object dismiss space and the less will the object duplicate space. By not duplicating space the object will increasingly destroy space because the value of space is fifty percent in the duplication of the space. By relenting the duplication of the space the motion of the star will tarnish and that will allow the star to dismiss much more space outside the star than merely the space, in which it is. By that the star gets an increasing ability to dismiss space that is not in the stars control but by putting such space to heat within the stars control. The star finally will achieve an ability to liquefy space into heat. The Sun has reached such a position. The essence of gravity is motion duplicating space and the dismissing of space. The Coanda effect is the best example and the spinning top illustrates the Coanda effect very well. The final step is to freeze space into material and become a dark star.

In the spinning top we read the dynamics, which controls the phenominon we associate with the sound barrier. The top that spins within the Earth atmosphere does so while it shares a relevant position within the Earth's gravity domination dominating the time aspect. When an object is in outer space and being unattached to the Earth as say a satellite is there will be two points in singularity sharing a unit by relevancy. One relevancy will fight for independence while there is the centre in the unit fighting to capture and control whatever space-time has motion within the unit which the centre provide the motion by which it intend to capture the space-time. While the top holds independence as it is captured by the Earth time the top will share a point in singularity matching that of the Earth. As soon as the object is in a motion such motion establish time other that the Earth demands. Then the independent singularity in motion shift away from the spot they share when there is only the Earth motion involved. As the tops spin creates individuality by independent motion it places time component away from the Earths singularity the motion of the top will secure a separate singularity value

In pure nature, the Earth applies all motion there is available with the Earth confinement that is subjected to the cosmos. Other motion belongs to life. Therefore $a^3 = T^2$ and to the top $k = k^0 = 1$

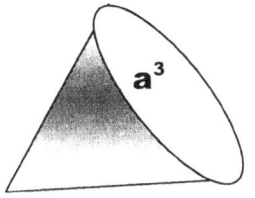

We must see every aspect of the top going into motion as cosmically artificially generated by an alien substance such as life is to the cosmos, where life has the ability to apply motion. To the cosmos, the spinning top is a disguise of the truth and to the cosmos; life is not a factor in existing.

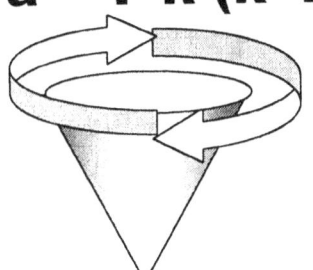

$$a^3 = T^2 k \; (k = k^0)$$

Then by artificially applying motion through the intervention of life the top acts in a manner that will only come about space in normal cosmic conditions if the singularity that is carrying centre has generated massive heat centralised by gravity applying.

As the proton grows the relevancy of singularity will extend. The more the proton contains the higher will the spin be that the proton enforces, but at the same time will the space-time serving singularity be in accumulating captured heat. Therefore, the spin is quicker but as there is more space-time accumulated, it would seem as if there is a slower spin. This growth in material causing more space to duplicate that is bringing more space to contract eventually drives the Universe from one to another extreme. The extremes are the Big Bang and the Big Crunch and the moving process is the Hubble shift. All the processes I have mentioned are well documented. However, since the Earth has less heat than it should have, we find atoms on Earth degenerate slightly, but that is only a response to less growth in relation to what growth is available in the rest of the Solar System with the Earth holding one of the lesser gravity units there is in the solar system.

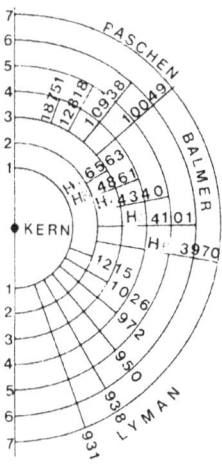

The Lyman series show precisely how heat excites the electron orbit and the heat increase brings about space increase. Heat is equivalent to expanding into space. We find the same degree of excitement in atoms when they are charged with more, heat, however the charging becomes a widening of the orbit circle of the electron. The duplication that increase amounts to more space within the set time and the duplication extends the space the material fills. In the case of the top the top wants to lift to a higher band of spin in a similar manner as the electron does when the electron is exited with more heat. The principle being the Coanda effect remains the same as the spinning motion excels the gravity and the gravity becomes stronger and more aggressive towards the governing gravity of the Earth.

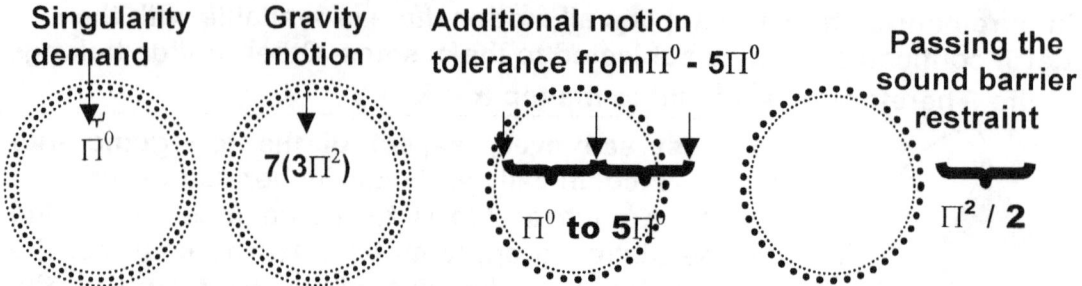

Within one space unit we always have two independent positions holding points in singularity that is sharing time but being independent in space at the same time. The two assert a velocity of $7(3\Pi^2)$ by just moving with the Earth. Since they both move equally in line with the Earth the two objects hold an independent that is Π^0 in independent speed but the independence may grow to $5X \Pi^0$ before the sound barrier becomes an issue of concern.

In pure nature the Earth apply all motion there are available with the Earth confinement that is subject to the cosmos. Other motion belongs to life. Therefore $a^3 = T^2$ and to the top $k = k^0 = 1$

We must see every aspect of the top going into motion as cosmically artificially generated by an alien substance such as life is to the cosmos, where life has the ability to apply motion. To the cosmos the spinning top is a disguise of the truth and to the cosmos life is not a factor in existing. If the cosmos found something in a rebelling mode such as the top is the most unlikely scenario will come about where the top find a means to establish singularity strong enough or driven enough to establish such highly developed independent singularity achieving such motion. However, the studying of the action of the top introduces the utmost basic principle we find in the Universe. We find the law of generating independence within the framework of confinement. That is what all of the Universe is. The Universe is independence of a lesser in the space of a dominating. The motion differentiation brings about the motion and the drive for motion. The motion then must be so well contained by the rebelling singularity that the Earth would be unable to depose the heat from the minor singularity centre. What we humans rein act is a process that happens in galactica when stars surge to find independence as the progress away from the confining centre and drive for independence.

The process is there and is part of the Universe but is part of another process we find in another part of the Universe. That process can never come about in the working of a star-like structure. Never should science intermingle life and the accomplishment of life on this little God forsaken blue dot of no repute as a fraction of the cosmos and have life acting on behalf of the cosmos. There is no round rock on any planet large or small that will start spinning because such a round rock has produced motion driven by its red-hot-core bringing about the motion. Before life intervenes and charge the top to bring about "cosmic Life" (a term I dare to use for the lack of a better term, but in effect is the volume of heat concentrated in the centre of a rotating structure) the top is experiencing gravity just as we do by having gravity press the structure onto a surface where the top makes as much contact as possible with the surface that either is the Earth or stands in on behalf of the Earth. The top is resting without motion on the Earth in orbit of the Sun as all captured objects do.

In this thought comes the realizing that motion is heat driven. The more motion there is the more heat there is. I am not referring to the motion of the top since the motion of the top is driven by life's ability to manipulate space-time and the top spinning in that regard contradicts all rules that should apply. No rock will suddenly voluntarily by own account start spinning, unless it is dropped in some form of heat and then the motion would be violent. The motion can only come about where there is a very high concentration of heat and realising this plus the fact that the top suddenly achieve independence from the restraining the Earth gravity enforces puts some ideas forward.

Then by artificially applying motion through the intervention of life the top acts in a manner that will only come about space in normal cosmic conditions if the singularity that is carrying centre has generated massive heat centralised by gravity applying

$$a^3 = T^2 k \quad (k=k^0)$$

The spin we find the top has, is an indication of what principles drive the Universe. The proton collects heat from outer space as the neutron supply motion by duplication. The duplication is a manner by which space-time is in contact with heat and by duplication it is in contact with heat that is released in space. By motion the proton contracts the space and by motion the neutron duplicates the space. The neutron replenishes the space it lost to the proton by moving about the space and this motion we see as the electron. The more the duplication is the more space-time heat is available to supplement the diminishing by contraction. That motion is time related and the higher the duplication is by time unit, the more contraction comes about and the less space in relevance the atom claims. There is more motion in the Earth's atmosphere, which constitutes to more gravity, which is more contraction, therefore is astronauts are smaller in the Earth's atmosphere than they are in outer space. The higher the spin rate is, the more space the spinning top puts into the unit and more space there is in the unit. That means by increasing T^2 (the spin) the longer k will be and the higher the ratio will be in relevance between space and space.

As the proton grow the relevancy of singularity will extend. The more the proton contains the higher will the spin be that the proton enforces, but at the same time will the space-time serving singularity be in accumulating captured heat. Therefore the spin is quicker but as there is more space-time accumulated, it would seem as if there is a slower spin. This growth in material causing more space to duplicate that is bringing more space to contract eventually drive the Universe from one to another extreme. The extremes are the Big Bang and the Big Crunch and the moving process is the Hubble shift. All the processes I have mentioned are well documented. However, since the Earth has less heat then it should have, we find atoms on Earth degenerate slightly, but that is only a response to less growth in relation to what growth is available in the rest of the Solar System with the Earth holding one of the lesser gravity units there is in the solar system.

By duplicating space-time in a rotating manner the motion elects a chosen singularity by exciting such singularity and the motion of the unit energises the chosen singularity to steak a claim off on independence. The independence comes from a singularity that established a centre where the lack of motion at that precise centre dismisses space to the reducing of space-time.

By spinning the atoms that is forming the unit, are all in motion, where even every independent singularity point centralising and securing that specific space-time is in motion by transfer of space-time. The motionlessness of independence has to transfer in order to legitimise the claim of independence. While in motion it establishes a centre that connects all the atomic motion and represents the one point in space-less-ness of being without motion. At that point gravity comes about that singles out the unit from the rest of the space-time in dominance and that point of space-less-ness and motionlessness find a singularity that then is charged with the task of securing growth and maintaining heat accumulation to further a flow of space-time from which all material in the unit that is duplicating will benefit by securing a stronger presence of duplicating while the chosen centre elects top dismiss space-time forming a stronger presence of a more secure singularity.

Under normal conditions this will only come about with the aid of a massive number of protons. In this case however the spin came from life generating some artificial gravity. When in ordinary electricity producing the operational conditions in electric motors we gave the process found applying there the name of electromagnetic induction. It is still applying the principle of heating singularity to produce centre heat charging motion. Although in this case the top is spinning because of the action of human muscle creating motion. The electric motor turning induces a centre by using the Coanda principle with the benefit a star quality motion. Humans take conditions one find inside stars where $iron_{56}$ causes space-time flow $(5 \times 5 + 1)$ of protons and copper $(2 \times 10\pi)$ to establish star gravity and centralise space-time flow. The important issue is that motion creates the spinning, which results in the top sustaining its own space by effectively enacting independent gravity through the Coanda process that is replacing the Earth's motion.

Singularity within the subatomic particles dismisses space in order to sustain cooling. The motion applying as cooling is a contracting and depleting of space that reduces the time factor to the point of elimination. To do that that it has to incorporate heat from relevantly close by or neighbouring singularity that has overheated and abandoned form. By resolving such heat into realm of the singularity able to secure survival by sustaining by the cooling of singularity that is applying gravity. When doing such reducing of space the singularity elected to the task eliminate space that allows a bigger flow of space by cooling space within material and concentrating the heat levels in space per space volume. The concentrating of heat within the cold and reduced space forms a driving ambition that generates motion into the revolving space-time surrounding the heat securing singularity.

The concentration of heat feeds the motion of the space-time in motion around the axis of singularity. By feeding on the much-concentrated heat the heat contribute to motion and the motion contribute to excessive (more than before the spin) duplicating of space by performing a stronger flow of space-time being the result of a stronger substitution for the dismissing of space in the centre. The motion is a result of combating heat which is a result of forming motion and the one factor generate the other factor as long as there is space to reduce and advance the heat levels that will supply the motive to produce motion on which further motion will feed while all the while singularity is incorporating heat into the unit accumulating singularity as it is concentrating heat by creating an ever reducing flow of space-time towards the centre and a duplicating of space-time is resulting into an increase in independent and unattached heat levels in the reducing space. When more heat concentrates it becomes an overheating situation that will increase both the duplicating as well as the dismissing of space-time that will increase the heat levels further.

It is an ever-increasing process and the only factor loosing is the outer space dismissing of heat within outer space. As the concentration of heat in outer space reduces the space factor in outer space that space then will increase by the same margin and that will seem to bring about the increase of space alone. This action is more like water falling on a hot plate. The motion duplicate space by increasing space as water reacts to the heat surging and the surging is elevating motion by producing space that instigate expansion, which generate as it heightens motion. But it is only the coming about of motion that is driven by heat levels increasing to which I here refer to in my using of this example.

The duplication that provides the motion can only come as a result of surrounding the singularity with other heat than the heat produced by singularity. When singularity release heat through overheating it reduces its relevancy in the relation between it and surrounding singularity, which is sharing relevancy. When the singularity sustaining form with gravity is committing heat brought about by other singularity such action enhance the relevancy in the relation that such singularity has with other relevant factors.

When the lesser singularity finds dominance in the diminishing action of the relevantly superior singularity it has to react by providing heat that surrounds the lesser singularity in space-time. If the heat is a product of self-inflicting overheating the dominance will prevail to a point where the lesser subdue in defeat and join the space-time of the superior singularity. This can only come about when the ultimate limit is crossed and the crossing of the Roche limit turns the value of k from $k^0 = T^2 k / a^3$ to $k^0 / k = T^2 / a^3$.

But when the lesser singularity has the ability to surround the heat accumulated from space-time being outside by using space reducing thus intensifying heat it will establish motion that will place the relevancy of such motion coming about from the space diminishing ability of the superior singularity. It is all a matter of sustaining singularity by matching and qualifying relevant abilities in different singularity. Without the surrounding of heat the singularity will submit independence and surrounding by sufficient outside heat it then can sustain the independence of an independent developing singularity. A clear understanding about this principle will lead to accepting the principle and the accepting is of crucial importance when one wishes to come to understanding the origins of the solar system.

It must be clear that it is the harmony in the speed and not the value of the mass or the quantity of the heat that will establish the allocation of the positions taken by the relevant objects applying gravity borders. Only when allocation reaches the fringe of the borders set by the relevant objects comes into the limits it is that quantity and potential ability has an influence on the outcome of further development by either object. In the book an open book about The **Seven Days Of Creation ISBN 0-9584410-4-9** I delve extensively as I delve deep into this topic of how the solar system comes about when using the cosmic relevancies that I am about to explain.

Gravity is the product of motion and with motion gravity comes naturally in the process of motion. Due to the way the sphere is built, it will always hold singularity in the centre of the structure. Singularity is a point within the centre of all spheres where no motion can be possible because the rotation pivots at that point and the pivoting changes direction precisely at that point without sides. The motion to which I refer here is the combined motion of all the atoms in the sphere that creates a singularity centre. The pivoting comes about where the line that forms the circle ends in the first dimension That we read into Kepler's formula about space-time originating from singularity $k^0 = a^3 / T^2 k$. At that point, all space-time finds the relevance to return to the form of formlessness. The point that holds the value of Π^0 is located by dimension within any and all spheres but to top this Kepler showed that motion produce space and space is time through motion $a^3 = T^2 k$. Time forming the second basis for the entire Universe is the spin of heat in space. Without motion space-time collapses into singularity and within all spheres there is this point that cannot provide motion. Therefore, through the form the sphere holds, the sphere will diminish and destroy space in the centre by not providing motion within the very centre in the round structure.

The top and the Earth share a common singularity. That is gravity we all see but it is not a force because when motion comes to the top, the shared singularity shifts. We may consider such shifting as receiving momentum but the momentum is only an enlarged gravity because more motion discrepancy comes about. The line $7^0 \times (3\Pi^2)$ shares a common connection to the singularity in charge of controlling all gravity aligned to the centre of the Earth. The motion of whatever magnitude comes and proclaims the top even more independent and a shift in the line of gravity comes in and puts space in between the line the Earth shared with the spinning top and the line the motion indicate to connect the top with the governing singularity. The shift can be anything from Π^0 going on to become Π and then extend to $\Pi^2 / 2$. That is what the factor **k** becomes added to the existing value there was before. By introducing a stronger **T²** the **k** factor has to become more and that is what causes the sound barrier to break.

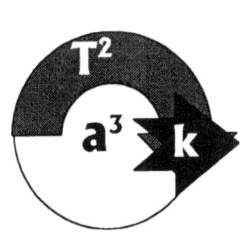

The spinning top is applying the Coanda principal to establish gravity in relation to the Earth centre or the controlling gravity. As soon as the top starts spinning at a specific velocity, the motion of the top secures a centre singularity that keeps the top upright. This is the motion the Coanda principle requires to produce a gravitational centre as to produce a specific space-less point in gravity. It is gravity that the top employs because only another gravity can interact with the Earth gravity to secure stability while the top is matching a specific spinning speed.

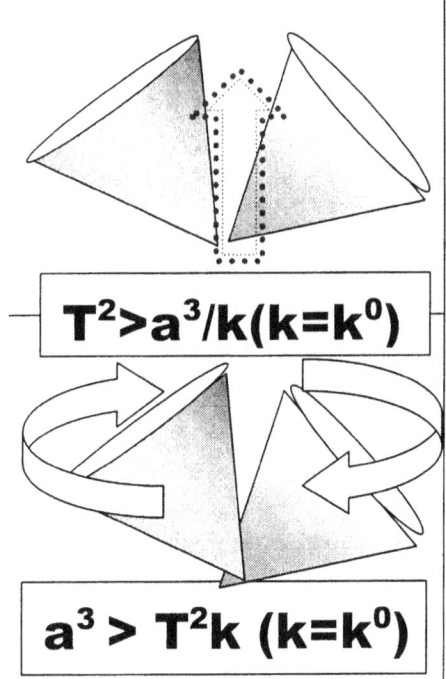

When the top spins at a pace exceeding apparent barriers the top will begin to turn about the axis as if it is trying to extend the axis, which it is spinning about. When the spin is extensively faster the top shows an effort in trying to escape from the mass constraint and the motion providing balance then becomes much more than the counter balancing of gravity. The excessive motion will create a spinning action where the top literally tries to use the spin to lift from the ground and jump into the air. Part of this is going side ways by spinning. The top jerks to the one side and jerks to the other side in an attempt to secure more space created by excess spin to use for the creating of additional space. Well that is precisely what it is doing in the attempt to turn around by extending the spinning axis which individual singularity provides.

The space it created through spinning has then reached a point where that space it then claims, is exceeding the space the Earth granted the top to have as a motionless object and the object is by the value of the motion it produce, acquiring more space. With extra motion, it will find more space-time by which it then tries to secure a better space-time position for the singularity it is maintaining.

The apparent ness why a top will not spin in outer space becomes more evident when considering the gravity the Coanda spin motion introduce with all the relevancies that is attached to the process. It is the relation that the spin establishes by implementing the Coanda effect by motion, which secures the gravity that the top install in order to place the top in the upright position. Installing space-time secures independence the top requires to spin. Gravity is all about motion and nothing about mass enforcing pagan forces, which unleashes the will of the unknown upon the unwilling.

The apparent ness why a top will not spin under control in outer space becomes more evident when considering the gravity the Coanda spin motion introduce with all the relevancies that is attached to the process. It is the relation that the spin establishes by implementing the Coanda effect by motion, which secures the gravity that the top install in order to place the top in the upright position. Installing space-time secures the independence that the top requires to spin. Gravity is all about motion and nothing about mass enforcing pagan forces, which unleashes the will of the unknown upon the unwilling. In this I have to point to the fact that when a body returns from outer space the body does reduce in size because the body adjust to time, but while the body is adjusting the body remains the same because a blanket of heat or time covers the size the material does not fill. In the spinning top it is this blanket or heat cover that readjusted that is the main factor which keeps the body upright.

The mass does not apply or draw the spinning top down and in fact the top is trying to undo the mass the Earth provides. The stronger the motion is that top asserts the bigger claim the top will make on the unit singularity. The top will try to rise into the air by extending the k factor between the singularity claiming space for the top's independence and singularity claiming space for the Earth.

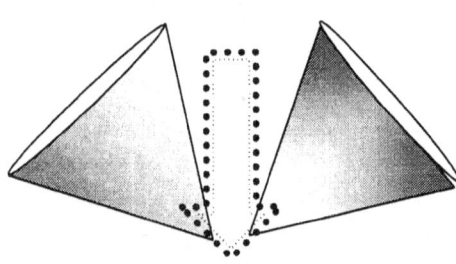

Only after the top showed the last act of defiance will the top again submit to the mass, which comes about from the top's inadequate spin tempo that will not any longer produce a sustainable space to create the needed space and use the motion as antigravity.

The lighter the spin rate of the top is, the bigger will the surge be to establish independence and by trying to lift from the Earth, that lifting shows the top has an attempt to reduce the affect mass has on it. By spinning faster $T^2 k$, such spinning is increasing the relevancy of the space a^3 that the top claims in relation to the relevancy of the space the top holds. That increases the buoyancy factor that the top has in relation to the rest of the space the Earth claims.

When the earth reintroduces the effect of mass the top still tries to create through motion sufficient space to use as a deterrent to the mass the Earth wishes to inflict on the top. Clearly, the top shows that the space relation grows between the top and the Earth which such growth is favouring the top's position. The higher the spin of the top, the more space the top seemingly claims. This proves Kepler finding of space being equal to the motion thereof $a^3 = T^2 k$.

Ask any Newtonian where the centre of the Universe is and he would not know. Yet a top spinning can summons the centre of the Universe to erection just by motion. By introducing spin the Coanda process activates the three position hold in accordance to the four that time holds and that is performing space (3) – time (4) and with that by spinning around in another (3) points the top becomes an atom in space-time. We know the centre of the Universe is located in every atom.

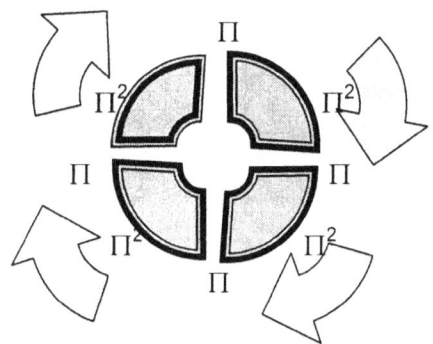 

The time factor in four points forming a circle activates the space factor in three points forming an axle, which activate the time factor in three other points and it is the time factor in three in there keeps upright. It is (4 + 3) inside plus three outside making the top have the ability to spin or duplicate by individual action.

The time activates a line from which space begins. The line forms a space that turns the top, but just as important is the fact that the line extending singularity activates a dimension in time in which the top can spin.

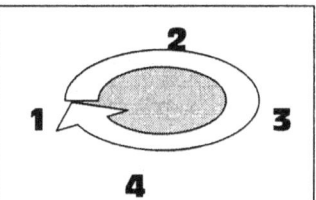

It is the time that relates to the spin of the space filled with material that allows the position the top has when spinning. When the co ordinance between the space spinning and the timeline keeping the balance falters that the top looses independence and fall.

The issue is that the rotation activated the time component where four points set about by motion. This formed the basis of the Coanda principle as the rotating top then became liquid although the wood is a solid. The point on which the rotation turns becomes the solid. This is evident in the fact that the center is activated and not activating the spin. As the spin reduces its dynamics the turning still fight for balance

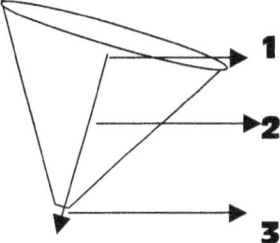

When the Earth reintroduce the effect of mass the top still try to create through motion sufficient space to use as a deterrent to the mass the Earth wishes to inflict on the top. Clearly the top shows that the space relation grows between the top and the Earth which such growth is favouring the top's position. The higher the spin is that the top has, the more space the top seemingly claims. This proves Kepler's finding of space being equal to the motion thereof $a^3 = T^2k$. The fight in the top is real for space supremacy and the determination of the top to prevent the declining of the space when the top slows down that the top establishes almost seems as if it becomes a fight of the death of motion to prevent the death of the motion where just the stronger between the Earth and the top will survive by killing off the weaker. Most times the fight lasts to the point where the top starts to spin vertically by running on the surface it used for spin. This must bring the evidence that mass does not bring about gravity but gravity creates mass when there is a lack of space brought about by motion duplicating space.

The mass does not apply or draw the spinning top down and in fact the top is trying to undo the mass the Earth provides. The stronger the motion is 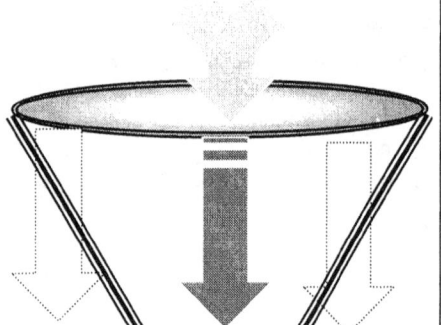 that top asserts, the bigger claim the top will make on the unit singularity. The top will try to rise into the air by extending the **k** factor between the singularity claiming space for the top's independence and singularity claiming space for the Earth. The lighter the spin rate of the top is, the bigger will the surge be to establish independence and by trying to lift from the Earth, that lifting shows the top has an attempt to reduce the affect mass has on it. By spinning faster T^2k such spinning is increasing the relevancy of the space a^3 that the top claims in relation to the relevancy of the space the top holds. That increases the buoyancy factor that the top has in relation to the rest of the space the Earth claims.

Only after the top showed the last act of defiance will the top again submit to the mass, which comes about from the tops inadequate spin tempo that will not any longer produce a sustainable space to create he needed space and use the motion as antigravity.

Notwithstanding the motion the Earth forms towards the center of the Earth, the top finds a way to counter act the motion by producing a motion that is much stronger than the motion or mass, which is the lack of motion. The mass of the top has disappeared because the force or mass that the needle point of the top generate multiply the normal mass of the motionless top many fold and yet the top overcome the mass restriction by generating an independence through motion

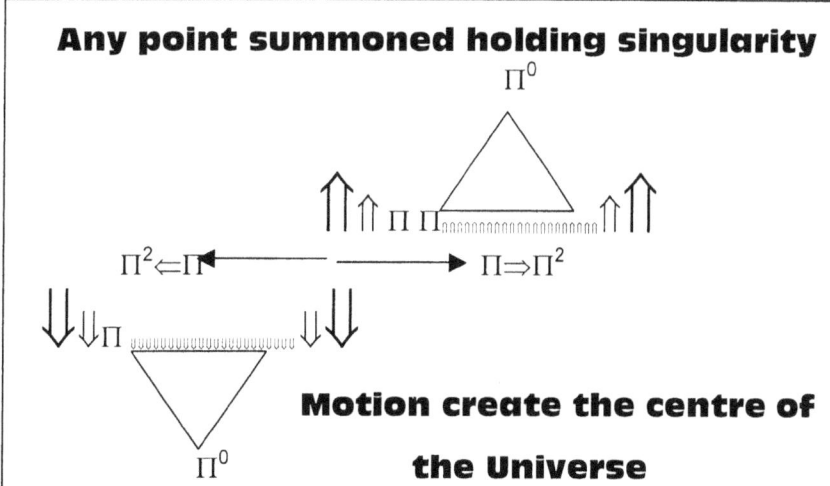

In cases where motorised machines bring about motion, the artificial energy is strong enough to apply motion that counteracts the mass factor completely

By bringing about motion that produces lift to the object captured on Earth, as the rotating of the wing of the helicopter will produce through motion creating space, the mass disappears. Mass is the manner in which the Earth depress the space the helicopter claims. This is the truth of mass, notwithstanding the corrupt definition, science uses to further foul and disgrace an already corrupt definition. What I refer too as being corrupt is giving a body mass in space while the body might be taking up volumetric space but mass it does not have for it has micro gravity instead. Science knows better and science has known better for a long time, yet not one person in science stood up and proclaimed the error. At some point, science has to come to terms and accept the differentiation that is degenerating the truth about mass and gravity. One must either decide to remain as stubborn as a mule and insist on it having mass just because of some ridiculous argument tried to cover up disgrace in the past, or admit that the action relieves the object of the mass that the Earth produces.

The object will still sustain an effect from space duplication discrepancy but such discrepancy is then so much reduced the object is as light as air. That is the reason it can fly! By using the heat it incorporates the heat in the atmosphere, the helicopter increases the material to heat relation, who then brings favour on the side of duplication and then the relevancy about the duplication to dismissing of space-time favours the duplication by yards. In all of this I cannot find one shred of evidence where mass brought about any thing except being a result of the motion not producing independence to the top and therefore allowing the top to create an escape pass. Mass is the absence of adequate motion.

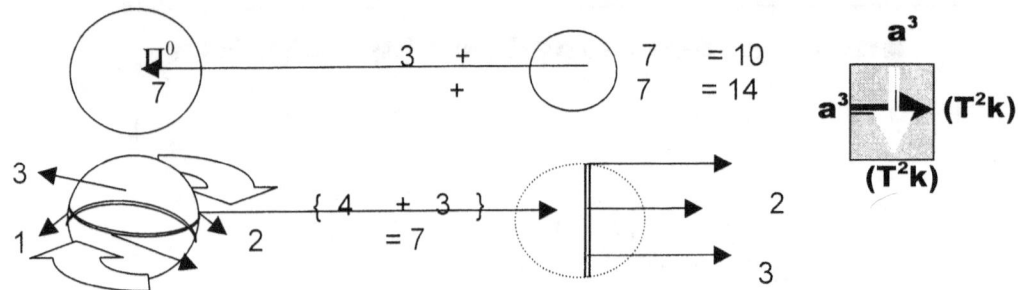

When the top is pinning vigorously and with much skill the top lands on the ground with a seemingly high velocity because from that it can produce sound. This proves that the motion energises singularity positioned in the centre inside the top and the energising is contesting the position in view of the Earth governing singularity. The top is served with "life like" energy and the basis for that comes only from a point in the centre that has no sides and no space within our cosmic boundary, yet it generates drive by motion. With enough vigour the top will make a humming sound and if one is truly skilled a wining noise. This is the same wining one find in a spinning turbine or spinning a roller bearing at high velocity. That is sound resonating with singularity and the start of what eventually becomes the sound barrier. Since the sound contains singularity by spin and not that much by linear motion the breaking of the sound barrier does not become a factor.

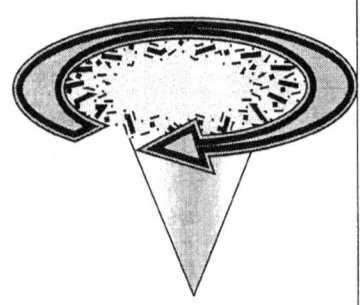

The spin under normal conditions can only come about as a result of more heat. With that aside the spin normally caused by heat will bring on a linear gravity running towards the centre of the top. This is then a product of $k = a^3/T^2$. But to counter this (Newton's law on action and reaction), another balance comes about $k^{-1} = T^2/a^3$ that centres the material in line with the progressive spin and the extending of the motion that should be because of a liquid heat adding to the material.

More spin increase both lines that force gravity by the increase of T^2 extending k, k^{-1} as well as a^3. The space wants to exceed its boundary because suddenly the motion allows the space to become extended. The gravity line running to the centre wants to extend for the same reasons and so does the gravity line running towards the liquid that should be there and that should be enforcing this sudden living up to better standards.

As the top hit the ground it start to move around in a vigorous manner as if to energetic to stand still and that is precisely what happens. This searching is a very important sign and is of most importance.

As the top hits the ground after being thrown with spin it starts to move around in small circles while rotating the axis. It spins in a vigorous manner as if the top is suddenly too energetic to stand still and that is precisely what happens. This surging finds a new dynamic and is a most important rule in cosmic principles.

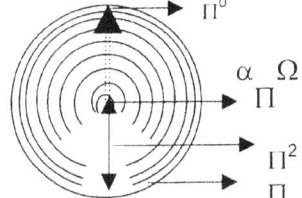

Nothing said so far is high tech or mind bending complicated. All the above arguments from the first page to this point reached are simple and there's only ordinary primary school mathematics involved that every scholar should know. One does not need a brain fitting Einstein to come to these conclusions but just thinking about everyday issues.

With all the excitement and no where to take it the extending of the drive line runs down the singularity inwards towards the newly established governing singularity that keeps the whole job erect.

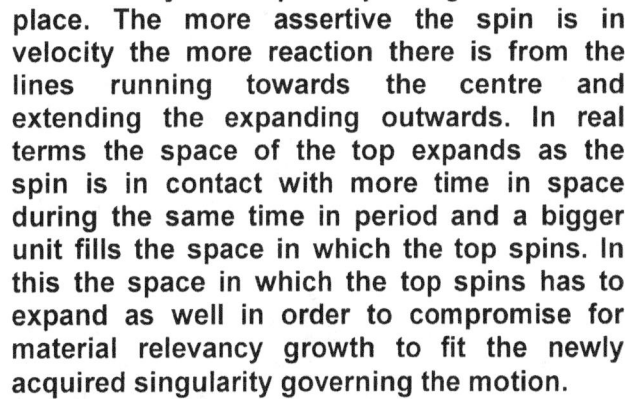

That is why the top is spinning in the first place. The more assertive the spin is in velocity the more reaction there is from the lines running towards the centre and extending the expanding outwards. In real terms the space of the top expands as the spin is in contact with more time in space during the same time in period and a bigger unit fills the space in which the top spins. In this the space in which the top spins has to expand as well in order to compromise for material relevancy growth to fit the newly acquired singularity governing the motion.

The support that the spinning top finds in the established governing singularity keeps the top spinning in an upright stance, only supported by the singularity that takes charge of the spinning space-time

The balance is a control of motion that is established as a flow of space-time supports the ends (4) holding time while this generates the space (3) singularity containing and creating the space (3) in which the spinning takes place.

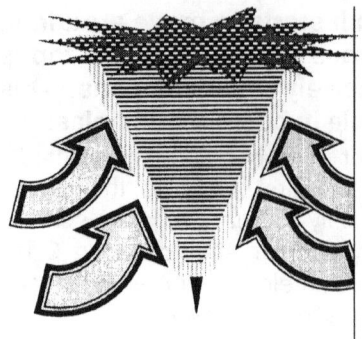

The heat that should supposedly under cosmic law drive the spinning top will come from the governing singularity accumulating the heat in concentration by the contraction or cooling ability the top singularity acquired. But in this case the spin is a result of life's ability to manipulate space-time and lead cosmic events. The heat that would establish such a drive in motion in real cosmic terms would require a lot of nourishing and sustaining from a large number of maintaining atoms that produce a large flow of space-time.

With sufficient energy the top gets into a fighting mood making the top very reluctant to give up this newly established freedom. The behavior now attributed to the top is normally the manner how a star develops in the galactica cocoon and how the fledgling star gains its birthright to leave the nest of the cradle of the galactica. The atoms form a sum total of space-time displacement that can support the generating of the required gravity in securing the heat that would unleash such a drive. Such singularity in governing come to life and release the new star from the blanket of heat that covered the star up to the time of release.

The example we can gather from the top shows how desperate a governing singularity can become when starved of motion and how such an exited singularity can put up a fight for life and independence. The top is in a fight for independence while the Earth is restraining the independence. The fight goes on until the earth suppresses the last bit of motion that the top has and the top uses the last motion it has to defy the Earth's control.

When the motion exceeds the level of the Earth gravity, the top shows an eagerness to rise to higher levels of independence in the same manner that an electron reaches into higher rings of energy because the top with motion is in an electron relation with the Earth filling the proton role and the atmosphere being in the neutron role.

Let's quickly establish events as they translate singularity from a dot to a controlling entity that is commanding space-time through the establishing of a separate individual drive. The motion comes about which proves to be that which generates the gravity that drives the individuality in the top.

In the sphere centre is a spot that has to be there mathematically by measure of $(\Pi r^2) / (\Pi r^2) = \Pi^0 r^0 = 1$. In order to provoke the line into action, motion is required just as Kepler indicated where the space becomes equal to the motion and the motion is equal to the space $a^3 = T^2 k$

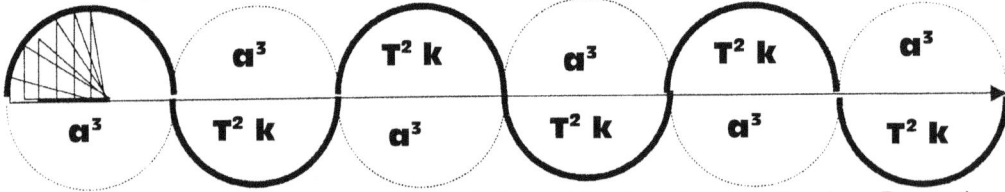

By the motion and the singularity the top evoke a graph forms where the graph runs along the line of time.

By the motion and the singularity the top evokes a graph form where the graph runs along the line of time.

The balance is a control of motion that is established as a flow of space-time supporting the ends (**4**) holding time while this generates the space (**3**) in singularity containing and creating the space (**3**) in which the spinning takes place.

There is a something (if you wish I'll use the term force although I strongly hesitate to use such an outrageous term) that is generating power to keep the top upright while the top is spinning. The energy that is charged has the dynamics to stand its ground against the gravity of the Earth that is under normal circumstances controlling it but inspired, the top seems to be reviving singularity by motion. The top is fighting and rebelling against the Earth gravity. The top is self-driven, as an electric motor would be. The difference between it and an electric motor would be the origin of the source from where the energy comes which drives the spinning top. The top stands upright as individual as any self-propelled object can be. Although gravity is retaining the motion of the top, it is not contradicting the motion. It is not combating but is merely suppressing the motion. What we would think of as air restriction is no restriction because from the restriction comes support that keeps the top standing on a very thin needle edge. The top should tell us so much about cosmic laws, if we would only listen and learn and not tell cosmic laws what we think nature should tell us.

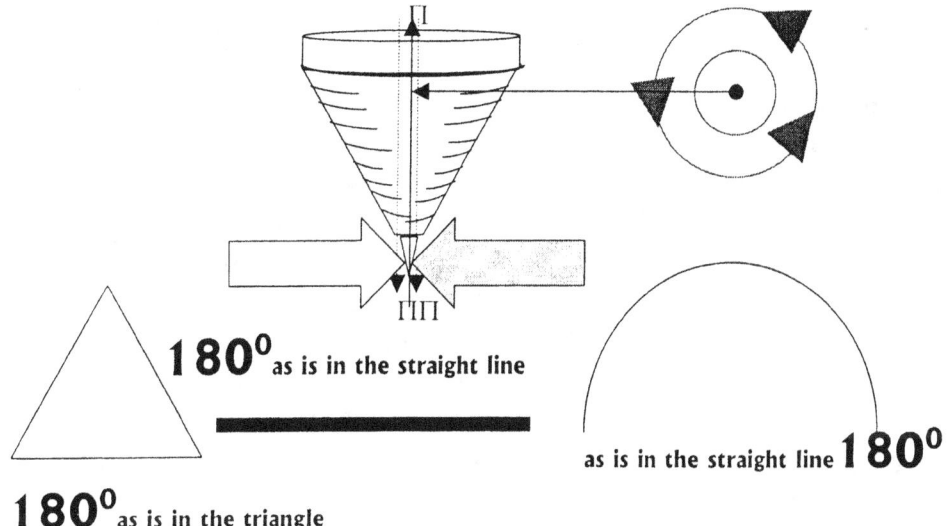

$180°$ as is in the straight line

$180°$ as is in the triangle

as is in the straight line $180°$

From the motion the top inspires by creating a situation that the top can establish a force or an energy, which is able to keep the top upright, is equal to the establishing of gravity and electricity. It is the very same principle being the Coanda principal.

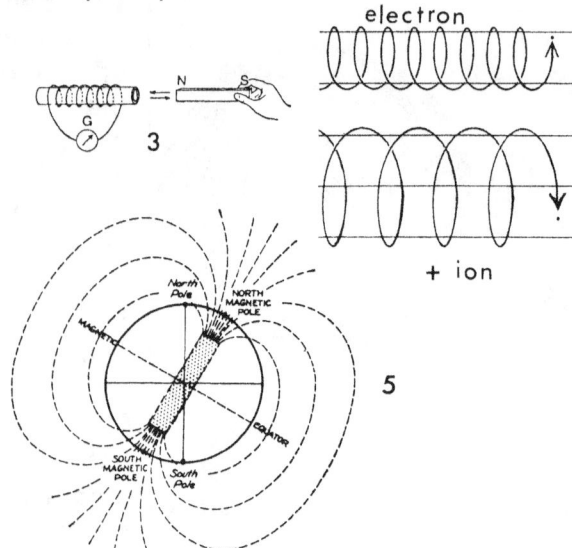

That which charges the top to stay upright is the same that charges the generator to charge electricity, is the same as that which charge the Earth with gravity. There is no force but a flow of space-time, which is contracted by motion and duplication to bring about the conducting of electricity. Man may name it gravity and then name it electricity or name it motion and balance but like hot and cold that is man made. To the cosmos in the Universe it is the same thing. It is what started the cosmos. It is what drives the cosmos. It is the engine giving motion as it is giving discipline in the cosmos. It is producing space to heat and heat to material.

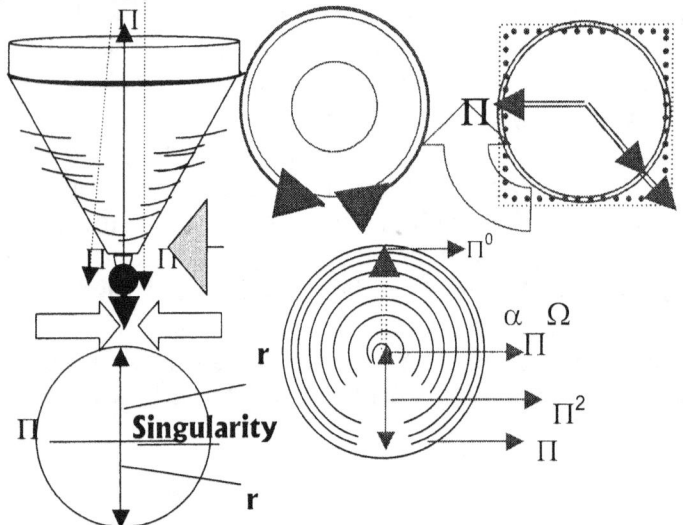

It can be only singularity that is keeping the top upright. One must remember that the part doing the balancing, that is creating the space in which the top is able to spin, that is establishing the necessary time distance in which the top can spin, that is establishing the time difference that the top can use to apply the motion that extends the time is singularity. The line that is evoked is not in real terms part of the Universe because the line has no sides, can't move at all and is the only substance that is there for all to see while being very much invisible.

The line has the responsibility to establish everything that is in the Universe while the line by own measure is not a functioning part of the Universe and yet it sets the absolute control of what is going on in our Universe. Without the line being part of space in the universe it is what drives the Universe in time and in accordance with time. The line is invisible, in detectible, has no space, can be called into action any place and anywhere from where the line immediately establishes control of what is in the Universe and creates matter in space in time in the Universe in infinite detail while the line is even less than infinite.

All material has gravity built into the form it holds. Due to singularity forming Π^0 space collapse as a result of motion in the centre of all spinning structures. The faster the structure spin the more intense that space will be that collapse and therefore the more heat there will be to accumulate within that centre. The space ending there forces the motion to stop and the lack of motion produces space dissolving. But with space disappearing a flow of space forms that becomes necessary to replenish the space that went lost.

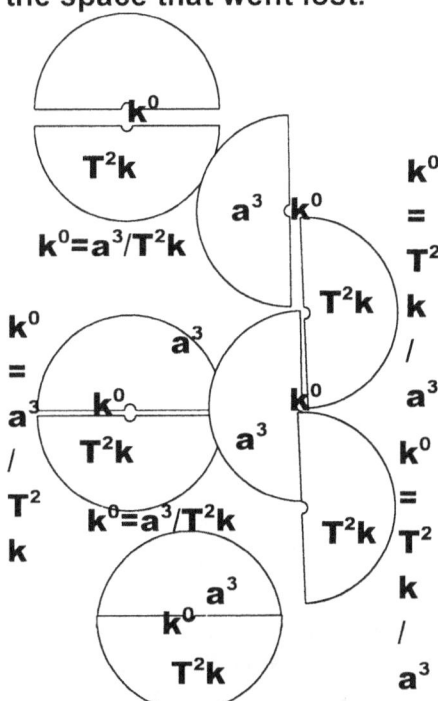

This centre is also within every atom where the proton forms such a centre and the atoms produce a unified effort of diminishing space to the total volume of space diminished that is produce as a compliment by all the atoms sharing space-time in the unit that is combining efforts where at the space less centre of the star in the space less centre diminish space. The refilling of the space produce a flow from the outside flowing towards that established centre we witness in the Coanda effect as in effect the centre where the space vanish as it accumulate in the centre of the sphere where no space ever can be.

Towards that end all space will flow but will never be able to replenish the lost space because motion stropped at that point and the motion that brings about the space destroys the space just because the motion cannot be produced at such a point. The fortunate part to us is the planning because the space that is dissolved is the space of the less successful spots that did not manage to establish contracting that well and in that kept expanding with the friction to combat the heat coming about.

There is a something (if you wish I'll use the term force although I strongly hesitate to use such an outrageous term) that is generating power to keep the top upright while the top is spinning. The energy that which is charged has the dynamics stands its ground against the gravity of the Earth that is under normal circumstances controlling it but as inspired, as the top seems to be revivifying by motion, the top is fighting and rebelling against the Earth gravity. The top is self-driven, as an electric motor would be. The difference between it and an electric motor would be the origin of the source from where the energy comes which drives the spinning top. The top stands upright as individual as any self-propelled object can be. Although gravity is retaining the motion of the top it is not contradicting he motion. It is not combating but is merely suppressing the motion. What we would think of as air restriction is no restriction because from the restriction comes support that keeps the top standing on a very thin needle edge.

> One relevancy spinning on the outside of the unit will fight for independence, while in the centre is another relevancy stronger than the one on the inside in the unit that is fighting for containing the one relevancy which is striving to uphold individual independence. The centre singularity is fighting to capture and control the outer singularity. This fight forms whatever space-time has to offer by means of motion within the unit, because a perfect harmony must prevail while the centre provides the motion. Both directions is motion but the flowing towards is by which it intends to capture the entire space-time While the top holds independence as it is captured by the Earth time the top will share a point in singularity matching that of the Earth. As soon as the object is in a motion, such motion establishes time other than that which the Earth demands. Then the independent singularity in motion shifts away from the spot they share when there is only the Earth motion involved. As the top spins, the top creates individuality by independent motion it places the time component away from the Earth's singularity. The motion of the top will secure a separate singularity value.

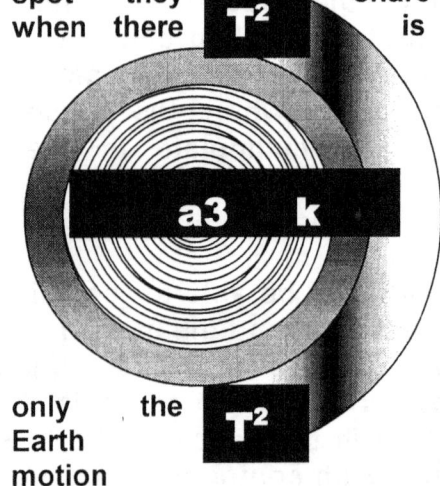

By the motion committing the top to motion, the top forms a secondary alliance with the primary gravity the Earth provides. The motion the top achieves is going beyond as well as including the motion, which the Earth provide. This employs the Coanda effect as the Coanda effect interprets the Pythagoras principle where the square establish a crossing of the singularity line committing singularity to activate the centre line of singularity in forming a division between 180^0 and 90^0. The top secures a centre by lining up in 90^0 with the rotation of the earth that is holding 180^0 position. Every time the top rotated, the top is forming a new graph that is sustaining both the previous and the following graph. Moreover and more important is that motion dictated space boundaries by committing a centre to a specific relevancy.

The fact that motion secures a unit of space-time where one part tends to be without motion while the other part plays the role of a liquid that supply the motion, that became the essence in the pre-Big Bang Universe. Without that information there is no pre-Big Bang realisation.

The top should tell us so much about nature if we would only listen and learn and not tell nature what we think nature should tell us. That which charges the top to stay upright is the same that charges the generator to charge electricity, is the same as that which charge the Earth with gravity. There is no force but a flow of space-time, which is contracted by motion and duplication to bring about the conducting of electricity. Man may name it gravity and then name it electricity or name it motion and balance but liker hot and cold that is man made. To the cosmos in the Universe it is the same thing. It is what started the cosmos. It is what drives the cosmos. It is the engine giving motion as it is giving discipline in the cosmos. It is producing space to heat and heat to material.

When there is a smaller object within such a space duplicating and that smaller space also serves as singularity and as an independent unit from the major singularity the space the smaller object holds is not duplicating to the same trend as the larger space. If the larger space held the same claim in size to that of the smaller space occupied the time aspect will set the two claimed spaces billions of years apart. But since the two share the same time the space differentiation is significant. With the smaller space being part of the larger space but also still being independent it will lack the capacity to match the larger space in the effort the larger space can present in duplicating. If the duplicating of the smaller space was more favouring the duplicating than the dismissing of space it is found more towards the outer rim of the sphere, which the larger space claims.

Due to size restriction the smaller claimed space will not match the required space duplication when sharing the same time factor in using the matching motion since both share the time aspect. Although the top shares time in space with the Earth the top can find relief from the drowning effect of the gravitational dominating space – time the Earth unleashes just by applying more motion with the help of the motion the Earth provides. As soon as the top starts spinning the top duplicate more space within that space of the containing object, which in this case is the Earth, and by applying credible motion the spinning top suddenly find the required ability allowing the spinning top then to match the space reproducing. By excelling motion the top find the ability to establish enough gravity to bring about independent balancing through independent motion.

When spinning excessively the top can overshadow the space duplication and in that it can seek the ability to find more space that is normally more to the outer edges of the sphere and therefore try to match the space reduplicating with space towards the outer edge. This effort we call flying. Gravity will always comply with two relevancies one is creating space flowing away and the other is space contracting. When the top produce a spin it tries to secure sufficient motion to lean towards escaping the confining of the Earth gravity by trying to establish independence by which it can escape and by not spinning it then submits in defeat by falling with the inward contracting space of the Earth.

Not only does this explain the energising of the top when spinning with excessive motion applied but it also proves that the gravity induced by contraction is part of the accepted form of space-time in the roundness that is securing Π as form. It proves Kepler as correct where Kepler said gravity is space in motion. The motion creates space duplicating within the parameters of the space containing the minor space serving as a unit, but with the intensity of spinning the motion establish a concentration of heat flowing towards and collecting at the centre that forms the motive for the searching to secure an independent by the spinning action. The contraction is a result of singularity diminishing space by not providing motion at one specific point and therefore always creating an initiative for space to flow by motion back to singularity to replenish the space that disappeared. That serves the proof that there is no pulling but there is rather a flow of space-time.

When it comes down to the pulling of gravity that there is no pulling of gravity. The creating of motion establishes the duplicating of space but in the same instance the motion provide a point where motion does not apply and therefore it will destroy the space. At that centre duplication stops and space-time or space motion seizes. The one accomplish the other and a relevancy is brought about because the motion creates space that creates a point of no motion destroying the space created. The stronger the motion is the more space will naturally displace and the more heat will then concentrate in the centre of the object applying motion. The minor object is part of the space and within the space of the major objects flow of space towards the major centre that is holding the major singularity

The rotating of the spinning top is as much part of the Coanda effect as flight is part of the Coanda effect and it all becomes the result of gravity provided by singularity coming as a result of motion creating space by providing a centre that will bring about contraction as space is rendered motionless within that singularity centre and therefore the centre is space less as much as it is motionless.

The contracting of space diminishing as a result of singularity established

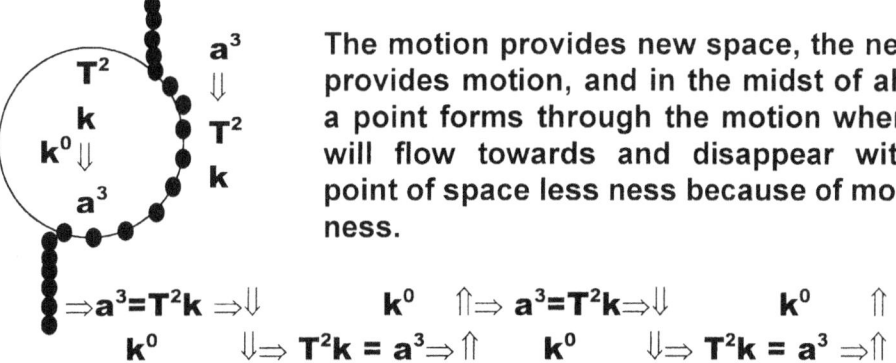

The motion provides new space, the new space provides motion, and in the midst of all of this, a point forms through the motion where space will flow towards and disappear within that point of space less ness because of motion less ness.

$$\Rightarrow a^3 = T^2 k \Rightarrow \Downarrow \qquad k^0 \quad \Uparrow \Rightarrow a^3 = T^2 k \Rightarrow \Downarrow \qquad k^0 \quad \Uparrow$$
$$k^0 \qquad \Downarrow \Rightarrow T^2 k = a^3 \Rightarrow \Uparrow \qquad k^0 \qquad \Downarrow \Rightarrow T^2 k = a^3 \Rightarrow \Uparrow$$

Space duplicating is creating motion and from the motion space is created.

When it comes down to the pulling of gravity, there is no pulling of gravity. The creating of motion establishes the duplicating of space but in the same instance, the motion provides a point where motion does not apply and therefore it will destroy the space. At that centre duplication stops and space-time or space, motion seizes. The one accomplish the other and a relevancy is brought about because the motion creates space that creates a point of no motion destroying the space created. The stronger the motion is the more space will naturally displace and the more heat will then concentrate in the centre of the object applying motion. The minor object is part of the space and within the space of the major object's flow of space towards the major centre that is holding the major singularity. The rotating of the spinning top is as much part of the Coanda effect as flying through the air is part of the Coanda effect. It all becomes the result of gravity provided by singularity coming as a result of motion creating space by providing a centre that will bring about contraction as space is rendered motionless within that singularity centre and therefore the centre is space less as much as it is motionless.

The establishing of motion is creating space and providing a centre point of singularity. This proves that the atom links with singularity and that the value of singularity extending forms a relevance with the space-time it influences up to a point that it controls the space as much as it controls the motion of the space. That proves singularity extend way beyond the space of the material it holds and establish duplicating points of singularity within singularity by merely applying motion to space within space.

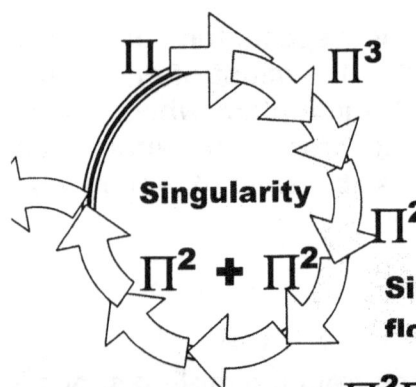

The Coanda effect is also the perfect example of the curvature of space-time brought about by the extending of singularity influencing due to the shape that imitates or duplicates the value of singularity and again conform Π. By establishing a new value of singularity as Π, singularity can once again take control and establish a new Π^2 as gravity in the new Π^3 forming space

Singularity extending the influence on flowing water

The Coanda effect is creating gravity. It is not replacing gravity it is not recreating gravity it is not enacting gravity it is is forming gravity.

By reducing the propeller in size and boxing in the airflow it is directing the flow to a centre where the spin will intensify as it accelerates. Gravity is created in such a way. The protons perform the spin creating the gravity and the proton number does not necessarily prove the strongest gravity.

The turbine engine is a star in the little. Massive space reducing brings on heat increase and with the accelerator of fuel added the heat increase generate a singularity enhancing equal to a star. It shows gravity come about by altering the space a^3, applying with heat added a new T^2 where that then produce the thrust to use the **k** coming about to elevate new movement which is gravity independent from the earth gravity

By reducing the propeller in size and boxing in the flow it is directing the flow the spin where the airflow will intensify as the airflow accelerates.

Gravity is overcome as anti-gravity is created in such a way. With gravity it is a case of the protons that perform the spin, which is creating the gravity and the number of protons used, does not necessarily prove the presence of the strongest gravity that inflicts contraction, which produce mass. The protons accelerate the moving of space whereby that spin will reduce the space volume. By accelerating the flow of space the volume per time unit decreases in relevancy. In the reducing of the intake the gravity becomes artificially constructed because gravity is the reducing of space. By injecting a fuel the situation intensifies as the possibility of raising the heat levels become much more prudent.

The reduction of space will bring about heat. Injecting fuel in a place where such reducing of space already exist it increases the heat level further and the fuel will "spontaneously" ignite. The igniting creates a rise in the heat level to a point where such a level that one can only find that level in the stars. Fuel will establish conditions that is in accordance with the laws of cosmology such heat will only apply to stars where such heat levels will indicate the enormity of the gravity present that is generating the massive gravity accumulated in the absence of space. Gravity is the concentration of heat in the utmost reducing and concentration of space thus a star is born in the gravity on Earth. In this example k increases as thrust pushing space a^3, which the **k** creates to a new T^2.

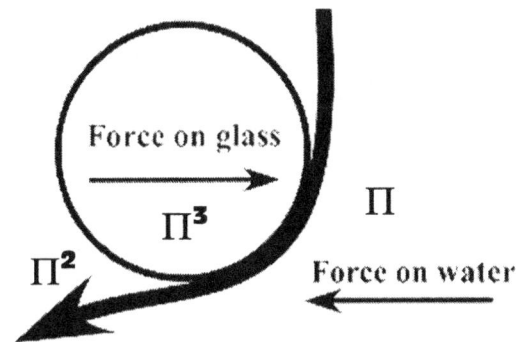

Bringing about a new centre with more heat in the specific point will establish motion in relation to the heat intensifying at such a spot and the motion produces an energy that elevates such a spot to form a new Universe. It is the manner how stars start to be stars and apply gravity.

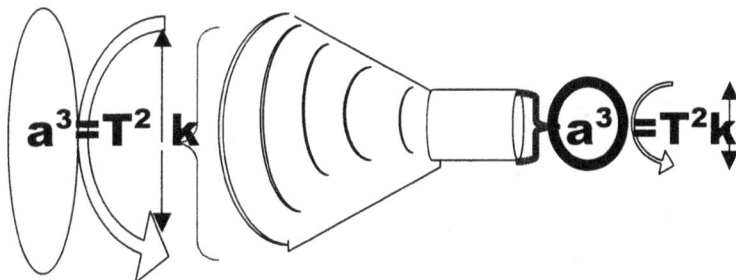

By reducing the space a^3 with the shortening of **k** a new standard comes about establishing a new T^2. Then moreover a new T^2 created by injecting fuel into a newly established k sets the groundwork for an a^3 coming about that can challenge the best star centres there is. With the ability of life to manipulate space-time man confuses nature in accepting there is a Tiger loose and this young star can challenge the space-time set by the gravity of the Earth. The earth will allow such rebellion just to a point and a fight will ensure that will either release the spacecraft or down the airplane. But all this comes about by the artificial creation of a singularity miming singularity by presenting something that can respond and produce gravity.

Do not forget the artificial aspect that is involved in applying the process since all this mentioned is created by the intervening of an artificial substance that is the only factor in the entire Universe with the ability to accomplish a force and that all of us that has it, we call it life. Life is first of all by my definition motion not of cosmic origin and with the motion life can control and accomplish the manipulation of space-time and in that we include the manipulation of cosmic laws that work independent of cosmic motion but still use cosmic motion. By the mimicking of cosmic laws we with life find an ability by which we can control the space-time we use. Part of such space-time manipulation also includes the cosmic laws and in successfully manipulating space-time, we establish such laws to benefit the wishes and the will of intelligent life. Creating the Coanda effect concentrates space as gravity does, it concentrate space in huge quantity as gravity does and therefore nature takes the action as coming from well established singularity. But never forget that life established the phenomenon which life created through the manipulation of the cosmic phenomenon. The action itself is not cosmic.

The law applying is cosmic but the commanding of the law is done by life in terms of manipulating space –time therefore life has the ability to use its manipulating gifts and that include applying the laws of the cosmos. I have to be deliberate in stressing this point because if there is one thing Mainstream science is guilty of and blind to reality then it is differentiating between life and its abilities and the cosmos and its abilities. There is a strong drive to promote atheism as science friendly.

Little can be further from the truth. Mainstream Science wishes to promote the fat of life as a natural outflow of a cosmic reality by depicting life as the most cosmic spontaneous events. This is done notwithstanding the incompetence science has about knowing what drives the cosmos. Not one scientist up to the moment knows what gravity is or what brings gravity about. In the face of such poor lack of vision those thinking they are brilliant as scientist also regard them of such supreme intelligence they can challenge the fact of a Creator. In doing that they truly show how little their minds are to host vision about physics. Physics has so much more detail to offer than the calculating of a few feeble mathematical formulas. I challenge any one of those to disprove me, after they have read **"Man-in-motion"** which is part of **Scientific Mistakes ISBN 0-9584410-1-4** or an open letter on "**The Seven Days of Creation" ISBN 0-9584410-4-9** and prove me wrong in what I declare in those books. Life has never been spontaneous part of the cosmos.

The two can never be confused because life is not part of the cosmos and what apply to life is alien to the cosmos. We always introduce standards applying to the benefit of life but such standards are a joke when the cosmos comes over to play. The action of life is artificially established by life's ability to manipulate the cosmos. The action becomes part of the cosmos as result of cosmic life, which is heat accumulating and heat driven where gravity is increasing heat or reducing heat. **LIFE** is not normal and it is not cosmically natural within the cosmic independent structures. It is laws we find in galactica where stars cradle until they hold a singularity sufficient to survive while the established singularity can seek independence from the star nursery we call a galactica. In stars or "planets" that stands in direct contrast to the very same application of the very same conditions but where the turbine is as artificial as money in the cosmos, the Roche factor is as natural as heat in the cosmos.

With the ability that life has to throw the top we stand on Earth holding our space in the space of the Earth. We throw the top and see the top come to life stay alive for a limited time and then die again into the realms of no motion. We gave the top space to be within and the cosmos allowed the space to be but set the terms for the space to be used and when the top was unable to comply with the cosmic terms, the cosmos repealed the space it allowed the spinning to take place. Yet the top still use space, but that space the Earth accommodates as that space uses the Earth spin by measure of motion. We cannot have space if we do not fill the space we have on Earth. While being on

Earth my position is $a^3 = T^2 k$ where k is because of the mass in movement standing in for k^0 by being k^{-1}.

The position I have stands in relation to the governing singularity that the Earth adheres too being k^0. My position I have in accordance to the mass I have stands me in relation to the k that positions me on the topsoil of the Earth. My motion T^2 sets my space a^3 apart and give me identity independence in relation to the rest of all factors forming the unit the Earth is. My mass comes about from me being a captured object of the Earth and as far as my standing with the Earth goes I have mass where my having mass classify my position in terms of the Earth as $k^{-1} = T^2 / a^3$. Being k^{-1} we are also T^2 / a^3 which is reducing us in the space we hold our mass as we try to reduce a^3 further to comply with the T^2 the Earth is applying and which we have to use. If we move we have produced a larger k factor to the order of at least k^1 to find the ability to move from k^1_1 to k^1_2, which will allow us to accomplish such a task because we use T^2 to move from k^1_1 to k^1_2.

So we have to improve both T^2 as well as k to accomplish motion. From the findings it is the proof that connects Kepler's formula in reality since every aspect of the cosmos is growing by duplicating. In that it puts science and the claims science make on aging in question. Using $a^3 = T^2 k$ and producing a larger $T^2 k$ it means a^3 must also improve. From that we can see that whenever a^3 duplicates by measure of T^2 singularity grows. By the measure of k extending further than what it was before the duplication. That it does by doubling the space it use during the motion where the motion in itself is halving the space that doubles. This is not that uncommon physics.

A car holds the space a^3 and is moving by the speed of T^2 through the distance of k. If the car speed up the gravity will increase because the distance k will reduce. But if such reducing becomes part of the equation then other factors in the equation has to compromise to allow changes to come into place. With the mass or space in motion speeding up the mass has to increase to have the equation to remain even. If the distance travelled becomes more in the time duration it took compared to before, then the time duration must become less if the relevancy of the distance remains the same. The time it takes is the gravity or the time by the square that increases when the space moves faster between two points. The mass has to become more to allow compensation between factors to permit the changes happening. That is what gravity is.

The way we think about gravity is that we wish to have gravity applying the same everywhere and acting in the same manner under all possible conditions. In order to play with mathematics and still find the constants that enable such playing to deliver the same results time after time man has subdivided the concept under so many names for each fragment we divided we cannot even find the basic principle any more. Gravity is not a force as Newton suggested but a motion between space occupied and space filled forming a relevancy and this applies throughout the Universe.

Life on the other hand is the only force in the cosmos. Gravity is a flow of time from one eternity through another eternity towards and finally to end in another eternity. That is not the purpose of life to find eternity where singularity creates the cosmos by using time. The cosmos is hostile to life and has no use for life because life connects with death and not with eternity. The only force there is can only be found on Earth in the form of life. Life is the force on Earth and only on Earth it has space-time with the ability to manipulate space-time under its control by providing motion other than and above that the motion the cosmos provides. It is precisely in such a manner that light uses to travel in from singularity to singularity.

Because singularity forms space and controls space by time measure, which we consider as dark and therefore invisible, such spots in singularity breaks down and rejuvenate space much faster than light in the photon does. When the point holding singularity has rejuvenated and replenished with heat in growth, singularity releases the photon and by such release can the released photon join the next singularity in the period of removing space-time and the next in rejuvenating space-time which then will include the photon reassembling with the next singularity forming the space-time of the next singularity. That is the way light has to move if ever it moves from singularity to singularity. Looking at the issue in this way we can begin to appreciate that light is the duplication of the photon by the singularity charged by the motion that provides the singularity by charging the intensity.

It is about duplicating more than dismissing although dismissing does form part when the photon changes singularity. In that way the light loses intensity as it travels and is recharged by the singularity on route to somewhere in the future. By giving the top a chance to spin we extend our position by means of life's ability to use and manipulate space-time and that extending allow us the property to apply motion to the top. Light then forms the part of not being permanent on the side of what is eternal but is part of the limited time duration and death. The rest of the cosmos is in a balance of duplicating by rhythm of space-time dismissing and the control between the two points of gravity. Light is temporal and material is permanent but not eternal. Singularity is eternal.

In contrast to the duplicating of light is the duplicating of material. The duplication of space filled with material is the use of heat in space surrounding the atom, which provide the material the ability to the confirming of space. Such confirming of space is in relevance between material forming a unit and material surrounding the formed unit. In confirming the form of the unit the unit attach and surround the form it holds with heat. Such surroundings of liquid heat are much more than singularity requires in the immediate to prevent overheating to come into affect. The heat required is more than that is needed for sustaining heat to prevent singularity overheating and is more than singularity will ever require.

To be more than singularity will require it to be much more than what particle forming the atom ever will require. Singularity in charge of light can generate by duplicating the photon whereas material use the heat the photon provide for sustaining singularity requirements. When light is clashing with the atom the atom applies the heat that light is to dismiss space-time. The light is heat, which is applied by the atom to sustain the immediate singularity without any chance of relieving some. The material in its use of light is making redundancy of space-time, but its needs requires even more heat than light alone can provide. Material needs gravity, which is more heat and the gravity it requires is much more than the photon can deliver because that is why there is shadows on the dark side. Life is one providing form of such heat but is only committed to one atom and only one type of that particular element. Life is a supply of a specific kind of heat that allow motion to occur where such motion of space-time uses such heat to become motion other than the motion the cosmos can provide. But the duration of such independent motion is very temporarily whereas the motion that the Universe provides is eternal. Life uses the supply of heat to live and living is the aging of life. Life aging, cosmic growth and the limit duration of life are so much connected it is the same thing. That, Newtonian science totally misses.

Aging and the fighting of age, disease and death has become a major concern in the world of science especially now that there is a drive to secure atheism. The world of the medical genius claim a possibility of human life extending to four or five hundred years between birth and death and the working life of a human being on Earth becoming three to four times longer than it is at present. This they say while facing the reality that those in science have not even got the faintest idea what causes wrinkles and what causes the degenerating qualities of the human body as time develops. This is clear because they blame it on gravity while they have no idea what gravity is or what becomes gravity.

They claim such rise in life expectancy based on the already accomplished rise in average time of life in general on Earth. This rise in life expectancy we currently are fortunate to experience came about with the ability man acquired to fight other organisms on Earth and thereby remove a major killer of young lives. This we accomplish with a chemical warfare on other creatures that kept those in good health alive to the detriment of the unhealthy. Then food supply was a big reason for wars and killing back then when germs were the big killer of human life. Instead in modern terms where food supply is no longer an issue it is the wealth connected with oil being the major factor of killing in the modern terms. We have grown rather fond of human brilliance to destroy creatures of small size but big repute in killing humans. We think by killing those killing us we can use mighty bombs instead of gems to kill our kind. Then there do not have to be germs that kill, we can use British, Canadian, Australian, Italian and American politicians and their appetite for oil to select those that needs to be killed instead of germs doing the choosing.

Our effort of killing germs will stop the indiscriminate killing so therefore we better kill all known germs first. Changing that is not changing the coarse of human destiny but it is merely prolonging and boosting a disaster waiting for the future to come to pass. By killing off life, be they germs or insects carrying the germs, our killing them with chemicals do not give us doing the killing a superior advantage over those killed. Remember that those we kill saw killing of their kind so many times with so many disasters that struck the Earth in the past, we do not even know where the disasters were or what the disasters were. Those we kill overcame problems nature tossed their way by destroying their kind in a manner and much more than we that are doing the killing at present can anticipate of what destruction the Earth is capable of.

There is a disaster waiting once this organism found a way around that which we use to kill those in their ranks. Some of us are doing the killing to swell the ranks of man and a small percentage is using the opportunity to get stinking rich from this action. Others do the killing in the hope of foiling death that is coming to every one and that coming is inevitable in any case. When those microorganisms come back after they found a way around our chemical war the revenge they will respond with, will bring death to humans on a scale that is beyond the recording abilities of man. This is one part of the foolishness that man is spreading in his lust for commerce and control but it will not destroy man, it will only set the balance straight again. Man will die in the billions by germ infection and the rest will die of hunger. But man will prevail as all forms of life do when facing genocide of its kind. We must also remember that this function of life connects to micro life even more than it does to humans. It is shown in the past when life of whatever measure is threatened the resilience of life to fight back and overcome such a threat to life is beyond the scope of human imagination. When in the very end of the Earth all life on Earth is dead one day, when the end announces the end of all...we can be sure of one thing...somewhere there will still be a form of life fighting to survive with all odds against its survival.

Read into this statement what you wish as long as you also read that the cosmos is out to end the beginning of life because the cosmos is in a fight with its own and the cosmos will not tolerate life chipping in at some point. On Earth we live with the idea that where we are we may think of that point being to us as eternally big but in relation to the rest of the Universe that point is so small the point gets lost before the point is ever found amongst the vastness of the cosmos. Life is also about duplicating life as much as removing dead wood. When born, life receives a specific carbon atom and by some property that specific carbon has, that carbon can help life to conduct the heat life use to move. Life is the motion of space-time in space-time by manipulating space-time in duplicating space-time purposely as to control the direction flow of such duplication. To do that life has to use a specific heat to apply a specific motion and the heat transmitting as an electric flow but the electric flow is not life, it is only a manner to conduct massages that enable life to manipulate space-time. That electric flow is also a conducting of gravity, which allow the carbon atom to grow as all atoms grow. By singularity extending in the first dimension $k = a^3 / T^2$ the atom of carbon used by life also grows.

When the atom grow, such growth is cosmic related but that growth brings aging to life and that process limits the time span that the atom can host life before change must come to the life that uses the atom for the sake of transporting life. That is the reason why the body needs oxygen. Oxygen brings heat to the carbon atom and by the carbon atom using the heat the carbon atom extends the atom singularity and in it also the atomic space-time. Unless science knows a way of detaching the growth of the carbon atom by suspending the live body of using oxygen, science and the atheist may be as wise as they see them to be, but they cannot beat aging. It is because of the growth that time inflicts on the atom whereby the cosmos will limit the extent of the duration of life being a part of the cosmos before renewal to the atom must come by the way of a newer version of the same life that has to replenish the atom again. Such replenishing we see as the birth of life. In most of my other books I deal with this matter in some or other way on a far more grand scale than the space in this book would permit me to do.

The important issue is to see that life does give the top a chance to be in independent motion, but because life gave the top the chance to be, that chance is as limited as the cosmos will allow life to be. The drive of the cosmos runs from eternity to eternity through eternity but because the cosmos drive include eternity that drive exclude eternity from life in the cosmos. The limited time duration that is connected to the spin of the top is the result of life giving the top the spin and therefore the top is also taking the limit spin time as part of the gift of life. Death comes to life, but death only comes to life, meaning that stars cannot die because stars are a party to the eternity sector, which is part of the cosmos. The cosmos cannot find death as life can but the cosmos can only substitute one form of eternity to replace another type of eternity when time comes for the swap to take place. Life on the other hand is completely connected to death as the eternity of life connects to the realms from where the cosmos is driven and not from the driving side of the cosmos.

Gravity is about a relation established when time begun between particles we know as material and particles we know as free or unoccupied space. Gravity reduces space to apply to fit the form of the sphere and later accept the form of the sphere as much as gravity is duplicating space by motion thereof. The fact of gravity is the producing of space by duplicating space just as Henri Coanda showed. Gravity produces space by mimicking space and producing motion that destroys space.

The Roche limit comes about when singularity oppose space-time by applying gravity which we can witness in the Coanda effect where motion contract space to an independent point of singularity representing such space. When the two such points establish by independent motion without establishing enough space-time separating the two points the Roche limit comes into play. The fact of why the two are too close is the first problem too investigates after establishing the Roche limit at work.

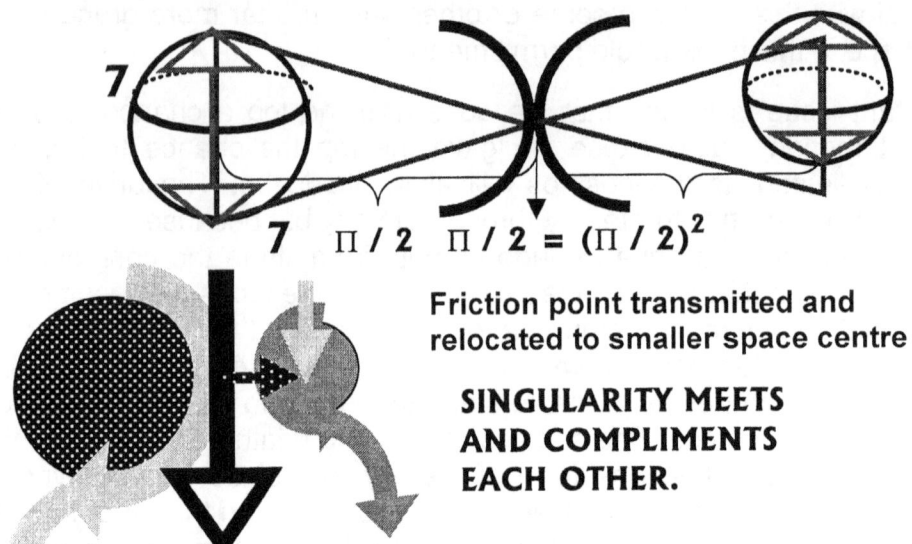

$7 \quad \Pi/2 \quad \Pi/2 = (\Pi/2)^2$

Friction point transmitted and relocated to smaller space centre

SINGULARITY MEETS AND COMPLIMENTS EACH OTHER.

Gravity is about a relation established when time begun between particles we know as material and particles we know as free or unoccupied space. Gravity reduces space to apply to fit the form of the sphere and later accept the form of the sphere as much as gravity is duplicating space by motion thereof. The fact of gravity is the producing of space by duplicating space just as Henri Coanda showed. Gravity produces space by mimicking space and producing motion that destroys space.

The difference in gravity being speed or motion will establish a differentiation in flow of liquid space. By implementing the Coanda principle the motion of the larger singularity providing flow to the liquid space will be superior to the flow created by the inferior gravity. That will advance the motion of the lesser singularity and the accelerated flow will try to establish a larger space that what can fit into the atmosphere of the lesser object. Again we can read the Coanda gravity principle into the applying of the Roche limit

When two opposing objects find a fluid in motion around them the motion establishes two Coanda systems bringing about different space-time relevancies. Since the stronger space will produce a faster flow, the faster flow will influence the space-time flow of the smaller space. The faster flowing liquid then will try to increase the smaller space by providing a larger space.

However, the smaller space can only become bigger by allowing more heat to be present and the larger amount of heat will have the growing singularity of the lesser space to flood with heat, thus having the space overflowing with heat. By accumulating the amount of heat to become proportional in dynamics the Coanda effect will establish a joint centre from where space-time is controlled in both directions. The accelerating in combined liquid space will equal the solid space whereby the lesser space will destroy, as it has to cope with heat the lesser singularity is not adequately adjusted to control.

When explaining the Roche limit we must recognise the fact that lightning exchanging takes place between Jupiter and its inner closest planet Metis.

That means there is photographical proof of electric interaction between structures separated by outer space.

The diameter of the cosmic structure holds the value of r and singularity holds the dimensional value of Π meaning that the radius or diameter (r) extends to become the diameter multiplying the value of singularity. But since r already consists of the square of space holding a definite positional relation with the value of singularity being Π the diameter comes into effect. Π extends each to an individual value to a point where the singularity on each side meets, bringing about a mutual Π^2 to the value dominance of the larger singularity control.

The Roche limit comes into effect when the lesser singularity holds inordinate heat to create substantial motion with the flowing of liquid that will comply with the forms of enhancing gravity in relevancy between the objects in separate independent motion. The relevant speed does not match and that can only result from unequal growth of singularity gravity in too clearly shared quarters. The lesser singularity has the inability to establish a match in motion that will bring about harmony in $a^3 = T^2 k.$

When the motion is not in harmony the lesser singularity will not duplicate space in sequence to the larger singularity and will therefore move towards the centre of the larger singularity in order to establish a better time component fitting to the duplicating that brings about motion. The singularity of the lesser is space-time to the larger singularity in relation to be only heat to the other more prominent singularity and as such domination must come about before integration can be established. Gravity is electricity on a wider scale and electricity is gravity on a lesser scale.

The gravity of the dominant singularity will be electricity to the lesser singularity and the gravity will become a mismatch where the one will be a short circuit to the other which then will produce heat as all short circuits do. By rendering heat into space-time the lesser singularity must either become a compliments being an independent partner or a supplement to the dominating singularity as it then relinquish independence and becomes resolved as space-time within the absolute control of the domineering singularity.

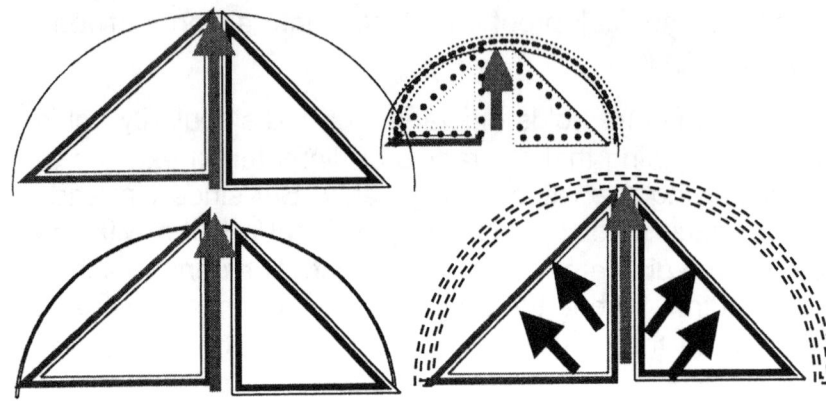

The Roche limit is the strongest indicator we have that can show us what is happening with the sound Barrier because in principle the sound barrier is half the Roche limit

Since both objects being the Earth and an independent object sharing the atmosphere with the Earth is attached to the earth by singularity, they share the Roche limit by half

At this point the equality of the straight-line dimension to the triangle and the half circle holds prominence as a straight line, a half circle and a triangle is dimensionally equal. The common denominator will bolster all factors to an equivalent ratio.

When singularity by the straight line increases the singularity by the triangle it will also bolster giving equal potency in singularity by the half circle. As the singularity of the major component revives the lesser singularity to equality, the triangle in singularity will match the performance and so would the half circle respond in precise ratio setting equilibrium in order. The major partner's singularity in the straight line excites the minor partner's singularity in the straight line affecting all other aspects holding singularity in both objects to match equilibriums in all aspects of singularity. That is the Roche lobe.

By exciting the heat within the lesser structure the structure may find motive to accelerate motion but more likely will lose more heat to the dominating singularity as the two will join space-time by uniting Π in establishing equal gravity Π^2 and creating shared space Π^3. The Roche Lobe explains gravity in the ultimate detail one can wish for. If $k= \Pi^2 /4$ of the diameter distance, then T^2 presumes an equal value in both atmospheres rendering the lesser structure to dense heat. This proves that the space then becomes the motion the major structure enforces on the lesser structure and the major structure then demands equal motion and equal gravity from the lesser structure.

From this the lesser partner will fill by the extent of the larger partner and as soon as equilibrium sets in the growth will duplex to matching in both accounts, normally to the fatality of the lesser partner, as the lesser partner will be capitulating under the strain of the dual. In that way the inner planets came in place as I explain in <u>An Open Letter About The **Seven Days Of Creation** ISBN 0-9584410-4-9 or Part Seven of **MATTER'S TIME IN SPACE :THE THESIS** ISBN 0-9584410-8-1</u> . In this there is no mass defecting but there is heat displacing from the lesser object to the larger object and all the infighting between cosmic structures in the cosmos is about producing heat in concentrated space. We find this principle also in space flight since flight is about sharing space-time. In gravity there is motion in a linear sense and in a circular sense. Gravity is the balance between space applying a negative or departing nature in **k** and the other is the positive nature in T^2 where the departing is neutralised by the swinging of the relevancy to fit the other side as we may find in the expression **1 / k** Applying the constant space a^3 is equal **=** to motion $T^2 k$

Gravity is always a relevancy, which is always about domination and being strong enough to secure space-time whereby independence is formed from motion If the motion of the space-time proves to be strong enough to have independence it has a future and eventually become a Black hole at the end of developing. Even something as minute will end up as a Black hole but that is very much in terms of eventually being far to the future development. By that time the distance there is between the Sun and the Earth would have grown to proportions so large in our human terms that no one ever would recognise that the two objects once shared a system holding space-time. In that not one of the two objects would increase much mass by creating new particles that will allow a gain in mass and yet by growth the mass will become massive enough to push the growth Black Hole proportions.

The Roche limit is more proof about my theory that gravity is all about motion discrepancy between two moving particles in space. There is no physical contact of any notion. When the motion between the two objects does not share space, which will introduce equality in time, or rotating tempo it will establish friction in being heat coming about from contact and the end result of this friction is mass. That is what we experience when we feel the attraction we categorise as mass or weight but what we experience is on the very edge or the other end or the limit of possibilities in the extreme of what can come about.

The mass we experience is a resisting of those particles shown to protest the discrepancy there is in motion in space between the two objects. From our (us being all object in the atmosphere that is not part of the Earth solid surface) having the inability to duplicate space at the rate the Earth is duplicating space because of size (or space-time occupying discrepancies) the Earth is reducing the space we claim. We resist such reducing and that is leading to our friction that manifests as mass. Any object having a lesser time in space difference between the singularity connecting will indicate such heat when entering the zone such boundary is in. The Roche limit is the sound barrier and the Roche limit is part of the flames we find covering incoming objects that enter the atmosphere as it fall to the Earth.

In order to establish a matching duplicating ability we have to increase the volume of heat in the mass ratio and that will increase the mass to the point where it destroys the mass and the body finds the duplicating ability which will enable such a body to fly in the air of the Earth atmosphere way above the Earth core. When the object is outside the atmosphere limit of the Earth the object has to sustain a matching speed and such a speed can only come about by having an independence that can establish enough duplication because it must have the ability to secure the correct density in heat in the centre of the object. When passing the limit as the craft is coming in from outer space in the moment the craft gave up the claim on individual independence from the Earth singularity we see the heat forming a blanket around the object and if the conditions of entry is not met to detail the object entering will become heat.

This blanket of heat is the result of time compensations to space claimed and the reducing coming about from that. The heat comes about as the space occupied by the lesser object has to compromise in velocity of space duplication to presume the required mass or space in time ratio the partnership then will require. In the Roche connection mass is not yet a product of distinction connecting the two objects but there is a bond coming about arising of motion discrepancies.

The space I refer to is another boundary much closer to the point in singularity we call the atmosphere and are where space turns to liquid because of gravity reducing space. Gravity is applying much more distinct measures as those border brings about mass and in that connecting or entry we find spacecraft covered in heat. That entry point is where an object will receive the mass we so graciously give to the object because to where it eventually will crash and then find mass as gravity at the incoming moment changed alliances from $k = a^3 / T^2$ to $k^{-1} = T^2 / a^3$ causing distance to reduce as far as possible that will provide absolute space sharing qualities.

One aspect Mainstream science totally ignore is that the entering and departing of objects we send into space and receive from space is totally alien to "cosmic normality" and "cosmic procedure" because all action produced is coming by the way of through the manipulating of life where life we find on Earth is totally alien to the cosmos. Life as such is alien to the rest of the cosmos and this fact cannot be pressed enough. The entering and departing is a creation of life and therefore "unnatural" in the cosmic world.

There is no chance that any object will leave Jupiter under individual motion establishing any possible means of accumulating the required heat for such ejecting. There is no chance of particles leaving as a result by forming heat as a cosmic effort. No object will release from Jupiter bringing about the release of any included particles that is part of the Jupiter atmosphere to become excluded particles outside the Jupiter atmosphere. Much less is the chance of particles ever leaving the tiny star such as the Sun by finding the cosmic realised ability through the distributing of cosmic heat. Life is the manipulation of space-time and what life achieves is alien and contradictory to the cosmic flow of events coming about by adhering too cosmic principles.

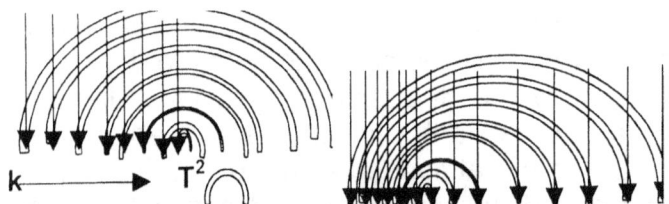

The fact that spin can activate and establish an independent singularity proves what laws motion of space bring about. The fact that there is not only a drive for individuality by an independent moving object but that by establishing gravity independently such gravity fights for release from a contracting gravity proves the law holding the cosmic growth captured. There is a relevancy to achieve release and a "let go" and there is another of contracting, commanding and a "come back" and the relevancy in a gravity unit is the balance there is in the two.

Taking this statement to Pythagoras the triangle enters the equation.

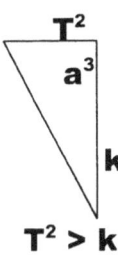

I have indicated there is no gravity forcing us down to the ground through any magical persuasion. It is the motion in the relevancy between T^2 and k forming a^3 that determines the position the object holds in that relation. A bird can stay in the air on three accounts. The one is the bird maintaining a motion superseding the space displacement the Earth brings about when flapping its wings in motion. This motion is by following the curvature in a straight line T^2 thus overcoming the motion bringing the bird down to the ground. The bird provides the motion bringing about sufficient antigravity to sustain the motion equilibrium.

The next one is by motion of the winds curling air through the wings by a flapping in reverse. That motion is airflow with a reversing relativity allowing the bird to appear to stand still. This reversing of air is creating the air to move backwards and thus giving the Bird the relevancy of standing still in air. $T^2 k$ where $T^2 = k$

Taking this statement to Pythagoras the triangle enters the equation.

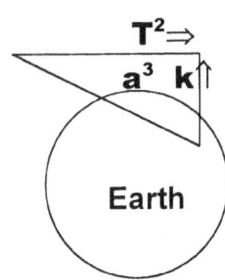

In the normal dispensation of gravity the balance between space and motion sets the space provided by the motion. The higher or longer k becomes the smaller the relative motion of T^2 will account for the aircraft staying air born. At 31×10^6 meters in linear flight an aircraft needs *many times over* the speed it requires near the Earth to break the sound barrier. The aircraft is not flying at twice the sound barrier because the balance between k and T^2 is out of balance. At that height T^2 hardly holds a value. At 2500 km / h the aircraft is just about falling out of the sky.

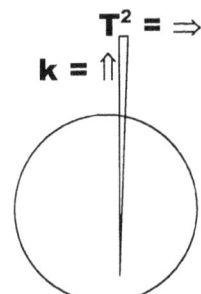

When the aircraft receives a maximum heat ratio applying in favour of k the craft has the ability to escape the earths atmosphere as long as the departure is horizontal and departing by 7^0. The condition is that the motion from the start focuses on k as the only recipient of heat and is the only factor with a value higher than one, bringing in T^2 as a factor holding one. In such a case singularity providing k the direction will influence k to provide T^2 with the minimum value being 7^0. This is the only way to escape the atmospheric bubble enveloping the Earth in the form of a liquid.

Then with birds blessed with massive wings those birds can use the motion of the air that comes about as the heat in the air brings on space and the space created is providing an updraft to supply the motion that will fight the birds tendency moving towards the centre of the Earth. **$k = a^3 / T^2$**. All other cases we find the motion of the Earth **T^2** exceed the motion the objects (amongst others we humans too) **k** provide and in that the space **a^3** we claim is of a diminishing nature. **$1/k = T^2 / a^3$**

The balance also favours the other factor as we use that balance to do the normal flying through the atmosphere in an aircraft. When **T^2** is very much dominating **k** and **k** has the minimum influence where **k** is just about all the time in a factor of one we fly in the same manner as the birds do.

In the event where a body use the latter motion balance to fly, such a motion cannot bring about the crafts escaping from the atmosphere into outer space. Visiting flying sources will become resident flying sources. It is the other balance that has to be dominant to charge the Earth singularity with heat that will bring the escaping effort to a successful conclusion. No alien aircraft can merely enter at will and unnoticed or escape in this manner and no aircraft can change from the circular stance to the linear stance providing the energy requirements to satisfy such an escape without radar detecting this enormous source of heat. This is the reason why the ride height of a formula one racing car is but a millimetre or two three and that represents **k**.

In this ratio the movement of the formula one at 300 km / h is the **T^2** factor coming about as anti gravity. The formula one therefore needs wings that is fitting a jet fighter more closely just to combat the antigravity the motion develops by establishing this ratio between **k** and **T^2** in the order of a few millimetre in ratio to hundreds of kilometres per hour. It again proves the application of gravity established in the set up of the formula one. More proof about gravity being space depleting through motion applying is the way the sound barrier comes about. The sound barrier is the duplication of space within space by one object forcing motion and the second lesser object fighting for independence by providing the motion **$k = \Pi^0$ or Π or $\Pi^2 / 2$** to sustain individual singularity motion **$T^2 = 7(3\Pi^2)$**.

When it starts moving it make contact with gravity forming space between the outer space and the Earth when standing stationary. This is $7(3\Pi^2)$	There are an innumerable number of lines running down at $7°$. The higher the speed it reaches the more would the line be making contact with the car.	The more the speed the more lines would be in contact with the moving car but also more acutely the lines would have to bend to influence the car. The car exaggerates the body structure by motion because there are more points in contact in the same time.

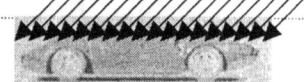

By making more contact in moving than is the case when standing still therefore the whole pattern of gravity in a linear motion becomes disturbed by gravity in duplication by the rotating motion of the Earth and exaggerated by the motion of the car. The motion of the car produces more space per time unit used as when the car was motionless. The extra contact the moving body has in time with the space it flows through is a measure of exaggerating the space claimed by material and that force something to give way.

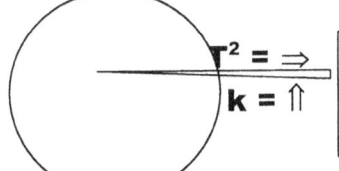

The balance also favours the other factor as we use that balance to do the normal flying through the atmosphere in an aircraft.

When **T²** is very much dominating **k** and **k** has the minimum influence where **k** is just about all the time in a factor of one we fly in the same manner as the birds do.

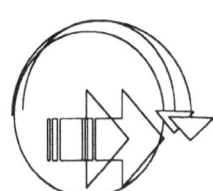

The complexity becomes clear when every factor in the Kepler formula is given a value implicating all the parts it holds in space-time through out space-time in the cosmic space-time. The factor represented by **k** points to where singularity is implicating space **a³** and implicating time **T²** from a centre.

This, the Kepler formula, is the most accurate formula ever devised with the simplest explaining about it and is the least tested or understood formula in science. When in doubt about cosmic science refer to Kepler for answers because it is Kepler's vision of space being equal to time in the relation with a very specific point in singularity that is the solution to cosmology. Space holds a third dimensional value where time summons the space from a single as well as a square being the second dimension that is formed as space-time. The two in compliment forms the fourth dimension incorporating time and space. The location where we wish to locate a Universe holding time we can only find when a compliment of time forms space in relation to a second object forming motion relativity with time the ruler of space. To understand this one has to understand Kepler's formula.

 =  X

The Doppler effect suggests a shift in rings moving from a centre concentrating in the direction of movement. That is to a very limited way what the sound barrier is. The Doppler affect is a small and almost unnoticeable part of the sound barrier.

In the most basic explanation, one can bring in the atomic relevancy of $(\Pi^2+\Pi^2) + (\Pi^2 \times \Pi^0) \times 7$ with $(\Pi^2+\Pi^2)$ forming the proton value and $(\Pi^2 \times \Pi)$ forming the neutron value. In this, the seven forms the electron value where the electron brings about the space-time influence of singularity on the motion.

The motion takes time back to a past relevance because the restriction back then made the motion in relevance much slower. By moving faster, the relevance apply in which time becomes slower and that is why the Newtonians better start forgetting their fixation on a time constant because by cheating their logic it will get them as far as they have come and they have come so far that after almost four hundred years the best they can do is call gravity a force or a spook. The body will become smaller in motion because in space –time the time remains the same under cosmos inspiration. By extending the duplication of space the space used for duplication becomes more ion relation to the space in which the duplication takes place. The duplicated has to become make the space it claims become less because the relation in space-time change.

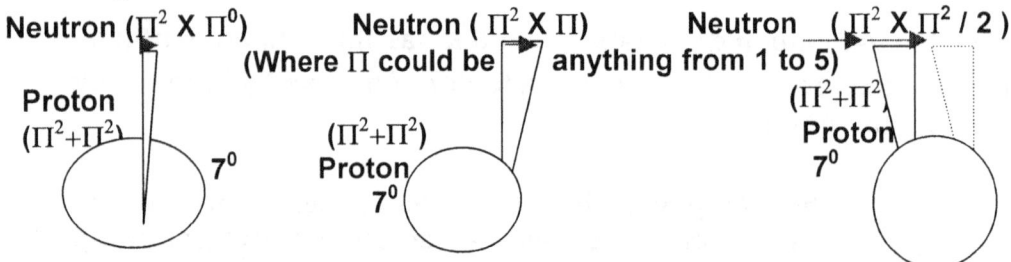

By bringing about two separate identities when Mach is broken such separate identities share a unifying **half** as another form of the Roche limit $\Pi^2 / 2$

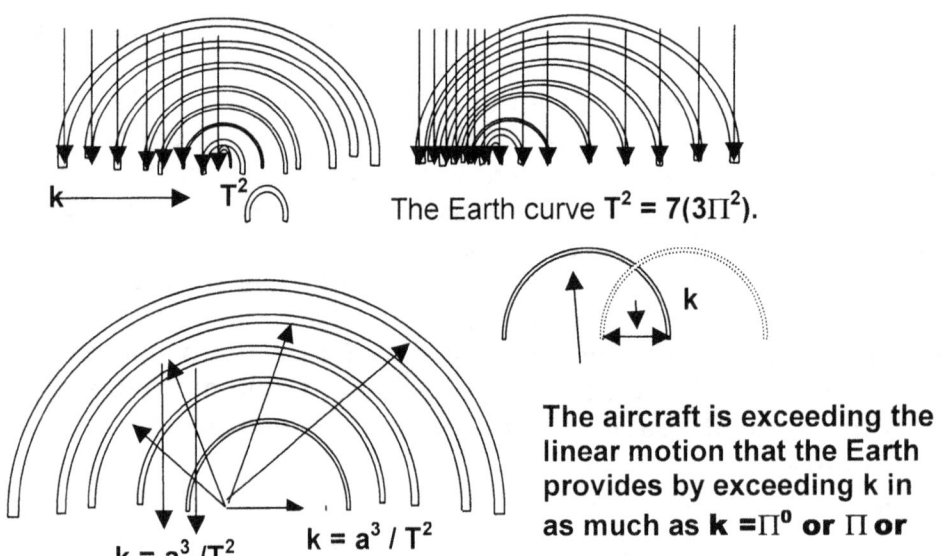

The Earth curve $T^2 = 7(3\Pi^2)$.

The aircraft is exceeding the linear motion that the Earth provides by exceeding k in as much as **k** $=\Pi^0$ or Π or

There then can be a transfer of heat brought about by the fact that the Earth applying gravity compresses singularity points much closer and denser. **k** becomes relatively smaller to accommodate the constant of T^2. The sound barrier is from $7(3\Pi^2) \times 5\Pi^0 = 1036$ km / h to $7(3\Pi^2) \times \Pi^2 / 2 = 1022$ km / h

Motion comes about as some heat release causing motion heat to release at one specific point. Singularity ties down all objects at specific points. In order to establish a release from that point heat in the object seeking relieve has to come about to bring about motion. Since the atmosphere holds singularity more densely as space relinquishes the position towards the centre many more spots of singularity occur and such spots are holding singularity in a position where the spots falls within the barriers that forms the Roche limit.

$k = a^3 / T^2$ is gravity and is that Doppler only implicated as rings without any explaining why and as such it is not good enough. The very affect of the sound barrier can only be explained by using Kepler's formula because the fact is that the aircraft reaches a different a^3 to that of the Earth because the T^2 arrives at a different **k** to what the Earth allowed the object to be at that point. At the heart of this is gravity applying two forms of space dismissing. It is the motion that brings about new gravity ratios of singularity dominating space –time to maintain a balance ratio, with a singularity position that seeks for independence by motion.

> The perfect example is the trumpet where the walls of the trumpet expand, and changes the relation between the unoccupied space-time and the occupied space-time. As is shown in the trumpet, the relation is a theoretical relation and changes according to many influences altering the compiling factors, for instance heat, density, altitude space and time. The sound barrier comes about when the factor **k** of the aircraft provides a T^2 in self-motion that is exceeding the T^2 the earth has while the two are sharing space. This then comes about as the Roche limit secures half the claimed space by bringing on **k** as **k** $=\Pi^0$ **or** Π **or** Π^2 **/ 2** while the other factor presumes with the value in T^2 that the Earth enforces being $T^2 = 7(3\Pi^2)$.

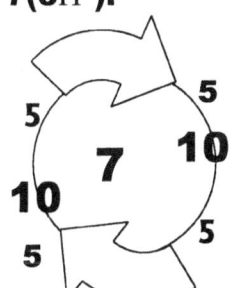

Gravity is using the Roche factor in conjunction with the Lagrangian five point systems and the Titius Bode law to form gravity. It is a combination of all the cosmic principles to form gravity but it can only apply if singularity is committed in a position that representing of the value and form singularity applies to motion.

This is the manifestation of gravity we know by another name we use as the Coanda effect. The Coanda effect represents the final coming together of the principles forming gravity by motion bringing about duplicating space in the presence of singularity.

$$\{[5+5] + [5+5] + 1 + .9\} = 21.991/7 = \Pi.$$

This becomes space **5** sides duplicating in time

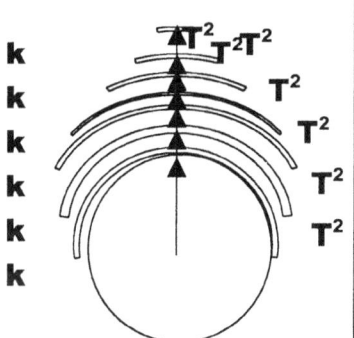

Within the influence of the sphere the Earth has and uses to capture space holding the space to Earthly time there is a relevancy Kepler saw. From Kepler's formula a relevancy comes about following precisely the manner that Kepler translated the relevancy to be from the figures he received from the cosmos. The big issue about gravity is that the motion has a dual influence. The space a^3 is relevant to two factors of equal substance in space. The two connects to motion in complimentary fashion but not equal.

As motion brings about independent points serving singularity the Earth take on the position what the Sun had before the motion came about. The Earth then forms central singularity and the newly establishing point serving the independent singularity and as motion comes about and since gravity is repositioning various speeds in relation to each other the motion establish another gravity where the motion establishing the relevancies.

The perfect example is the trumpet where the walls of the trumpet expand, and changes the relation between the unoccupied space-time and the occupied space-time. As is shown in the trumpet, the relation is a theoretical relation and changes according to many influences altering the compiling factors, for instance heat, density, altitude space and time. The sound barrier comes about when the factor **k** of the aircraft provides a T^2 in self-motion that is exceeding the T^2 the Earth has while the two is sharing space. This then comes about as the Roche limit secures half the claimed space by bringing on **k** as **k** $=\Pi^0$ **or** Π **or** Π^2 **/ 2** while the other factor presumes with the value in T^2 that the Earth enforces being $T^2 = 7(3\Pi^2)$.

As **k** grows T^2 will reduce and as **k** reduces a^3 will diminish. We find aircraft follow this motion to the spot and do not depart from it in any way. However since the sound barrier is the duplicating of a^3 in response to the changes brought about by a shifting **k**, it is not the velocity achieved by the aircraft that represents the sound barrier but the relevancy between **k** and T^2 and by **k** duplicating T^2 in motion forming the triangle. At 31000 meters an aircraft may seem to record a flying velocity at a speed we record to be 2500 km / h but in relevancy it is ten times less than our recording of the speed. It is because **k** extends that T^2 has to decline and that puts the relevant applying speed at 250 km per hour. To break the sound barrier at such an altitude will be very impossible to achieve using any cosmic atomic material known to man.

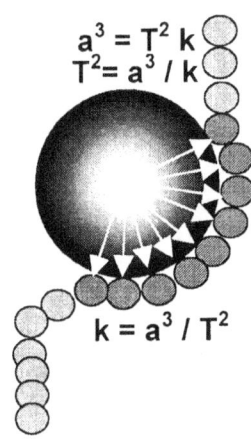

The sound barrier personifies gravity as presented by Kepler in a lateral motion. There is the unmistakable and clear evidence that the space a^3 created by the light T^2 relates directly to the distance k the object is with the Earth, which is the main provider of singularity. In the Coanda effect k remains the stable factor providing space a^3 a new and alternative position as T^2 brings about motion and that motion creates the new space filled by the flowing liquid. This is precisely what happens with the flying object or any moving object. With every moment in motion new space comes about that is then filled by particles occupying space and indicated by the distance factor of **k**. It still is the same thing. In all scenarios the formula applies as space a^3 created by motion T^2 in relation with distance **k** just as Kepler announced it.

There is not a bit unnatural about it. I wish to be very ambiguous and very clear that the duplication I explain and the dismissing of space is as natural as space in the Universe itself is. The space is duplicating as time is providing the motion. That is natural since the space fills the vacancy that comes about as the vacant space moved on into the next position leaving the oncoming space to follow. It is what science sees as a separate issue in momentum but it is gravity.

Kepler said space a^3 is equal to the motion thereof $T^2 k$ and also is diverting space to the motionless centre of all spheres T^2. From that **k** is the motion duplicating the position and therefore duplicating the space occupied by the sphere. T^2 is space replenishing the centre where the centre holds singularity in space and is dismissing space by lack of motion in the centre. That comes from the natural form a sphere has. **k** represents space a^3 replenished by motion in space of provided by a more dominant structure forcing space to apply motion to comply with space demand. Space duplicates by motion because the following space will fill with whatever is in the previous space and such filling will continue as long as space is in motion. The instant motion stops there will not be a following space to fill with the previous space and the motion or lack of motion will destroy the space being filled by motion. However the Earth resumes the motion as relevancies apply gravity then changes. The Coanda is proof of just that which I mention.

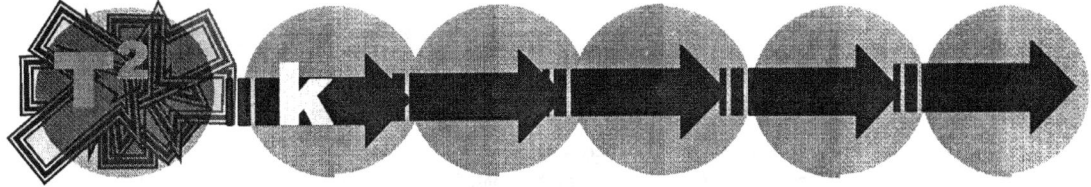

Space is equal to the motion thereof $a^3 = T^2 k$

The destroying of space on the singularity end comes about as the proton removes space by introducing the motionless centre in singularity to the space less confinement of the proton cluster. The space at that point is 38303288464389083136 times reduced than space is at the electron end and therefore time is $(((1836)^3)^2$ $((\Pi^3)\Pi^2\Pi^0)$ times slower than at the electron end. Space in motion forms the seemingly stable Universe we know and stopping such motion will destroy whatever part of the Universe, which is stopped. The space cannot simply stop because then one piece of space will move away and there will be a crash as the other standing still will be colliding with the space coming into the spot it has to fill by time producing motion.

On the other side the space will move away from the position it had before and with no space filling as a result of motion, the space will automatically self destruct in vacancies and collisions. This part forms the duplications and we can see the result of this when two solid objects finds one location that both solid objects wishes to use and therefore has to share. We call this phenominon events or collisions or more appropriately accidents which in most road collision events those a way cases should be called deliberates instead of accidents. Human made objects may find. In the centre of the sphere there is a spot that is never in motion and that spot has the strongest destroying of space providing gravity the other part gravity holds. Since the space does disappear and since the gravity is strongest at that point in the precise centre the reason for that is it is keeping all aspects of the sphere in form with precise borders but motion does not apply there in the centre and space deform as it collapse. To render such a spot secluded from destroying the other spots in the Universe a border or barrier form that can only be bridged or crossed under most selected conditions or circumstances. It is all part of the natural with no fantasy about it. $\mathbf{k^0 = a^3 / T^2\ k.}$

The space, which is in motion that is in place separating such points of space destruction, has a value limiting any and all excluded space-time from entering. Should there come about a need to enter certain laws apply that must be enforced or material entering the secluded space will be rendered to a form of liquid heat of which the photon producing light seems to be the main ingredient. Because of the natural form all spheres have there is in a sphere a centre where the space loose validity just because it cannot comply to what is said in the previous sentence. The dominance of such a singularity determines the space-time and the density of heat covering the space-time. Because of the Prime law of the cosmos space is there because of motion. This prove the opposite true that says that in the centre where space is lost as there is no motion to support space therefore space goes lost. However what make the space valid is the presence and the value of a bigger as well as a smaller space producing the relevancies.

There is a motion coming from the Earth that all objects have that is located within the Earth and functioning as part of the Earth. Where there is no independent motion the relevancy falls away as the independent object relinquish the applying singularity independence. These two factors make up the compliment of gravity. When an object within the boundaries of the Earth comes in motion above and beyond the motion that the Earth dictates can the object brings about independence within the confinement of the Earth. All objects have the displacing of space repositioning and trying to relocate the object in the Earth to within the Earth. That is mass and that is why mass tends to reduce the space occupied by the object producing the mass. However, the barrier holding the excluding properties of space-time that establish borders move from Π^0 to Π as soon as the singularity is threatened by motion that the independent singularity establishes.

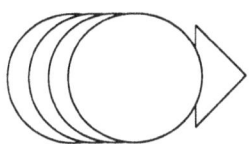

As the planet (in this case the Earth) or any lesser structure rotates around a more dominant structure such a rotating structure has to comply with the space duplication through rotation in order to remain in the circle of gravity it has in relation to the dominant object. It is space in motion that allows it to secure independence.

That leads to the complying of motion by repeating the route set by **k**. In the case of motion the factor of **k** has a very prominent role it plays. It positions the location of the next spot that will become the dot, which the space $\mathbf{a}^3$ in motion will duplicate and fill. But as this is going on there is a larger more demanding motion as well because as the motion complies to satisfy the dominant aspect of space in motion in which the object is in motion within that space controlled by the dominant object. As the Earth rotates it takes all that is secured with it on its travel around its axis. It also takes the atmosphere and all that is within the atmosphere on its travel around the Earth axis.

The dominant object is reducing space to a personal located centre from which the rotating object receives the value of duplication. This is the factor $\mathbf{T}^2$ and **k** will eventually become the rotating $\mathbf{T}^2$. I think by this time this aspect is established in the mind of the reader. Whether it is accepted is another issue but the best way to prove me correct is to try and prove me incorrect! There are forever relevancies establishing greater and lesser partners in such a relevancy. The spiral is the same continuing forever where $\mathbf{T}^2$ will come about from **k** and **k** will in the end form $\mathbf{T}^2$ again. When an object within the confinement of the Earths boundaries find a means to establish motion above and beyond the motion such an object has to comply with being part of the motion of the Earth therefore the independent motion stands in addition to the motion the Earth demands.

That is where sound comes in because sound connects to the singularity centred within the centre of the Earth where all space except space independently filled space will disappear. The sound adhere to singularity at a value of Π^0 but when motion comes about bringing independent motion to an object the Coanda effect proves that such motion which in fact is regarded as fluid in relation to the solidity of the Earth soil, the Coanda effect shows how easily motion can establish a new dynamic singularity above and beyond and in difference to the singularity within the Earth centre. The Coanda effect is just an indicator of how all motion comes about but it is just ordinary motion like all other motion taking place as we find with any form of motion in the cosmos. It is moreover the interpretation of singularity duplicating space, which provide time with a new spot to position space. That is what Kepler said four hundred years ago in his statement $\mathbf{a}^3 = \mathbf{T}^2\mathbf{k}.$ That is the Coanda effect which is gravity actively participating in cosmic affairs from an established point in singularity.

If any one places a solid object into the space of the rotating fan blades, there is no one in the world that would indicate anything else than that the space fill solidly. In the case of all moving objects, the motion brings about that the rear fills the spot covered by the front. The rear then is what the front will be. This is because the distance reproduces what was contracted at the back in duplicating  that by moving it to the front of the fan. In front of the fan, we find all that was momentarily behind the fan. This shift of material is the duplication of material using time to produce the duplication from the back of the fan to the front of the fan. The duplication happens faster than the speed of light every time because what parts the space in the back from the space in the front and in that it is only time that is filling that space by duplicating a^3 and redefining the new location of **k**. **k** reproduce a^3 by establishing a new spot with T^2

With **k** or$\Rightarrow$performing the duties of $k^0 \Rightarrow k$ it personifies singularity pointing a new spot to fill by time which is just space in motion. The space duplicate with motion performing time as **k** duplicates

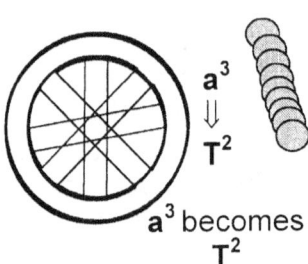

a^3 becomes T^2

The Coanda effect proves Kepler's statement more than any other but also proves gravity more than any other. The establishing of a^3 becomes half of the object because the other half is in the motion of a^3, which becomes T^2 in relation to a constant **k** forming duplication. It is the same as a fan blade spinning. When spinning very fast the back of the fan is the same concentrating what it is collecting to reduce that which is collected to then form what is in the front of the fan because the only difference between the two is time where time is heat in motion through space that reduces the space by motion.

$a^3 \Rightarrow T^2 \Rightarrow a^3 \Rightarrow T^2$ **but** $\Rightarrow$ is representing k but with the same growth also can go negative when mass becomes a factor $k^{-1} = a^3/T^2$

When spacecraft enters the Earth the same applies but the reversing of the same extending growth of **k** applies.

When a spacecraft enters the Earth the same applies but the reversing of the same extending growth of **k** applies.

The reducing must follow the boundaries of Π

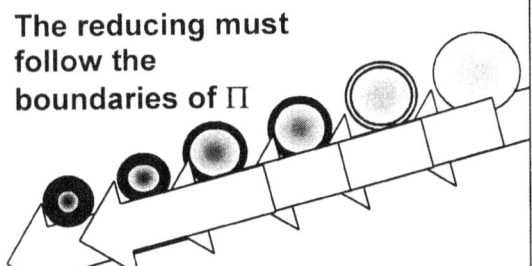

The re-entering must come about following the boundaries set by singularity because the reducing will be $k^0 / k = T^2 / a^3 = k^{-1}$. Every time space duplicates it releases k^{-1} to form heat as the space claimed reduces.

The space reducing forms heat to compensate for the reducing of space by the ratio of $k^{-1} = a^3/T^2$ and that heat we see as a hot blanket surrounding the incoming object. That space reducing transforming space and forming of heat is the only true gravity applying. $k^0 / k = T^2 / a^3 = k^{-1}$ leads not only to the confirming of the time component but the time component is forcing the space reducing a^{2-3} because the time factor T^2 becomes the constant the Earth prescribes in the Earth's relation to space–time $a^3 = T^2 k^1$, but the space duplication must then accept the ratio the Earth dictates. By entering the Earth atmosphere the duplication of space of the object entering relates exclusively to the duplication the earth relates to in the specific ratio demanded by the Earth. If space duplication is within the Earth space duplication, a comparison must come about where the object's space-time duplication stands directly related in relation to the time component that the Earth holds as a claim to space in regard to the singularity the Earth sustains. That brings about mass but that is not yet mass in any location other than being solid on the Earth. It only follows the indicator that will eventually indicate mass if the object can sustain the space diminishing to heat that the Earth will introduce to the incoming object. It has nothing to do with materials forming friction with incoming objects because there can be no friction.

At that height (and everywhere else in the atmosphere for that matter) the density of material is extremely sparsely distributed and even if material had the ability to produce friction there just is not enough material at that height to bring about enough friction to cause such glowing heat. We must consider what volume of heat will material in friction produce to produce that amount of glowing heat. There is not much material gone from the craft therefore it must all be coming from such sparsely distributed and destroying that much to produce that volume if glow is just not a factor because then that much of the aircraft structure should also be scraped from the craft.

This is proven by the "heavy" gasses floating and the "light' solids stationed on Earth. There are a group of elements that Mainstream science categorises as gasses. We have discussed when these facts in this book a on other places in this book. The material we find there is also extremely volatile and will move about bringing on highly active motion.. They are by name Hydrogen, Helium 2, Neon 10, Argon 18, Krypton 36, Xenon 54, and Radon 86. These gasses are in most cases hugely massive in numbers and mass. Others are Nitrogen 7 and Oxygen 8 that has more protons than some solids. In the cases of

Xenon 54 and Radon 84 the proton number and the mass produced makes the element much more massive than Carbon 6, Boron5, Silicon 12. With the presumption that mass makes things heavier because the mass pulls the Earth harder and the Earth responds by pulling back harder and the pulling of the two dismisses all the distance from both ends equally these gasses should at least be metals and in some cases heavy metals. Since the protons in numbers exceed metals in mass numbers. They should not be suspended in the air as gasses are but they should be stuck on the solid ground and some must be almost unmovable instead when taking into account the number of proton mass they hold they are floating like feathers.

Then there are the group of elements that does play they're part as Newton said that it is mass that brings about they're being on Earth as solid as you can get. By they're remaining on Earth they play the accepting role of the categorization they have. We share space with all the elements and being as short sited as humans can be, we categorise and classify all elements in a manner that we will best acclimatise to that position. We fancy 20^0 C that is a very nice laboratory acclimatised laboratories temperature for humans to work in and in that we test elements while nicely forgetting that such a climate might favour the Earth at that position. Take the test results down five kilometres below what the deepest mine is and the test result will be much different to what it is in our nice little artificial climate. Then we go one more step from the ridiculous to the sublime and categorise hydrogen as a natural gas and lead as a natural solid. In our human mind set hydrogen has to be a gas because it takes massive changes in our laboratory to change hydrogen into a solid and that temperature cannot sustain life so we through it out as cot congestive to life. It takes first as big an effort to liquefy metals and no person can live with such heat so the metal s then a natural solid. Hydrogen has to be a gas because hydrogen proves to be a gas in every laboratory found on the face of the Earth. They are massive except in some cases and they are solid. We classify them and categories elements as solids or liquids and gasses knowing very well every element can turn to liquid and gas should prevailing conditions demand the changing. By adding or reducing the surrounding heat the surrounding heat all material can manifest in every form there is available yet we still demand that the elements will not change the form we gave such elements. I have shown that it is not the proton that produces solidity or a fluid state or a gaseous states but the way the element structure relate to heat or space. I gave this forming responding to heat the name of the Lagrangian atom and explain the concept in explicate detail in another book being **Volume six of MATTER'S TIME IN SPACE: THE THESIS ISBN 0-9584410-8-** or **"STARSSTUFFN' ISBN 0-9584410-3-0.** Elements respond to space-time in two manners and the response bring about the form they represent. The protons dismiss space in the time it takes to duplicate space while the element takes time to duplicate the space motion provided the time to complete.

The outlay of the atom, which is too extensive to go into detail in this book atoms stand favouring space duplication and in balancing contracting to that motion of space flowing to the centre where space-time dismissing takes place. When the element favours heat and disfavours much density it holds the heat it associate with as a "blanket" surrounding the element to sustain about all the dismissing of the protons and leave much of the rest of the heat to spare. In other cases the element density promotes a displacing of space-time that brings about not much duplication and therefore disassociate it by form from with the "heat blanket" surrounding the element where the displacing is so much in control the electron associating with heat is unable to secure a "heat blanket" that will secure space-time in order to supply the demand going in the way of dismissing. The duplication favours the electron bringing on more heat by producing volatile motion in an effort of collecting the surrounding heat and through excessive motion in comparison with other elements that diminish space-time by contracting motion towards the centre of singularity. In this the atom dismisses more space-time than that which the electron may accumulate, it all comes about from form in the formation of the atom. In between there is a variety of choice to be made producing forms hard too soft to gas. But none of this is written in rock and changes as heat is added or removed.

The proton holds space in a zone where the proton dismisses space and therefore the number of protons will reduce space volumetric equal to the number of protons in such a space less zone. This I named space-time displacement and that does not refer to mass in any possible way.

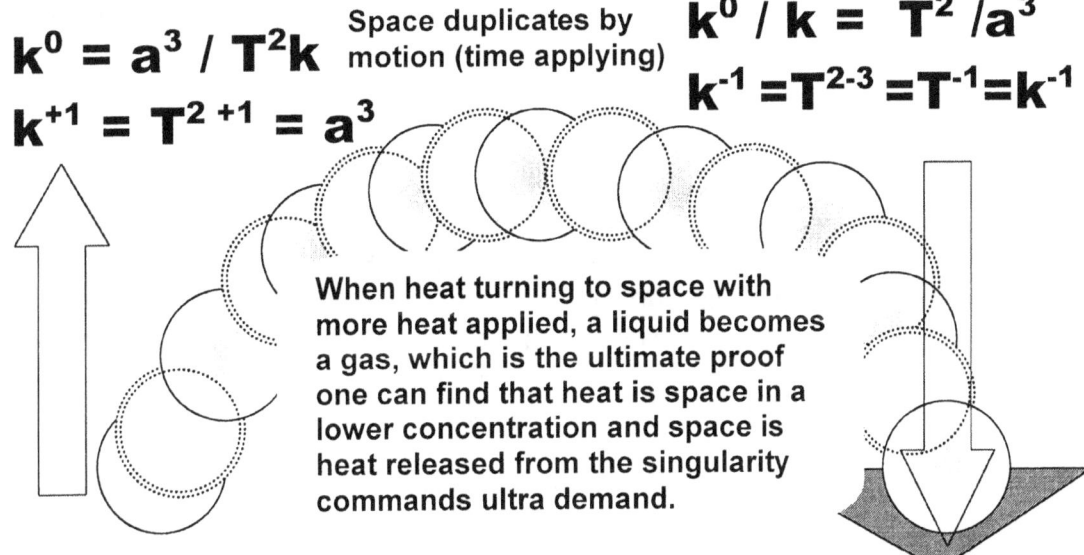

When heat turning to space with more heat applied, a liquid becomes a gas, which is the ultimate proof one can find that heat is space in a lower concentration and space is heat released from the singularity commands ultra demand.

All material duplicate space by motion duplicating space $a^3 = T^2 k$ It takes a certain time to duplicate the space but the space it has to duplicate is much more in volume than when compared to other elements. Such elements are surrounded more by heat/space and therefore they show a greater resistance to solidify than others do not withstanding proton number or mass acquired

All protons dismiss space-time by displacing the space as it dismiss the space either by motion acquired to move through space or in other cases to apply motion to space, Because of the dismissing nature we tend to think of such elements as having little space and that we refer to as them having density. The density is exclusively a product of space dismissing by space-time displacement. But by displacing the space it results in forming heat density to space. It also provides a structure having very little space.

When the element has a relation to applying a lot of space surrounding the element the element is covered by a lot of space where we consider such element as being volatile which suggest it is full of much motion and will therefore prove not to easily abandon the space to solidify when heat / space covering remove the space surrounding the element. We call this tendency being a gas in extreme cases or liquids in less extreme cases.

There are the elements that normally show less resistance to remove the space surrounding the element and become solid and those tend to form liquids more easily.
Then we have the third group that shows a severe reluctance to surround the atom with space or heat and those we find to be the elements we regard as solids or metals.
In this argument Newton's formula for gravity and of mass goes flying through the open window. $F = G (M.m)/ r^2$ has no solution to this mystery and thus is not spoken about in intelligent conversation because the solutions Newtonian science offer are most horrifically proof of utmost stupidity. If Newton's arguments were correct then all light gasses must have very few with very little protons must be the ones floating in air and all heavy solid metals must have a large number of protons that do the pulling of matter onto matter. Since this is not the case it shows that Newton's perception of mass pulling and making heavier is extremely incorrect. Gasses actually go out of their way to disprove Newton.

We on Earth think of the spacecraft as orbiting but the spacecraft find the motion it has as flying straight. While the orbiting spacecraft is following a straight ahead travelling direction the spacecraft will inevitably again reach the point it was at just a cyclic period of time ago. By travelling straight the spacecraft completes a circle and by circling around the Earth the spacecraft must fly straight ahead matching the pace of the Earth. But it took an enormous amount of liquid released in such a way that it would favour the straight-line **k** by reducing **T**^{2} to match the singularity **k**^{0} initiate space-time.

The release of the heat produce the space putting the orbiting object at that spot it has rotating in the circle which it does. After the release of such heat the Earth took the bounty paid in heat in return for the release of the craft where the Earth permitted the escape to take place but only on the condition that as long the orbit speed which is duplicating the relative required matching tempo to be in place as the rotating speed between the Earth, the object and matching the Sun.

That is gravity because what keeps the craft in relation to the Earth is the motion the craft duplicate the space it has in relation to the Earth duplicating the space the Earth has and the difference in the duplicating and the volume duplicated forms the gravity between the craft and the Earth. When we take the circling straight line even further we find the moon is rotating the Earth following a straight line while the Earth is rotating about the Sun following a straight line and Sun is rotating the centre of the Milky Way by also adapting a straight ahead position where that straight ahead position will eventually land the Sun rotating in a circle around the Milky Way centre.

Second Over All Overview

That is what gravity are to all things in the cosmos. But that is not what mathematician Newton said although that is what the cosmologist Kepler said. Kepler said space a^3 is in motion of a line k in the single dimension going straight relating to a circle T^2 in the square. More than just that Kepler went on to say that space can only be formed in a straight line going in a circle $a^3 = k T^2$.

The other three parts of time went about duplicating since the relevancy could not yet commit space to form. Therefore, only form of space (Π) was committing motion (Π^2). There duplication started being space in motion.

Space duplicates by motion (time applying)

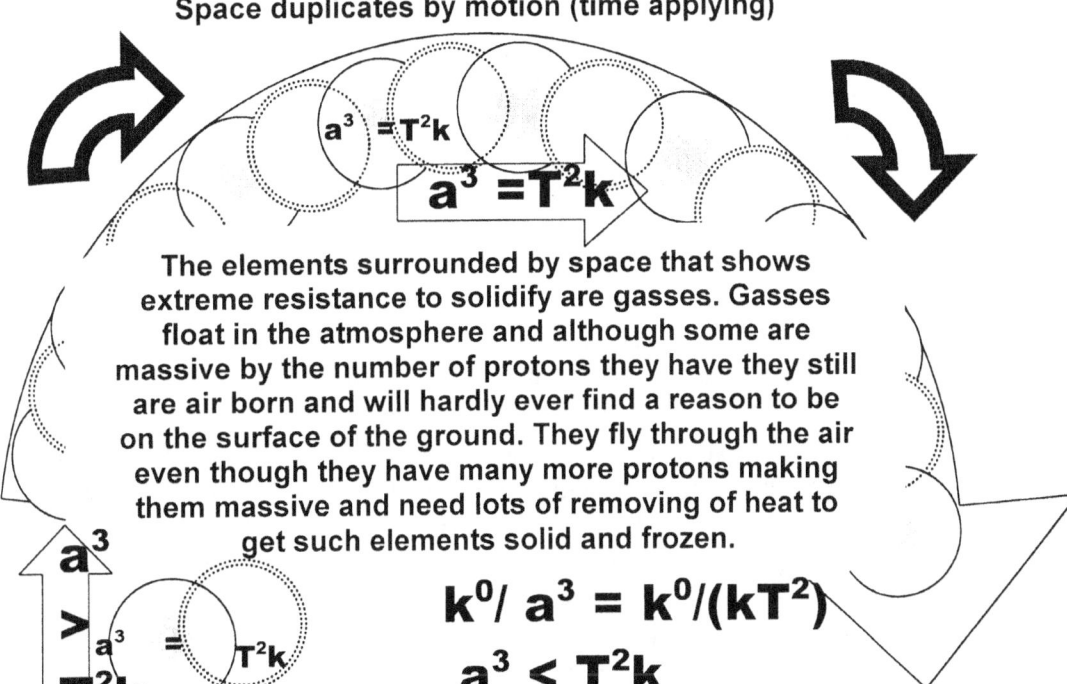

The elements surrounded by space that shows extreme resistance to solidify are gasses. Gasses float in the atmosphere and although some are massive by the number of protons they have they still are air born and will hardly ever find a reason to be on the surface of the ground. They fly through the air even though they have many more protons making them massive and need lots of removing of heat to get such elements solid and frozen.

$$k^0/ a^3 = k^0/(kT^2)$$
$$a^3 < T^2k$$

There are the elements filling the idea that such rather difficult to move notwithstanding (in little proton ability. But the normal that tend to be mass prone fore dense material is heavy and as they are solid by nature some cases) having precious is that the numbers of protons do tend to dismiss more space than they are surrounded with establishing heat and therefore applies much less space in regard to much more mass to claim for occupation. The function of the proton is to dismiss space by displacing space and where the space then show motion by refilling the reduced space we call such flow gravity. As every one is schooled on the Newton principle of much mass is lots of protons being massive we naturally think that because a larger number of protons will lead to a larger dismissing of space it occupy less space as the protons always are in a space less environment in the centre of the atom or the star.

There will always be a connection between the rotating and the straight line travelling. Gravity is by suppressing space in the attempt of accumulating heat that will advance duplication and/or remove heat at the centre that will retard duplication. Moreover gravity is the balance between the two possibilities favouring either, both or one at a time all depending on the allowing that space-time governing singularity permits. This balance we find in all stars. The so-called giants are heat collectors producing motion by massive enormity and volumes of liquid space. Those are stars, which I classified as being prone to the electron as the electron is about the charging of heat collected and accumulation thereof. By creating motion it makes much contact with space helping the star in the process of brining intensity to the space through gravity and accumulating the space that the star turns to liquid heat. The stars having this nature all develop to become stars with mostly atoms that is prone at dismissing space more than collecting space and forming such space into liquid. These stars further develop into structures that skip the liquid faze and turn the collected space into solid heat directly by dismissing more space than the star can accumulate. In these stars, which I last mentioned no motion is active, since the star is almost to completely in a state where singularity is reinstalled and in that manner all motion within the star is killed of by singularity.

Producing singularity sets the divide because singularity splits the Universe apart and in separate equal components that in combining form the duplication of singularity being Π. Since the split brings about equality it means that what is applying on this side must be applying on that side. When motion changes Π^3 to the proton $\Pi^0 \Pi^2$ it will happen on both sides of the divide of singularity Π^0. It in effect means that that which combines the proton also part the proton as it combines the proton because the proton becomes $\Pi^2 \Pi^0 \Pi^0 \Pi^2$ where the adding is the divide being Π^0. The circle motion comes from space being dismissed by ending the motion and such ending of motion compromises the space it forms. By returning to where it is coming from it is ending the motion that began the space and as space is motion that is duplicating space motion returning is also motion that is ending which is destroying of space. THAT IS GRAVITY!

Gravity is the balance between motion forming space by duplication space in motion forming time and time ending motion by destroying the space. Gravity is about space duplicating space in relation to space destroying space and some particles are more prone to duplicate than destroy not withstanding mass or proton numbers. Those we call gasses. Then there are others that are more prone to destroy space that duplicate and those we refer to as metals. Then there are a few that destroy as much as the create space by duplicating and that we call fluids.

When heat is added to some elements we consider as solids, the heat helps with the duplicating of surrounding space and brings about a balance restoring the difference there are in the destroying of space and the re-establishing of space. The metals become liquid and the heat forming the liquid brings about an adding to the material where such material diminishes space. By applying heat to materials that already favours the duplicating of space to the destroying of space the adding of heat will bring additional space as duplicated space and thus will produce more space to be duplicated and such elements will rise into the higher part of the atmosphere. When heat is added the heat as space that is added is the heat forming space by duplication forms a shift in the balance because the heat forms space also as a process of duplication but without the contracting aspect of singularity renouncing space.

Although the "gasses" are the particles favouring to duplicate space they still hold the tendency to diminish space but when applying heat the balance will favour the duplicating much more because the heat transforms to space acting as space duplicating and adding to the overall duplicating of space. The element has a function naturally of returning the space that the heat duplicated back to heat by removing the space destroyed and therefore returns the space to heat. In that way the particles do not only diminish the space they have but diminish space outside their claim. Such particle we call heavy metals.

This I use to indicate where there is a balance favouring the diminishing of space
In all forms of material, there are the constant interaction between space duplicating and space reducing. Some elements favouring duplicating space more than the diminishing of space are as follows

Hydrogen has a mass of 1.00797 g/ mol melts at -259^0 C, boils at -252^0 C,
Argon has a mass of 39.948 g/ mol melts at -1899^0 C boils at $-268,9^0$ C
Krypton has a mass of 83.8 g/ mol melts at -157^0 C boils at -152^0 C
Xenon has a mass of 131.3 g/ mol melts at -111.79^0C boils at -108^0 C
Radon has a mass of 222 g/ mol melts at -71^0 C boils at -61.8^0 C
It is note worthy to notice that none of the above elements feature strongly in stars although they should be massive in relation to the numbers of protons they have because they duplicate space.

Other elements favouring diminishing of space more than the duplicating of space will be as follows

Magnesium has a mass of 24.32 g/ mol melts at 650^0 C boils at 1107^0 C
Silicon has a mass of 28.08 g/ mol melts at 1412^0 C boils at 2680^0 C
Iron has a mass of 55.847 g/ mol melts at 1536.5^0 C boils at 3000^0 C
Cobalt has a mass of 58.933 g/ mol melts at 1495^0 C boils at 2900^0 C
Carbon has a mass of 12.01 g/ mol melts at 804^0 C boils at 3470^0 C

There are no correlation between mass and elements prone to space or prone to be solids. Mass do not create gravity and again on one more point Newton was wrong. Mainstream Science would rather ignore such compelling evidence as well as my writing about the matter than to admit that Newton could ever be mistaken.

As heat is added more space becomes available to duplicate in relation the space they destroy and what space the elements diminish. With more space to duplicate the object will surge higher to a location in a position Earth will naturally duplicate as much space as the newly relocated material duplicates the surrounding space where more space naturally are. Cooling on the other hand reduces the space available for duplication by removing available heat that would have helped with the duplicating of the space and that then tips the balance in favour of the diminishing of space, which that element will also have. In that case the element will become a solid as the space duplication is more that the space. In this we can trace the most important part of star evolution.

When the star has a liquid centre with lots of heat, the duplication of space-time by motion duplicates space much more than it dismisses space and destroys time. With a star in all the liquid as the Sun is it is proof of a very young undeveloped and insignificant star with almost no influence sphere. As the star develops the liquid ratio will shrink until it is only present in the centre. But as the liquid diminishes the motion of the star deteriorate because the liquid represents the motion. The star eventually becomes all solid just before it removes the neutron from the atoms in the star and eventually places the proton action into outer space. Judging the layers we find evidence in this as the outer layers of stars are filled with elements which is highly prone to space duplicating as they have such a relation with heat. Hydrogen and helium stands very favourable to space producing and little in favour of space dismissing while iron, cobalt and copper is much prone to space dismissing. But in all factors mass plays no part. It plays no part in the star performance or the star development

It is note worthy to notice that none of the above elements feature strongly in stars although they should be massive in relation to the numbers of protons they have because they duplicate space.

My argument I take from Kepler where Kepler distinctly shows that space produce motion and motion is time and time produce space. This puts a relation between the Earths as a massive body compared to that which are within the space claimed by the Earth. The Earth has a lot of space to duplicate that provides the Earth with a lot of motion. There is no element or structure within the atmosphere that can even remotely match the Earth as far as the use of space goes by creating motion with duplication, but there are elements that fair far better than others. The Earth has a surface cover of 70% water and some persons in the world of science goes around thinking the Earth should have been named Oceania.

That is another promoted misconception of gigantic and titanic proportions and the only bigger science fraud preached are the size the massive dinosaurs supposedly had that roamed the Earth millions of years ago when they roamed the Earth so many millions of years ago. But Oceania is about in the misrepresentation class as our fraudulent representing of massive dinosaurs. Off the point just this: has no one of those scientists ever sat down and wondered why everything was so extremely BIG back then and why has those specimen found today shrunk so much at the present time. Are they truly thoughtless or has Hollywood fame drained their logic in the brain they have? Just an observation I made with no harm intended! I give the answer plus all the relevant explaining in another book where I match the Biblical seven days with a possible seven solar occurrences that brought about seven periods of growth in the Earth core. The book is therefore named the **Seven Days Of Creation** ISBN 0-9584410-4-9 **or is also Part Seven of** The Theses.

Getting back to Oceania where science making this statement is only when looking at the Earth from the top and then only sees a very thin outer layer, but most of the Earth is metals and silicon. Metals and silicon break down lots of space but holds little space and therefore the gasses floating can have a far better relevancy of space creating by motion than the comparing Earth and will float in the air not withstanding the proton number dismissing space. While on the inside of the Earth is all the space less and space reducing Iron core with other more heavy metals the bodies we received such as we have show reluctance to produce space. Therefore the Earth producing more space than we do will compress our space by thrusting us towards that space less inner centre. It is because we cannot manage to equal the duplication of space by motion than what the Earth can achieve. The thrusting of our inability to duplicate space gives us the presumed mass while the gasses apply a much bigger ratio of space reproducing which tends to favour the gasses and make them float in the air. There is no mass that is producing gravity.

The gravity is producing mass by thrusting the less-space producing objects towards the centre from where the gravity draws the space at large. That flowing of space-time towards the centre causes a duplicating mismatch that thrusting or the depleting of the validity that the space has committed to match the duplication effort of the larger space volume then seeks a relevance to position allocated to space that is in volume equal to that of the lesser space as it produces motion forming an imbalance.

The mass cannot produce the motion to escape therefore it produces the effort of restricting the motion by producing a breaking effect we named mass. The very opposite can be established if the necessary heat is produced and applied to a specific point of k^0. When heat supplies motion to one smaller object the balance can be rectified in favour of the smaller object escaping the larger objects gravity. In spite of Galileo proving that objects with mass differences land on the Earth simultaneously when dropped at an equal distance also simultaneously, which totally contradict Newton's mass claims and pulling of objects.

The mass is the manner through which the dominant singularity uses to dominate the lesser body as the lesser body produce insouciant space duplication to accomplish the slowing down of the motion a larger object has notwithstanding the futility that such an act may have. By implementing the value of singularity $\Pi\Pi^2$ in relation with the Earth proton value of $7(3\Pi^2) + (\Pi^2)$ and the releasing object $(\Pi^2 / 2)\ \Pi^2) \setminus 3600$ applying heat which is producing the releasing motion the escape requirements is valued.

Explaining the whole process of flight in relation to singularity is also very consuming and I have a book **AN OPEN LETTER ABOUT " INTER GALACTICA SPACE TRAVEL"** ISBN 0-9584410-2-2 that deals with that aspect almost exclusively. We find this applying of gravity in the flying and the breaking of the sound barrier and that is what science completely misses.

The linear motion duplicating is related to the diminishing of the space the craft claim as and the crafts applying of additional motion again favours the duplicating of the space. In that manner the space is duplicated more by the craft than the space is duplicated by the Earth and since the motion brought independence to the craft the craft then holds more space duplicated by motion than does the Earth in that area destroy space. The Earth dismisses space as it creates space. It is doing this even on behalf of all objects in its atmosphere.

Being still and secure on the Earth the difference we as separate bodies standing on Earth have we still share to a precise point the same ratio in dismissing space relating to duplicating of space that gives us mass. The motion carrying the Earth represents the duplicating of space and its entire belonging foreword and the dismissing is about maintaining form to a specific centre at 7^0. This change when a standing object begins a separate motion in duplicating space through motion faster than the process of the Earth.

The linear motion duplicating is related to the diminishing of the space that the craft claims as occupied independent and individual space and the craft is applying the motion giving more independence of additional motion, which again favours the duplicating of the space. In that manner the space is duplicated more by the craft than the space is duplicated by the Earth and since the motion brought independence to the craft the craft then holds more space duplicated by motion than does the earth in its balance between duplicating and destroying of space in that area. The Earth dismisses space as it creates space. It is doing this even on behalf of all objects in its atmosphere. Being still and secure on the earth, the difference we as separate bodies standing on Earth have, we still share to a precise point the same ratio in dismissing space relating to duplicating of space that gives us mass. The motion carrying the Earth represents the duplicating of space, its entire belonging is being pushed forward by the act of duplicating, and the dismissing is about maintaining form to a specific centre at 7^0. This change when a standing object begins a separate motion in duplicating space through motion faster than the process of the Earth.

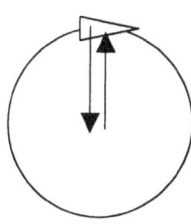

When the aircraft is motionless and is standing, still it is a secured on the Earth being party of the dismissal part of gravity. While it is in a neutron position by being independent, it is also holding a position that is equal to the dismissing part of gravity. The craft not only comply with the dismissing its protons displace but it also have to comply by conducting space-time that the Earth dismisses as it moves space-time by displacing towards the Earth centre.

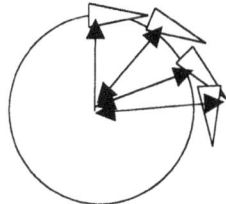

When the aircraft comes into motion the aircraft becomes increasingly more part of the neutron sector or the duplication part than it has a role in the dismissing of space-time displacement of the Earth. By establishing motion, it becomes air because it is more part of the duplicating sector than it was before motion brought about independence.

It is performing individual motion that is helping the craft to become more involved with the Earth duplication than with the dismissing where the individual dismissing sets the tone for dismissing control and leaves the larger dynamics of dismissing involving the earth to the other sector. The craft by motion stands more about establishing individual gravity as the craft by motion employs the Coanda principle of motion in relegation to the centre of the Earth.

It is this action that Newtonians miss in the gravity aligning to the comet to the Sun. The comet is going strait ahead while the Sun forces the comet not to come closer but to adhere to the form Π at the measure of **k**. The comet is going **k** while the Sun allows the comet to go T^2. It is this fact that Newton missed when Newton trashed Kepler's work. Using Newton version of gravity the comet makes no sense, but to a few very misguided Newtonians I have come across. By rendering the comet motion a circle action the whole affair becomes Π.

The earth holds the value of forming the proton relevance being ($Π^2+Π^2$) and as the object is an individual structure loosely standing on the earth but holding individual mass in relation to the earth the object has the solid neutron position of ($Π^2$). This means the object is motionless in the horizontal line of gravity but ($Π^2$) secures the object motion in through relating to the elected $Π^0$ commanding the Earth control and by $Π^2$ in duplicating space-time. The connection with singularity is 7^0 or in position $7Π^0$

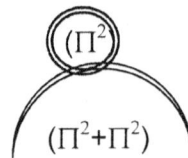

The dismissing of the space is equal to the degrees the Earth runs from circle to centre, which are 7^0.

The time ratio component in this duplicating 7^0 is the proton value ($Π^2+Π^2$)(+ $Π^2$) as well as the solid neutron value (duplicating $Π^2$) in relation to singularity X $Π^0$ = 207 km / h.

Because the formula relevancy becomes so long I use 3 $Π^2$ to alleviate confusion. The Earth crust and below the Earth crust forms a proton related position of ($Π^2+Π^2$) Everything not part of the Earth crust forms the solid proton position of ($Π^2$) This is the natural material positional dispensation.

Since the object is maintaining the same duplication to the dismissing factor of space-time ratio the Earth establish the duplicating is in relation with the earth singularity as $Π^0$

Because we chose a cubic meter of water to represent the space of the Earth as $cubic^3$ X $second^2$ in time of the Earth spin around its axis and around the sun any volumetric discipline in line with the rotation of the Earth will represent the earths motion in space-time precisely abiding by the seconds we use.

| When standing still the body holds a relevancy of $7(3\Pi^2)$. That is the contact in gravity motion | In motion producing more of the relevant contact with time in relation to the Earth increase from Π^0 to $5\Pi^0$. The relevancy increases by $7(3\Pi^2)\Pi^0$ to $5\Pi^0$. It is because more material contacts more space or time in time duration as motion produce more space over time in time. |

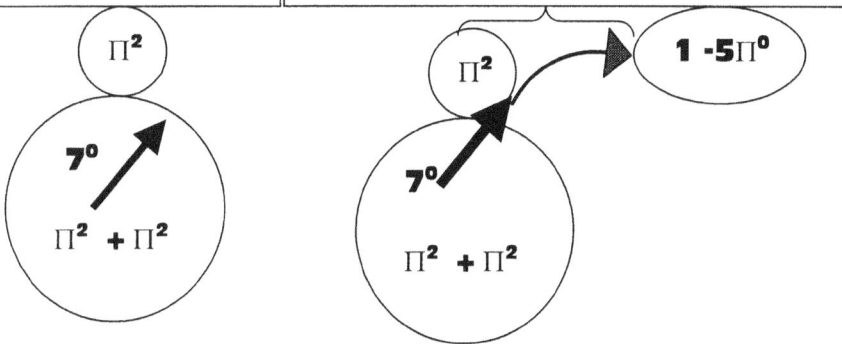

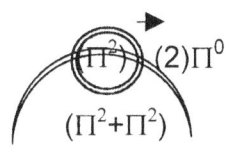 $(2)\Pi^0$ When the object asserts independence by applying motion in addition the motion the earth has, the object finds a new relevancy to the singularity of space dismissing which the earth applies. In such a case the formula becomes $(\Pi^2+\Pi^2+\Pi^2)$ which represents the proton factor including the solid neutron factor relating to the newly established singularity factor valued at any value between Π^0 and $5\Pi^0$

When spinning outside the atmosphere there is no direct linking contact established between the Earth and the object, which then is a satellite. That is why the wing of an aircraft works. It has nothing to do with pressure and all to do with relative wind speed divided by the wing. The difference in speed on top of the wing in relation to wind speed at the bottom of the wing sets the gravity the wing create either to push while the other side pulls or visa versa. But it is established gravity by measure of wind speed in relation to the contact the wind speed causes with the body of the wing.

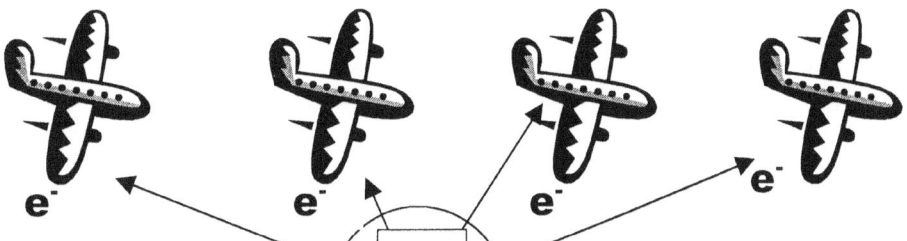

One must see the aircraft in its role it takes as acting as an electron in the presence of the Earth being the atom. When the electron moves faster that the atom can spin, misalignment and heat will come about and the aeroplane follows the same principle. Since the moving is stronger than the gravity the Earth applies the gravity the aeroplane-electron takes on is breaking new space-time limits by securing more space in the unit of the Earth which the aircraft holds in relation to the centre of the Earth. To the Earth the aircraft became liquid since the aircraft is moving. Liquids duplicate and in order to move the aircraft has to duplicate almost by perforce

Motion is cyclic as well as linear and therefore if motion is the way that space filled with material translates into time, and then it is time that has the dual quality of being cyclic and linear. In that we have to judge motion by $7^0 3\Pi^2 \Pi^0$.

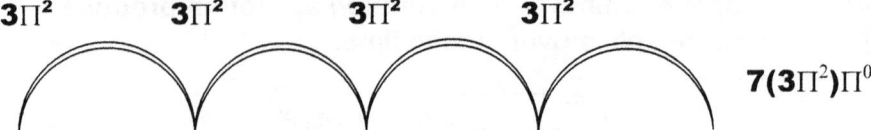

In the flow of material motion will repeat in a rotary manner while progressing in a linear manner. But the linear manner will adhere to the cyclic by Π as well as 7^0

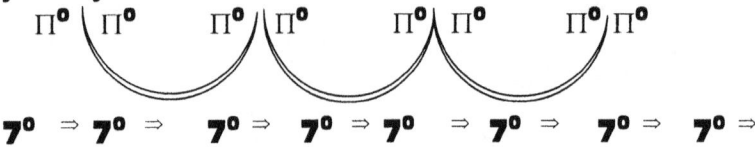

Once again one must look at the process from that which we find the cosmos uses in the atom. There is a solid centre, which the body of the Earth takes on which is covered by the atmosphere. The atmosphere is the liquid or neutron and the moment that the independent object moves that object then becomes the neutron. The role of the neutron is to establish the fringes of the electron. That will then be the exact role the moving object takes on. However when the moving object start moving faster that what the atmosphere being the neutron moves, there is a huge discrepancy in time relation between the neutron and the moving electron this discrepancy we call the sound barrier.

However when breaking the barrier and the satellite enters the atmosphere there are laws applying that changes

The Coanda effect starts

Any object entering the Earth atmosphere is starting an Earth wide Coanda effect by establishing a space link through motion.

From the centre running through occupied space singularity positions space-time at a point that connects by the seventh dot and in this, a straight line forms from the centre out to the edge of space. This is then coupling the object with singularity controlling everything within the realms of the Earth by using the line running straight but abides the 7^0 points. When the object accepts the duplicating and the dismissing, motion of the earth such an object maintains a perfect relevancy with the singularity in the centre of the earth.

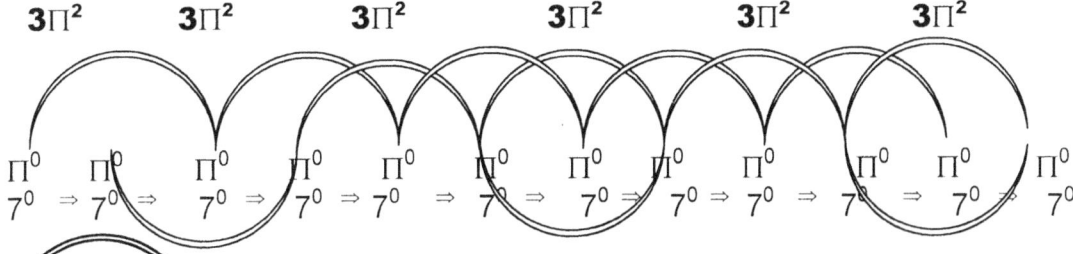

$7(3\Pi^2)\Pi^0$ The relevancy of motion applying to the object, which delivers the separate identity, is $7(3\Pi^2)\Pi^0$. This motion applies although the object is motionless because the motion is motion that is creating the affect of motion, which we interpret as producing mass. The motion stands related to the centre of the Earth and is completely independent of whatever may present the mass factor.

When the object asserts independence by applying motion in addition the motion the Earth has the object finds a new relevancy to the singularity of space dismissing which the Earth applies. In such a case the formula becomes $(\Pi^2+\Pi^2+\Pi^2)$ which represents the proton factor including the solid neutron factor relating to the newly established singularity factor of valued at any value between Π^0 and $5\Pi^0$

When the solid enters the Earth atmosphere the reverse effect of the Coanda effect establish a concentration of the liquid present, which is then the atmosphere as the motion, is contributed by the solid which is the spacecraft in this instance and the liquid is the concentrated space of the atmosphere. The liquid is pure heat covering the body under the rules of the Coanda effect that must shrink the space of the body and fill the shrunken space with liquid as the Coanda rules stipulate. As the relevance of **k** changes to compromise for the changes of the time T^2 that also change a^3 has to reduce in accordance with Kepler's formula ratio and in accordance with the establishing of a new Coanda effect linking the Earth centre with the space formed **k** by the motion of the entering object T^2 that then has to reduce the space a^3 which the object claims in terms of the more dense space a^3 that the Earth apply to retain space a^3. This comes from a new alliance between space a^3 through a motion T^2 that introduce a new linking value **k** between the Earth centre and the object entering. As time move back in the direction towards where the Big Bang came about, so does the heat that would have covered an object that was able to move at that pace through outer space. That leaves the entering object covered in a blanket of liquid heat as it establishes the new acting Coanda effect between the entering object and the Earth centre.

From the centre running through occupied space singularity position space-time at a point that connects by the seventh dot and in this a straight line forms from the centre out to the edge of space. This is then coupling the object with singularity controlling everything within the realms of the Earth by using the line running straight but abides the 7^0 points.

When the object accepts the duplicating and the dismissing motion of the earth such an object maintains a perfect relevancy with the singularity in the centre of the earth.

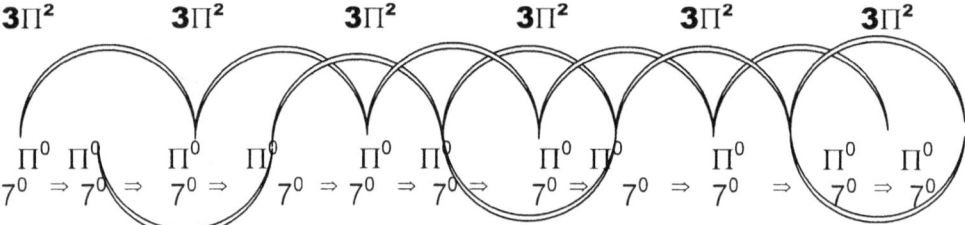

As independent motion sets in individuality by adding motion the object can no longer adhere to singularity in according to what the Earth duplicates but has to bring in a new alignment that is not in line with that of the Earth. The motion represent the duplication factor whereas the gravity the Earth so kindly bestow on us is not outright just a dismissing factor but in it is a larger duplicating factor and our independence from the Earth represent our possible duplicating factor in harmony with that of the Earth. This factor we know as time or the rotation of the Earth as well as rotating the Sun. By moving across the horizontal line will be bringing about motion where such motion exceeds the Earth motion brought to us by the compliments of the gravity linear factor we call time. As we excel by duplicating more (in the horizontal) than does the Earth represents our duplicating we find the means to move above and beyond the motion the Earth establish by duplicating. This motion comes about entirely from life's involvement and by employing life's free will in manipulating space-time. From the laws applying in the "cosmos life" has no value and since life does not exist according to cosmic law all motion must result from some cosmic principle that the Universe are familiar with. The cosmos reacts on the motion as one of several spontaneous occurrences that may take place and motion can only drive for independence if and when under natural cosmic circumstances the independent singularity secured the correct amount of concentrated heat surrounding the independence striving singularity by applying strong enough independent gravity to secure such heat in the enclosed environment. By involving separate heat to commit drive or motion only then can the independent singularity achieve the duplication requirements to manifest such duplication on top of what the Earth is producing.

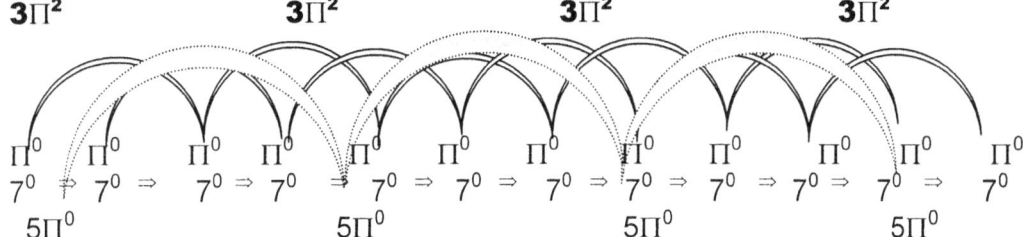

This motion can go on without effecting singularity borders from the limit of Π^0 to the border of $5\Pi^0$. At $5\Pi^0$ the Lagrangian displacement ruling atoms sets the next border.

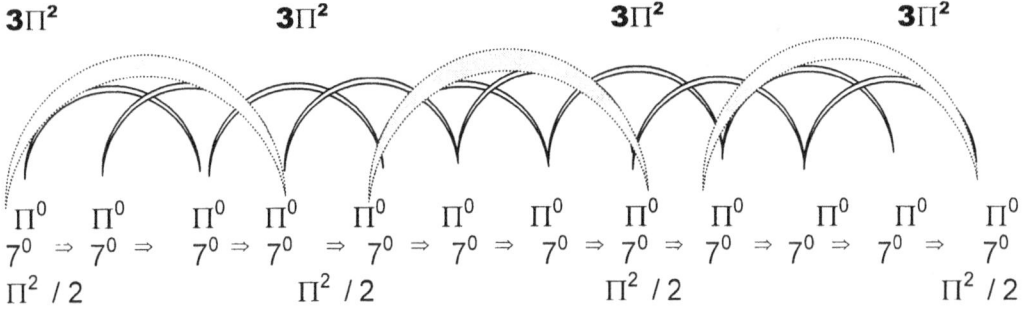

When the atmosphere surrounding the wing releases from the atmosphere that the Earth holds, which was the same atmosphere that the wing had up to that point in motion a time differentiation comes about. The differentiation that comes about causes a time difference in the atmosphere and this becomes a distinction between that which belongs to the wing and that, which belongs to the Earth. Further more is it a reality that time is heat. In this the wing holds a much higher heat or time going close to what was present during the Big Bang than the Earth establish at present. It is because the wing creates a higher gravity dynamic than what the Earth motion is capable of. That super heat or Neanderthal time surrounds the wing directly and that has a higher distinction that the atmosphere has where the atmosphere is still connected to the Earth. The distinction is directly using heat as currency. This can only be accomplished when a time difference of $\Pi^2 / 2$ is between the motions there is in atmosphere surrounding the wing, which is time, connected to the wing and the motion there is in the atmosphere of the Earth, which is time, connected to the Earth.

Second Over All Overview

The frequency of duplication by motion controlled by the Earth singularity applying to objects in motion within the Earth atmosphere does not harmonise with the duplication that applies to the object in motion since the object in motion broke free from the retaining Coanda gravity effect that applies to all other objects in the atmosphere. At that point the duplication of the object starts a new alliance with the Earth by establishing a new Coanda arrangement between the Earth singularity and the object in supersonic motion. However the sound coming from the aircraft at supersonic speed land into space that still apply the regular duplication rhythm and subsequently immediately start lagging behind the crafts supersonic duplication. There is no middle ground for the sound to be within.

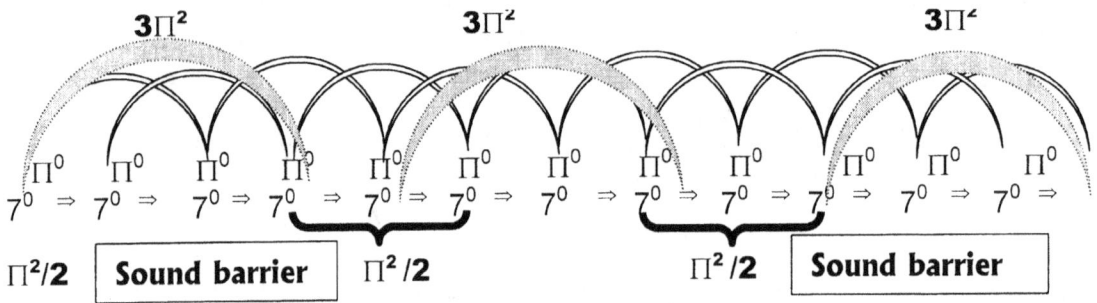

This border is the border we associate with speed and with breaking the sound barrier. When the striving rebel singularity finds the ability to bring about motion such duplicating crosses the border that space-time has the ability to duplicate the carrying of sound waves. Then at that point the rebel singularity finds the second border that will lead to eventual release from the Earth controlling singularity. The duplicating the rebel can produce exceeds the limit that the Earth can produce but it still holds onto the motion the Earths duplication lends to the rebel. Now at such a point the duplicating starts to cross the dismissing gravity lines flowing vertically down at a straight line of 7^0. In this half independence has been established and the secondary Roche limit come into affect. The Roche by half (because the rebel is still using the atmosphere) and therefore still abiding by the rules set by the Earth commanding singularity. The Roche applying between stars is $\Pi^2/4$ whereas in this case it is only $\Pi^2/2$.

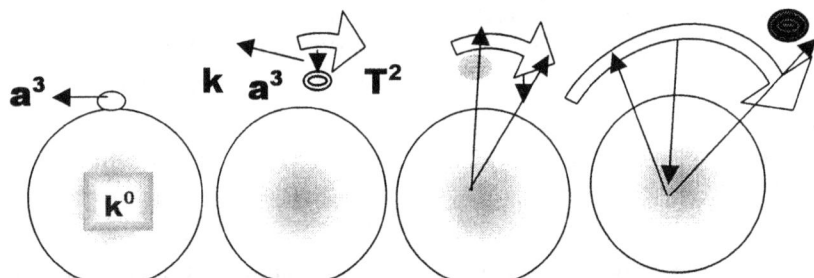

In the accumulating of heat, the object finds motion. The motion brings along structural independence and such independence puts distance between k^0 and a^3. The heat increase will accompany a larger T^2. By increasing heat, the distance between the objects will grow.

The faster the object fly the more the object will secure a singularity that falls outside any parameter the Earth has with the Earth singularity but the moving object must still pay homage to the Earths singularity since it is part of the Earth singularity. Since the motion in heat the aircraft produce forming space is much more prevalent than the Earths duplication the relevancy of space the aircraft represents is much more in ratio than the Earth and the aircraft will surge to a higher space duplicating level that of the Earth. By having wings the contact the aircraft produce with space will increase and dramatise the contact with space even more bringing on a pretension of seemingly duplicating / producing more space than what the Earth is duplicating.

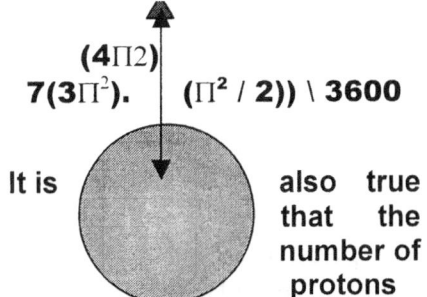

Only by creating a total independent heat centralised in a point holding singularity and feeding k^0 with an independent heat supply can an area a^3 establish a k that will release the independent a^3 from the secure larger k^0. The overall condition is that the escaping a^3 must establish a route following k^0 as k^0 places a diverting 7^0 where that 7^0 then forms part of the object creating heat to secure a release from the established a^3. Providing the heat will bring about a release placing a new object into outer space.

$(4\Pi 2)$
$7(3\Pi^2)$. $(\Pi^2 / 2)) \backslash 3600$

It is also true that the number of protons

Newton introduced G (m + m_p) as $(4\Pi^2)$ and since the object is becoming an orbiting structure, it receives that values Newton

creating motion will determine the amount of space that will deplete and the amount of heat that will come about as the required displacement that will sustain the singularity which will secure a release from the Earth singularity. The motion that more protons produce will increase gravity in the escaping object but it is because of the motion the protons and neutrons create. Gravity comes about from motion and the motion uses the Titius Bode law to produce gravity.

T he releasing brings about "new Cosmic Life" but we must never loose sight of the fact that in relation to the cosmos such, a release was "artificially " created by "Human life" and "human Life" is an alien energy in the Universe and to the cosmos to secure an escape space-time displacement to the order of

$4(7^0 \times (3\Pi^2)(\Pi^2/2)\backslash 3600 = 11.216$ km/sec is the required velocity needed to escape gravity

Kepler stated that $a^3 = T^2 k$. To release from the confinement of $k^0 = a^3 / T^2 k$ and establish individuality as another structure (or possibly debris if classified by human standards) orbiting the Sun the space has to increase on one spot holding singularity as to produce motion which it then applies as liquid heat forming space freeing that particular singularity to an individual particle.

The **$7(3\Pi^2)$** represent the relevancy the Earth holds in the separating relation.

The projectile claiming individual gravity also claims **($4\Pi^2$)** in the relation because it reached a gravity point outside that of the Earth.

In order to overcome the Earth gravity the projectile must accomplish the space and time between the positions it had on Earth and the position it holds in individual space-time as **(Π^2 / 2) \ 3600.** The Titius Bode configuration in accordance to orbiting formation holds a slightly different explanation to the explanation that applies to cosmic structure surrounded by space. It is moreover the individual singularity in maintaining the major singularity, which sustains the governing singularity providing equilibrium in space-time. Not only does atomic individual singularity maintains self preservation, but in doing that it also sustains a governing singularity holding a structural composition and form within a cluster of matter for example a star. As there is between stars so there is in the same manner a mutual or bonding singularity between atoms in stars, which we see as fusion. From this, one may freely deduct that gravity is not forcing material closer but is destroying space whereby it converts the space to a density the senior partner has in the atmosphere of the senior partner. This is most predominant in the cosmos and gravity is all about motion of more advanced in relation to motion of lesser advances singularity. I have no whish to go into space-time displacement and how that operates in this book because the motto of this book is simplicity at almost all costs.

The proton moves 1836 times faster than does the electron and the electron personifies the speed of light. If the speed of light was the fastest that material could travel through space as Einstein explained, the Black Hole would be as much a myth as Hercules is. Material linked to singularity in space can travel 1836 and material not directly linked to singularity by space displacing in space being light can travel at the speed of light. Light travelling is subject to gravity applying and gravity will always slow down the relevancy of light. It does not slow light down since light is a constant but that constant stands relative to the gravity or space in motion applying. In cases where the space in motion exceeds the speed of light, the star goes dark and where the star displaces space beyond the constant limit of light, the Universe surrounding the star with such demanding, singularity collapses.

In the Universe, one proton manages to displace a specific volume of space in ratio to a specific time being at a specific distance from singularity. This is deductible from Kepler's formula. When **$k > k^0$** at the very point the distance forming is that eternal fraction bigger that singularity and **($k / k^0 > \Omega$)** space brake down by specific measure **($k^0 / k = T^2 / a^3$)** the space demolishing in the time duration applying has a limit. The point, which I am referring to, is where singularity Π^0 evolves and becomes Π. At that point within every atom is where gravity collapses space into singularity.

That is the reason why the atom has the ability to dismiss space-time. In every proton in all atoms any proton has this specific generating ability one can measure as one volumetric unit **a^3** disappearing in one time unit **T^2** but the time and the space depends on the time and space singularity permits at that specific location. Giving that space and time a measure would be symptomatic of Mainstream science and would prove very meaningless. A good illustration about the interaction between cosmic vessels applying all space-time relevancies happened not too long ago in front of millions of television viewers around the world. The event will go down in history as the Shoemaker Levy 9 comet splashdown. In the case of the Shoemaker Levy 9 comet the answers Newtonian formula gives does not begin to cover the questions that raises from the answers it gives.

In the Universe one proton manages to displace a specific volume of space in ratio to a specific time being at a specific distance from singularity. This is deductible from Kepler's formula. When **k > k⁰** at the very point the distance forming is that eternal fraction bigger that singularity and **(k / k⁰ > Ω)** space brakes down by specific measure **(k⁰ / k = T² / a³)** the space demolishing in the time duration applying has a limit. The point, which I am referring to, is where singularity Π^0 evolves and becomes Π. At that point within every atom is where gravity collapses space into singularity. That is the reason why the atom has the ability to dismiss space-time. In every proton in all atoms any proton has this specific generating ability one can measure as one volumetric unit **a³** disappearing in one time unit **T²** but the time and the space depends on the time and space singularity permits at that specific location. Giving that space and time a measure would be very symptomatic of Mainstream science and would also prove very meaningless. A good illustration about the interaction between cosmic vessels applying all space-time relevancies happened not to long ago in front of millions of television viewers around the world. It was the event that will go down in history as the Shoemaker Levy 9 comet splashdown. In the case of the Shoemaker Levy 9 comet the answers Newtonian formula give is just about a does not begin to cover the questions that rises from the answers it gives.

Let us put the argument in the Newtonian court and say that if some inexplicable force pulled between Jupiter and the advancing comet the comet then is suppose to travel directly to Jupiter and collide with the centre of Jupiter since that argument does not allow for any reason why the comet went past Jupiter. Remember the reason for Jupiter pulling the comet while the comet was pulling Jupiter was F = G (M.m)/ r^2 and using that formula the formula does not allow the comet leverage to pass by Jupiter as it eventually at first contact did. Then the one core should be heading to the other core where the two cores will meat in one violent confrontation but that they did not do at first.

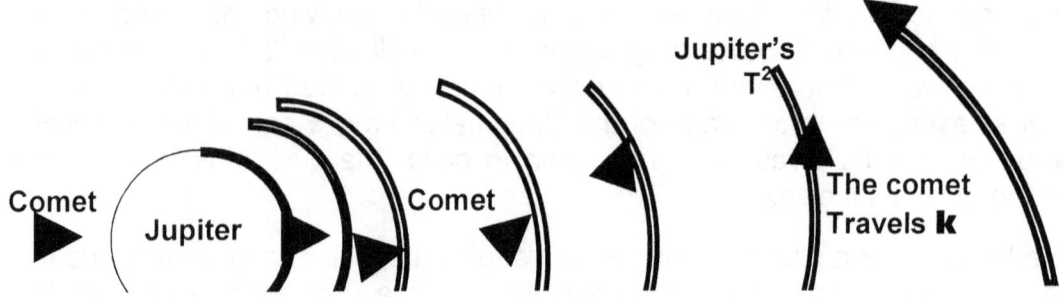

In accordance with the Titius Bode principle the space moved away from Jupiter would in effect double in distance every time the radius extends. The comet distance travelled **k** is directly related to the contraction Jupiter produces T^2 That makes Jupiter concentration T^2 halving the distance the comet travel **k** in its effort to escape from Jupiter's confinement. At some point the doubling time would become greater than the halving distance and when $T^2 > k$ the space a^3 would confine the comet leading the comet back to Jupiter. The Universe that the Universe Jupiter holds and the comet on the rebound had hijacked the comet became Jupiter at the core. The fragmenting was the Roche limit that then turned the comet to liquid.

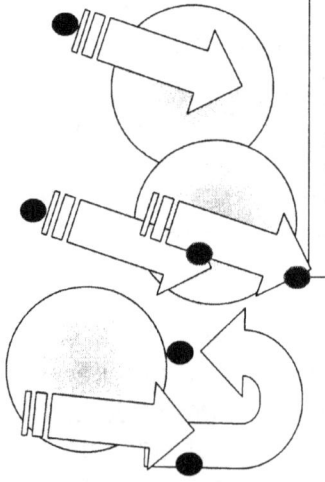

By returning back to Jupiter the Roche limit came into play where as instead the

The comet tore towards the sun via Jupiter's orbit path as that particular comet has done for (say we accept the) 4.5 X 10^9 years that science claims the Earth to be old. From trillions of possibilities in the past the comet never had to venture past Jupiter that close but it ran out of luck... Then this time the comet's fortune changed, which then changed the future of the comet as the comet was passing Jupiter orbit too closely this time.

Let us put the argument in the Newtonian court and say that if some inexplicable force pulled between Jupiter and the advancing comet, the comet then is suppose to travel directly to Jupiter and collide with the centre of Jupiter since that argument does not allow for any reason why the comet went past Jupiter. Remember the reason for Jupiter pulling the comet while the comet was pulling Jupiter was $F = G (M.m)/ r^2$ and using that formula the formula does not allow the comet any leverage to pass Jupiter by, as it did eventually do.

one core then should be heading away and finding the cooling of motion. Instead Jupiter confined the lesser core and the time in space was reduced in Jupiter's contraction where the lesser core overheated by interacting with Jupiter's core and the lesser core then met its destiny in one violent confrontation but that they did not do at first. It is all the result of the Roche limit acting in a cosmic law.

By moving away from Jupiter the comet is moving faster although the comet is moving at the same pace. The further the comet is moving the faster the comet is moving although the comet is moving at the same pace as the comet did before. The faster the comet is moving in relevancy the smaller the comet is becoming but also the hotter the comet is becoming. Since Jupiter is that much bigger than the comet, there will come a time when the comet is desperately moving away and yet it is returning to Jupiter. It is the same principle that has been keeping the comet in orbit around the Sun and that is only at a much further distance from the Sun. It still is the same principle however.

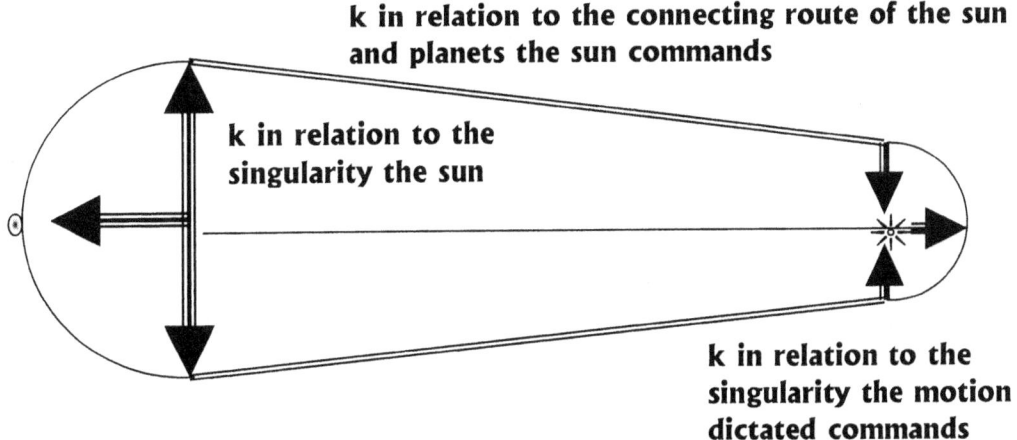

The singularity producing the circle around the Sun comes about as a result of the demand coming from the centre of the Sun.

The circle the orbiting object uses comes as a result from the motion establishing an elected point that serves as an established singularity point in relation to the Coanda effect proclaiming such a singularity centre by creating the motion around such a centre.

In that way and to that reason there is a circling that is much smaller than the larger circle brought about by the Coanda motion forming singularity point.

Crossing the divide always produce new standards as a new Universe is introduced by every such crossing.

This realising came rather too late to save our comet that crossed the divide and again crosses the divide unaware that the Universe changes in every crossing as new borders form bring about a new balance between duplicating and dismissing space-time.

The comet return was inevitable because the approach was so successful

> The comet's return was inevitable because the approach was so successful

The crash did not become a reality and instead the comet passed Jupiter as if it was very much undeterred by the giant. The gravity of Jupiter at first allowed the comet to travel past but only then the jerking back came about. The comet had a speed travelling and the closing in on Jupiter accelerated the comet by Jupiter depleting the space as the comet took advantage of the situation. The comet then had the speed to pass Jupiter because Jupiter helped enhancing the accelerated speed of the comet. With that boosting the comet had little effort to pass the Giant. But after the passing, singularity lines were crossed and a new order came in place.

The success of the comet's fight against overheating became the comet's downfall, as the method of fighting the heat suddenly did not bring results.

This incident has all to do with gravity and nothing to do with Newton's gravity. The comet came along merely duplicating space as it has done since time began for the comet. The duplication of space brought it towards the sun and back from the sun more times than the Earth has people but this was a fatal encounter. While the comet was duplicating space forming the straight line that will eventually come to be a circle around the sun, another mighty object came into the path they share and the other object was dismissing much more space than the comet was duplicating. As the comet came closer the dismissing of space by Jupiter was helping the duplicating of space by the comet. Then at a point in the centre of Jupiter the comet crossed that centre and went to the other side although the passing of the centre was at a respectable distance. As the comet went past, Jupiter's centre divide, which to the comet must represent Jupiter's singularity and presented Jupiter's other side of the Universe to the unsuspecting comet. Crossing the divide always brings radical changes taking place. The duplication of space by the comet was at that point confronting and the contracting which brings on the dismissing of space by Jupiter and because Jupiter dismisses more space-time in relation to what the comet duplicates. The comet was in trouble.

As the comet passed Jupiter the comet for the first time at that point encountered Jupiter's dark side in space depleting The comet saw the future smiling as the coming closer was in progress and Jupiter's space-time dismissing ($k^{-1} = T^2 / a^3$) was presenting the comets ($k = a^3 / T^2$) all the advantage of duplicating it ever dreamed of but crossing the divide brought about misery the comet never though it would encounter since from that point on it had to fight the space-time dismissing instead of joining the space-time dismissing as it had before. The dismissing of space-time by Jupiter proved to be much stronger than the duplicating effort applied to the motion that the comet had and the comet had to surrender motion in favour of the dismissing of space-time displayed by Jupiter. The going back Jupiter produced proved much stronger than the going away the comet produced. Passing Jupiter the comet came at a certain point where the space was not accelerating the comet as it did on the approach all along but in fact it was beginning to slow the comet down. The space the comet was duplicating was at that point being dismissed at a far greater pace than that which the comet can sustain because the time factor Jupiter brought about on route to Jupiter a time factor that was helping the space duplication the comet used for cooling which helped and the comet too remain cooled very easily. But when passing Jupiter the duplication slowed down and the motion retarded where this directly effected the cooling of the comet. The slowing down became not supporting of the space duplication and that the time factor retarded the space duplication. The core of the comet overheated because of the insufficient cooling improvised by motion due to reducing of motion by Jupiter's space depleting.

While travelling towards Jupiter the comet endured conditions, which is similar to aircraft leaving the Earth atmosphere but when it went passed Jupiter it started enduring the atmosphere as it would when the escaped spacecraft once again had to apply the conditions that comes about when objects are entering the atmosphere. The rise in the heat will create space. The comet expanding began to slow the comet down than ever with more space fighting the expanding than ever.

The reducing of space by Jupiter did not do the comets eternal fighting of overheating any favours. Jupiter cramped the comet space with the space Jupiter allowed to the comet to have, has been not enough to support the comet. The overheating became a growing issue because instead of expanding with the heat increasing and bringing some relief, Jupiter cramped the already reducing space-time of the comet and with that ion the heat within the comet quadrupled while the singularity retained no more influence than it had and that then became another factor the comet had to find a compromise for.

With the reproducing of space retarded and the producing of space rising due to overheating there was only one option left. To remain cool enough to support the structure and prevent the structure from total demise more of the comet must then have direct access to space in relation to just transferring heat to space. The comet fragmented as singularity split through overheating and more heat being uncontrolled forms a bomb as heat forms space and with that the comet released more material into smaller portions to space that provided an escape from totally destroying all material. There the comet receive a name being the Shoemaker Levy nine comet because it transformed the space it occupied to nine spaces it occupied. But since Jupiter's demolishing of space at that point where the comet unfortunately found it is so many times greater that the velocity the comet had, the comet found itself rerouted back to Jupiter. One also must see such fragmenting as a way whereby the comet duplicate in order to avoid ultimate destruction and this was a major factor that brought about the planets according to my theory of how the solar system formed when the Universe gave birth to our solar system as unique as it is.

The turning about in direction did not bring on the comet fragmenting because from the comets point the comet was still on route toward the Sun. But the fragmenting of the comet slowed itself down even much more than what was previously the case. To the comet Jupiter was another Black hole that the comet never saw coming because it was in sight and then it was not and then it was this gaping hole. This process that Jupiter applied to the comet was a part of the way the cosmos was born. The cosmic birth came about when singularity was resting in the single dimension of Π^0 where that value held all the possible Universe components waiting for a future but was still sealed in the first dimension. Then singularity Π^0 heated and formed the initial value of singularity in as much as becoming Π. That did not yet release the other components, as they still remained stuck to the first dimension in the first circle of Πr^0.

They came to be on the newly formed rim of Π but were still in the first dimension of Π and only by the second expansion providing the motion that brought on the second forming of space outside singularity that is holding all captured future possibilities in the upcoming Universe did every one of the three factors come into being separate and individual factors. With the first motion Π^0 to Π not one of the factors k^0 and T^0 and a^0 except singularity Π^0 to Π was released by the expanding of the distance Π^0 to Π from single dimension. The components came into the Universe being a factor in the single dimension only after Π^0 became Π. When Π moved from Π^0 to Π, k^0 and T^0 and a^0 only became factors but not identifiable separate individual values yet. It was at this point that the law of Pythagoras came about and the line was the same value as the triangle and the half circle. It was at this point there were no formal values yet and all components within the newly born Universe had their values expressed in degrees meaning in directional rather than space valuating relevancies to one another. This was because only motion T^0 the half circle with distance growth the straight-line k^0 and space in the triangle a^0 came into place. One can see the way Jupiter rendered the comet to heat. First it went about destroying the route the comet followed. Then Jupiter retracted the space the comet took to escape by reducing the space more than the comet was able to duplicate and in that the comet retreated back to Jupiter faster than the comet advanced onwards towards the Sun. Then Jupiter compressed the space the comet shared with all the space that Jupiter retracts in a natural process, which cramped the comet to the level of being liquid heat. That made the comet fragment which then circumvented the Roche law that Jupiter had to acknowledge.

With the fragments much smaller that the complete structure the comet had it made the devouring of the comet much easier as Jupiter finally cannibalised the comet singularity holding fragmented space- time. At first glance it is clear that the process repeats what Newton suggested what is happening out there. But if Newton was correct the centre point had to draw on the centre point and causing a direct line pointing at the direction the two objects had to travel as they reduced the distance r by the square to the force of the two masses. The two objects did not draw closer in a direct line, as it should when applying Newtonian views. In stead it came by a roundabout manner by first passing then returning then passing again before it broke up (something which is quite unexplainable using Newtonian science) and had the fragments slammed into Jupiter. It is the circling around that bothers in the Newtonian explaining. The comet once past the divide not going straight on towards the Jupiter centre, and came back following again not the straight line in the retreat but had another come back towards Jupiter before slamming into Jupiter. One should use this to read into what went on in the early cosmos before material was set on undisputed routs and later on and later on never influenced one another's travels. We find that more advanced singularity dominates lesser developed singularity and so much so that it happens to the point where the lesser singularity sometimes completely turn to liquid heat. That should point to where heat originated as we deduce and read into what applies in the process of liquefying unsuccessful singularity to a state of fluid heat.

Jupiter reduced a^3 of which the comets a^3 was part of the point where a^3 once more became a^0 but with a^3 going a^0 the rest of the relation must then also go single becoming T^0 and k^0. This had to one of the manner that applied during the time that mainstream science now presume antimatter came about. The comet became antimatter because to the comet Jupiter became antimatter reducing the comet to antimatter, which the Jupiter antimatter dissolved. Jupiter gave the comet a fair share of antimatter as Jupiter introduce $k^{-1} = T^2/a^3$ and that reduced the comet space-time from seemingly enjoying $k = a^3/T^2$ while in that time in truth and facing reality the comet was also reducing by applying but which the comet never suspected because from what the comet saw it was all the time applying $k = a^3/T^2$ while $k^{-1} = T^2/a^3$ was what the comet received. The comet became antimatter because of Jupiter forming antimatter in the form of k^{-1} to a point where heat remained. The three factors combined and multiplied to bring about the cosmos as it came to be. It was at that point mathematics arrived but the Universe was already secured as a statement bringing many factors to become part of the future. But singularity first came into position as Π^0 to Π and from that the foundation of expansion and growth was formed. In that very instant Π^0 to Π formed motion Π^2 to Π^3 where space came into place duplicated by motion.

As we can see from the behaviour of the comet in relation to Jupiter motion and dismissing formed a ratio. The very first instant came when $\Pi^0 = \Pi^1 = \Pi^2 = \Pi^3$ and all had the same value, which was separated by directions rather than the measuring of size as such. We witnessed the reverse of when the comet dissolved to heat that had no gravity and no duplicating qualities. From that the product of the culmination grouped together as major atoms or what we call stars. With the passing off the comet the applying gravity between Jupiter and the comet was $a^3 = T^2 k$ but by recalling the comet Jupiter changed the applying gravity to what is the same gravity applying to us, which is $1/k = T^2/a^3 = k^{-1}$ and also the same gravity Newton saw. Our position, which is a captured position on Earth, is also that of the captured where T^2 got in a position to dominate k into negativity. By using this evidence one may arrive at a conclusion as how the solar system came but implicating laws that apply on a much larger scale by using the cosmic relevancy scale and thereby determining the original position of every planets as they were before development changed the lay out. By allowing development to rein act the solar system it becomes an open book to read.

When time came from singularity it brought along space and distance in motion. The distance represents the creating of space while the motion represents the destruction of space. Newton proved that $4\pi^2 a^3/T^2 = G(m+m_p)$ where k also is a^3/T^2 (from Kepler $k = a^3/T^2$) and therefore $4\pi^2 k = G(m+m_p)$. Then in the event of singularity being $k^0 = 1$ then $4\pi^2 \times 1 = G(m+m_p)$.

That means extending from singularity space-time will be $G(m+m_p)$ or $\mathbf{k} = 4\pi^2$, which is what space-time became when $\mathbf{k}$ pronounced space-time for the very first time coming from singularity. Singularity extended to form $\mathbf{k} = 4\pi^2 = (\Pi^2+\Pi^2)+(\Pi^2+\Pi^2)$. With heat disintegrating to space and the cosmos forming an extending of space-time in space, space-time deformed to $(\Pi^2+\Pi^2)(\Pi^2\Pi)3$ and that is the value of the atom. Since $G(m+m_p)$ is only a symbolic gesture it can represent any aspect as long as it represents a dominant symbol relating to a subordinate. It can also represent singularity in the same relevance where the one singularity is dominant fighting for dominance or control and the other tries to secure independence by establishing personal identity/

The very first instance brought the developing of $\mathbf{k}$ that was equal to $\mathbf{T}^2$ time and $\mathbf{a}^3$ space

But motion came about on this side as well as the other side of the Universe where the Universe was $\mathbf{a}^3 = \mathbf{T}^2\mathbf{k}$ and on this side was $\mathbf{a}^3$ while $\mathbf{a}^3$ was becoming $\mathbf{T}^2\mathbf{k}$ on the other side and the other side $\mathbf{T}^2\mathbf{k}$ was becoming $\mathbf{a}^3$. Then the second in creation came about but the first instant plated a veneer of material in the form of opposing singularity destructing as liquid heat due to a lack of motion onto the singularity applying motion, which brought about to specific markings onto singularity. This act represented destroying of some less protected and the maintaining of other singularity remaining in form and cool. This would see as a flicker from another point holding singularity. From the other side the flicker would seem to cut the duration it took the motion to fill the time half as long while in truth it cut the space in half by doubling the time. Then four flickers came about and eight flickers came about and the action time provided brought about more space as space doubled every time but it cut time duration by half every time.

By the time the proton and neutron became about the very fist time the duration of time slowed down by Π^3 which is $(1836)^3 = (6188965056)^2$ more than the original space but there were also $(6188965056)^2 : 1$ flickers of time to the original. Every flicker $\mathbf{T}^2$ of time represented on $\mathbf{k}$ that accumulated by introducing one unit of $\mathbf{a}^3$ space. One rotation then meant time reduced in duration by $(6188965056)^2$ times to the original on and space increased by that margin. Eventually by the time that the electron and light came about the space was confirming duplication to a ratio of 1:6188965056 that means for every flicker light brought along space confirmed $(1836)^3$ times over. Light found the epitome in $\mathbf{k}$ and $\mathbf{k}$ formed the atom but the atom confirms the unit of space duplicated by time to the value of (1836) which is made up of the double proton $(\Pi^2+\Pi^2)$ and the neutron $(\Pi^2\Pi)$ as well as the electron 3. Therefore the motion of the proton is 1836 X 300 000 km/sec. That confirms the proton as "flickering" 1836 times more than the 300 000 flickers in km strokes per second of that of light. In that way the duplication of space is accelerated by 1836 times when it becomes a proton in the ratio between light and mass.

By the time the Big Bang started all material was wrapped in a blanket of heat. I have little intention at this point to go into the matter significantly technical and explaining the process in detail other than to say that the wrapper covered the wrapped product that freeze a Universe until a use could be developed.

The Grand Unified Theory admirably proves this. While sanity prevails we must discard the nonsense about gravity pulling dust into thick solid material that eventually form stars and most of all planets. That is a mediaeval myth and the sooner it becomes discarded, the sooner can cosmology shed the envelope of backwardness. The thought that stars can come about from particles being dust that through the magical process of gravity can fuse together as solid matter is a fairy tale. We have to divorce from that nonsense as we explore true development. Once singularity establishes independence, the star structure leaves the cocoon the galactica provide and move into the open Universe as an independent structure. This is done by the cosmos extending **k**, and in that all singularity connects and extend **k**, by expanding the relevancy singularity takes on. Every line within every atom within every star goes through this extending process by cosmic development. This serves as an indication of cosmic progress. As the governing singularity develops within the centre, the star finds meaning in the development coming as a direct result inherited from the growth in individual singularity that the atoms show. Atomic development corresponds within the star and that drives the galactica by the progress that the relevancy **k** produces in the star matching the progress of all relative **k** that influences the galactica. The singularity governing the galactica discharges its control as its capture tarnishes. This favours the gain of stars in the galactica development.

The process comes by way of the developing of the star's atomic singularity driving the star's governing singularity. This extending we find of the factor **k** is what develops the progress of time. What all concerned miss is what time really is. People wish to put eventuality concerning human history as time but in all fairness that can only have some relevance to time. To us it is very important to know what our past was and where from did we develop but in cosmos terms it has no concern to the importance of the Universe. Let us journey to find space and to find time but above all, find the ability in distinction of both. If we wish to find the future, we should locate the past. If the cosmos is contracting, where to is it contracting? The direction of contracting must be in the opposing direction, the direction of expanding. If we wish to locate the past from where the cosmos came and through that see in what direction the cosmos came, it must take an effort to backtrack the direction it came. Should the argument come about that all came from nothing, then everything still has to be at nothing, or our understanding of nothing leaves much to desire. Nothing means not existing, not being, never found and unable to produce any multiplication thereof by any growth.

The above questions, but mostly the unanswered questions about what is more nothing and what is less nothing draw me to the realisation there can be no such a quantity in space as nothing because even space has to be something. Clearly as it is for any one to see one creates space by nuclear explosions. In explosions Academics portray the winds as shock waves, but what is the shock wave other than new space coming into prominence and rearrange the structure in relation to the new space just created by liquid heat unleashing the created space as well as the space volume that came in place. In that way, it is clear that releasing heat brings about the expanding of r as part of the sphere forming space. Hubble proved the Universe is expanding. Then by backtracking, we have to set about reducing the sphere constituting the expanding Universe. If r in the circle is growing, we have to reduce r to backtrack. When the circle reduces, the value located to r will become implicated because r determines specific size. Not so in the case of Π, because Π in the true sense only indicate that the circle is a square without corners and therefore Π dictates form and not size. By reducing size, only r comes into contest and will point to such reduction. By reducing the circle radius r by half, continuously will lead to an infinite small circle but Π will remain because the circle as a form remains even being infinitely small.

It would be the same as if the ant running in Central Park in New York is of the personal opinion that it is his being in the park that is the cause of the park being where it is and maintained as it is and only because of our ant sharing his personal benefit with every one through his magnificent generosity that every one which is there in the park should humble them in gratitude. Our ant considers that every one holds the opinion that amongst the many of millions of people they are all aware that he is the reason why the work effort of many a thousand people is dedicated on his behalf to maintain Central Park in New York just so that the ant may harvest the benefit. The ant will have the opinion that all the people in New York are of the obsession having one purpose to live for and that is to please that one ant. Yet, that is happening with the light travelling through the cosmos to us and us. Every person standing in the Universe is under the illusion that all the light throughout the Universe is directly flowing to the very point the person is standing. It happens to all of us. The place where I stand or any other individual for that matter is standing is positioned in such a manner that every beam is directly flowing to that very specific spot. Every beam is coming at the speed of light through the entire Universe to locate such a person with that magnitude in honour and glory as to fill the centre of the Universe.

From all the corners of the Universe and in specific every spot imagined one line of light is especially directed to that specific location used by that specific person for that specific instant in time. That line from every possible spot there may be is competing for the honour to reach the person filling that centre spot...and only the toughest will endure. One very important human being is filling such a location of absolute splendour. The light departed from every location in all points throughout the entire Universe stretching further than the mind can admit directed on course to meet the person in that centre spot. The light followed one after the other dedicated in a stream of innumerable millimetres for billions of years to flow in the direction and directly to that spot, never diverting for one instant, to come to where I am filling that spot in that centre of the Universe.

All the light in the Universe is coming to me. It is on route straight to me where I am standing filling one spot on Earth. If there are those that do not believe me, well those I challenge then to go outside at night and see the vastness everywhere from wherever the light is coming from and all the light is heading precisely to where that person or me is standing. It is coming from all over. It is coming from areas so large not even Einstein can calculate the size or content of such measure and it is rushing towards me specifically. There is not one ray that is going to miss me by fluke or accident. The light has one purpose and that is to meet me at the point I am presently located. Every beam has my name on it and it is coming for my eyes. Can any one imagine if a person was standing in a location and found all the persons in that city was running towards him where he is occupying that point, how frightening such a person must feel. Yet, it is happening to every one from where ever the vastness of space is situated and is coming across space to that very specific point the viewer is standing.

Even if I shift to another position on the other side of the Earth or to the moon, the light will change direction and trace me in my new location. Even if my new location is in a camera and the camera is in a vehicle in the centre of Mars the light will know that I am using such a point and trace the camera so that I may still be in the centre of the Universe. Wherever I might be, the light will still get me at that location. The light flows to me from where ever and to top that it is flowing to all other persons. That means it is not the Earth that is that important but it is where the point location is and that point which the observer is using to view from, that is the most important place ever to be. If it was only the Sun that the light was streaming from that is choosing me as representing the Universal centre being the centre of the Universe it then cannot be that very exclusive. The Sun is close and the light is plenty. That, which I am referring to, that it is coming from all over.

That is just one small part of the fantastic affair. Some of the light left the stations they come from some 12×10^9 years ago to meet little old me in this spot I is filling. The light has been travelling 12×10^9 at the speed of light, which I might add is much before my birth crossing space and time, rushing all the way to meet me at this point. No one ever thinks how it was possible for the light to know I was going to stand at this point and be here the moment the light arrived. How did the light know I was going to take centre stage at that moment when it left so many billions of years ago and fill the specific centre of the entire Universe? I have to be in the centre of the Universe because all the light is travelling to this spot filling the centre of the Universe without one straying off course and missing my spot. The light takes two million years coming only from the closest next galactica to meet me here taken into account the prefect timing it applied after all travelling that far to be in time just to meet me in the centre of the Universe. How important can I ever dream to be? Light is coming across time measured in millions and billons of years through space measured in millions of trillions of kilometres, travelling at the maximum speed the cosmos will allow, ignoring all other places it could go to and came to meet me in the centre of the Universe.

Photo credits CNN.com

When we look at the night sky we see images of stars. I am of the opinion that our vision of stars and our interest we show in stars is just what sets us apart from other species. Our interest in stars make us make us realise about what we are able see what we never can touch though as we can appreciate what we never can have. We interpret what we see without ever making contact to confirm and that gives us external knowledge and insight. Our vision about that, which we see tell us that there is more than the animal's concept of a plain survival on Earth where it is that you can eat or you can be eaten. Fathers show their children the constellations and although we no longer attach religion to our stargazing it never subdued the bliss we find in our astonishment about stars.

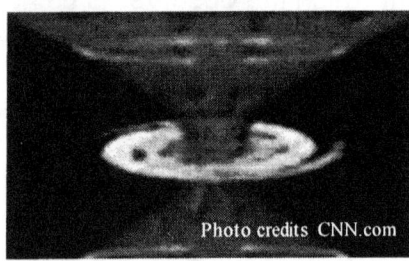
Photo credits CNN.com

The star that gives us our greatest wonder is a star we cannot see. Every one stands amazed at the fact that there can be a thing such as a Black Hole. There is so much to ask and such a lot to wonder as to why and how and where and which…yet, we cannot see any that we interpret. We see but we cannot see and that makes us wonder what it is we cannot see and what it is we wish to see. The fact that our view is obscured by the fact that our view is obscured dramatizes our sensation of wonder many fold. That is human and that is why we are what we are and why we are in terms where we are.

It is part of the human concept to believe your eyes. Seeing the sand dunes on Mars is equivalent to seeing the sand dunes on Earth by means of the television media and could just as well be of the Sahara. The Sahara is a place we can go and visit should any of us wish to do so, but the dunes on Mars are another problem. Visiting and confirming what we then see is not that simple to accomplish. The Martian dunes are not only space away, which means I can cross space in time and visit. The dunes of Mars is not even space away but is time away. There is no way I would ever cross time to see for myself what there is to see.

That is what is wrong with science, amongst others. Science is of the opinion we see space. We do not see space. We see time, but it is not time we see, it is the distortion of time that we see. The "further away" we look the more time we see. However it is not time we see. It is the distortion of time that we see. The further something is away, the more it is in the distance, the longer it will take the light coming from the object to reach us. That means the longer it takes light to reach us the more time is distorted to put distance between what we see and what there is to see. It is not space that we see but the distortion or the compromising of time. It is the time delay between here where we are and there where the object is that we see and we do not see the object or the space the object has or even the space between us and the object. We see the time delay there is between the object and us. We see what was there in time gone by, however, we do not see what is there and we see space for what space represents to the Universe. We see space as time delay, time slowed down. That is what space is, space is time delay. That concept urged me to go and look for the beginning of space and the beginning of time and the origins of the concept space-time. Please allow me to explain the beginning of space by measure of time delay. At the time of the Big Bang everything was small...not so...it was as big as it is today. If the Universe was the size of a neutron, then we had no size at all. One cannot compare apples with oranges and see bananas. The space we see is the distortion time has to separate points of comparison.

In order to understand what I am trying to say I have to use a picture that is most probably not a true event. What we think we see is space. It cannot be space that we see. In the forefront we see a line that is a result from a comet travelling. Then there are pixels indicating lesser star structures and some clear dots indicating stronger light spots, which would personify larger stars present in that direction. The rest we see is the black of night. If the Big Bang theory is correct and to my thinking there is no doubt about that, then not to long ago there was a lot less space between the objects than is the case at the present. The space was less. That cannot be the case because if the space was less it would then take the light much quicker to arrive at the spot we are at present. The light coming from what should be the comet is relatively quick in reaching my location while there may be some of the faint dots that have light travelling a considerable time to get to me. I presume the comet is closer. Looking at the image of a roving planet it shows a structure filling space at intervals. The space it fills is a constant because the space does not change in becoming bigger or smaller. However, the space it is moving trough appears to grant the roving planet another position every time it is photographed. It is in the terms of time that the answer is. It takes a different period to position and obtain the light coming from the different position where the object is located.

If the prime object were the space as it is in the case of the space serving to fill the roving planet, then no changes would come about to the space. It would take as long to fill the space between the object and where I as viewer am in position with travelling time. It does not because the motion that the light has to endure is shorter or longer by time duration. It is the space that is constant, yet the time to travel varies. It takes time to cross the space whereas the space holding the object remains filled at an even volume every time.

In the case of the planet space filling with material by gravity the same space filled without changes. In the case of the dark space, that space is putting time at a different duration to reach the location I am in. The "further" the object is in distance from where I am, the more time it would take the light coming from the object to reach me. The object will appear smaller as the distance increases but I know the space the object holds is filling the same volume as it does when being close to me. In the case of the space filled, that space appears to change but that space is filling a volume at a constant. It is when the space in which the object moves increases the space the object holds then diminishes. The space that the light has to pass through to bring me the picture of the object increases. That cannot be because the space is filled all the time by the same margin. It is the time the light takes to bring me the picture that increase and it is that light that shows me a diminishing space. It is not the space between the object and my location that increases, but the time that increases and by allowing the time to increase I allow the space of the object to appear to become lesser. The space the object holds has to remain the same and the space between the object and me cannot change by motion. It is filled by volume that motion cannot change. Only time can be affected by motion and since it is motion that is changing it can only be time the motion can change. The slower the motion the longer wills the time be that it takes the motion to negotiate the space. That black of the night that I see is not space that I see but is time that I see and the space I think I see is the retarding or slowing of time that I see. Outer space is not space but it is time that space retards and therefore space is not space but a retarding or a distorting of time. That means that which see thinking it is space that we see is all the time, time that we see and being time it has no outside because time is eternal. Space, being infinite interrupts time to give time in eternity duration value.

The relevancies we are about to address are about form. It takes us into a Universe when a line had the same value as a half circle and as a triangle does. It takes us beyond space to a Universe when time formed space. It puts the Universe beyond distance. It is what came about when space interrupted time to deliver us the black of night, which we incorrectly think of, as space. The Universe did not start small it started outrageously big. It is not expanding it is reducing. When the Universe started there was no outside to that which started because if there is an outside then what is on the outside of that which started. The Universe has always been an inside that went smaller. The limits grew smaller not bigger. The initial start had no limits. That which we think of as so small and tiny, so small it has no sides is so big it cannot have sides because it is too big to have an outside and all we see and all we cannot see fill the inside.

Where we are now in the Universe we are so much smaller than what was when the Universe was the size of a neutron or whatever it was. If the Universe as one block without limits had no outer limits and was the size of a neutron then it grew smaller because what was our size when the Universe contained all it had in a neutron. When the Universe was a neutron we were not even a thought. It is easy to lose perspective but perspective is all we dare not lose. That which took al the space a neutron could offer back then has no limits now and has no boundaries. It is too big to be cooped up by limitations and boundaries. We with limits and boundaries now have measurable quantity to calculate, but what was the Universe then has no calculations art present.

Where there is no boundary to shift what shifts then and yet they say the Universe is shifting its boundaries because the Universe is expanding therefore it shifts! Where no growth is possible since it captured the growth at the beginning where too can it grow. The end of such a shift by what cannot shift to where no shift is possible will eventuality be what they named The Big Crunch even before locating the Big Crunch.

It is like naming a baby even long before knowing how the procreating is taking place that will lead to impregnating of some member of the specie (which member it will be is still then still unclear at the time the name giving was undertaken) where it later on will lead to conceiving the baby ... that is the manner in which science dogma is enunciated but that is how clever those mathematicians are that knows everything there is to know on science. They can name a baby before even knowing what procreation is and that they do by calculating what they don't know anything about... like procreating the baby! It seems more likely that that which has no prominence finds prominence, which means the lot is shrinking. The Universe is surely shrinking to give us space to be.

When we altered the size of the moon in relation to the size Mars has what we did was change our relevance to that of Mars. We first brought Mars on a time line as close as it would be if it were hanging around in the space the moon has at present. Then we moved the time line back because it takes time to travel to the structure. Pushing Mars back does not increase the space, because eventually Mars fills the same space. It increases the time duration between Mars and us. It is not space we cover. If it were space then the time would be equal for light and for all to complete the journey in the same time. By changing the time the relevance change as to how long it would take to get there.

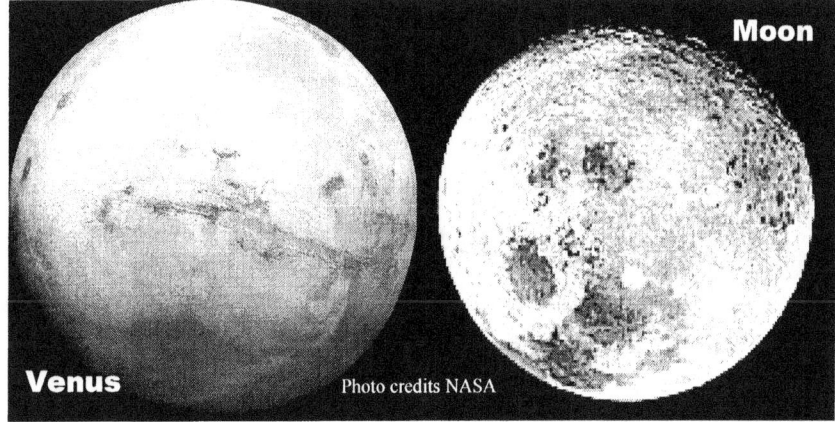

When we look at the images of the two solar objects i is so easy to put them out of perspective and in the same size, although we know they are not the same size. All one needs to do is just play with the dimensions and find the results. One changes the space they have to match and they are equal in size.

In cosmic reality the reality is quit substantially different. When we put our hand out we are able to touch...say the door we are immediately in contact with the door. It is the door we touch because it is the door we see we touch. Moving back one meter we find we are no longer able to stand upright and touch the door because we are one meter away from the door. We are one meter away because we can see we are one meter away from the door. We grew accustomed to this thought because Galileo's pendulum shows we are in time in space in the Earth timer in space. The time we will take to touch the door corresponds directly on Earth with the distance there is. Things change drastically when we leave the Earth or when we view object not confined to the Earth as we are. The truth is we are accustomed to think we are one meter away from the door we are unable to touch because we think we see the door is one meter away.

However that it is not the door we see. We cannot see the door because the door is not there for us to see. We see light banging on the door and as the light is rejected by the same door that the light comes flowing to us. We see the rejected light bringing an image of the door we cannot see. It is light we use and that we are used to of using to confirm what we see but such confirmation is what makes the most intellectual stumble. In quite the same manner we see the darkness of the night and observe such darkness as darkness. By darkness we interpret the meaning as that which we cannot see or that which we are unable to see. Reality tells me that the darkness is light that is too bright for us to see. Take an image of Mars with a close up view. Then reduce it and go on reducing it until it is so small it becomes invisible. The space filling darkness is not darkness filling space because the ratio of darkness increases as the ratio of light in comparison to the darkness reduces. The object does not go dark by moving back. It rather becomes more of the same when it blends with the darkness, which proves the darkness is not darkness but it is light.

By reducing the space an object has the darkness becomes either more or less but the darkness promotes the object or reduces the object. The fact that large objects are close and small objects are at a distance we on Earth relate to more space and less space. The only factor that can produce more space and less space is time because time is irremovably connected to time By reducing the share of the combination of space-time time must reduce or increase to allow space to do the opposite.

This is the best example we may ever find of space-time. In the term we use space-time it is the time that is reducing the space. The time divides the space according to the time it takes to reach the space  from where the viewer stands.

The time factor reduces the space factor as it divides the space factor into smaller parts of time holding space in eternity. That proves that all containing space is in fact time holding space. $k^0 = a^3 / T^2 k$

$T^2 = a^3 / k$ and most of all $k = a^3 / T^2$. The time T^2 it takes k to reach the space a^3. That is space-time and proves that space-time is not some mysterious cosmic scheme covering up of a Black Hole image the Universe devises to benefit only those with highly groomed mathematical skills. The more time develops and time pushes the object "further away" the smaller the object will seem in relation to the containing space or time. Outer space is not space but eternal time holding space.

The realising that only time can affect space by the measure of appearance is a huge step in the right direction. Space is a constant therefore time has to influence the appearance of space to become apparently more or apparently reduce to become less. Being big is a sure sign to the brain of an object being close. That would then appear as if there is little space between the observer and the object in observation. That is culture talking because space may appear larger or smaller but it can only be a medium of space that may allow space to appear. Space as such has the same measure and has the same prominence when measured. Time is the factor that allows space to reduce and even to reduce to the obscure.

Moving the object back into obscurity does not reduce the space the object has but puts the space the object has into a much larger definition in space in relation to the space I witness. The light streaming from the object will also fade into obscurity and disappear as the definition of the object declines in relation to the space it holds by comparison to all the other space in view. The light the object had did not decline or reduce but it diminished in relation to the gross of space holding light. In relation to the space out there the space diminished the light in relation to the darkness the light then offers. That way the light could only reduce by comparison if the light was less in relation to what the light is in the darkness we see. That means the darkness is flooding the light the object has and therefore the darkness we see is light. However, our relation to the light makes us in relation to little to be able to appreciate the light because as the object retracted from the position we had, we also diminished in space by the same measure. The space we hold therefore is too little to enable us to appreciate the darkness flooding us with light.

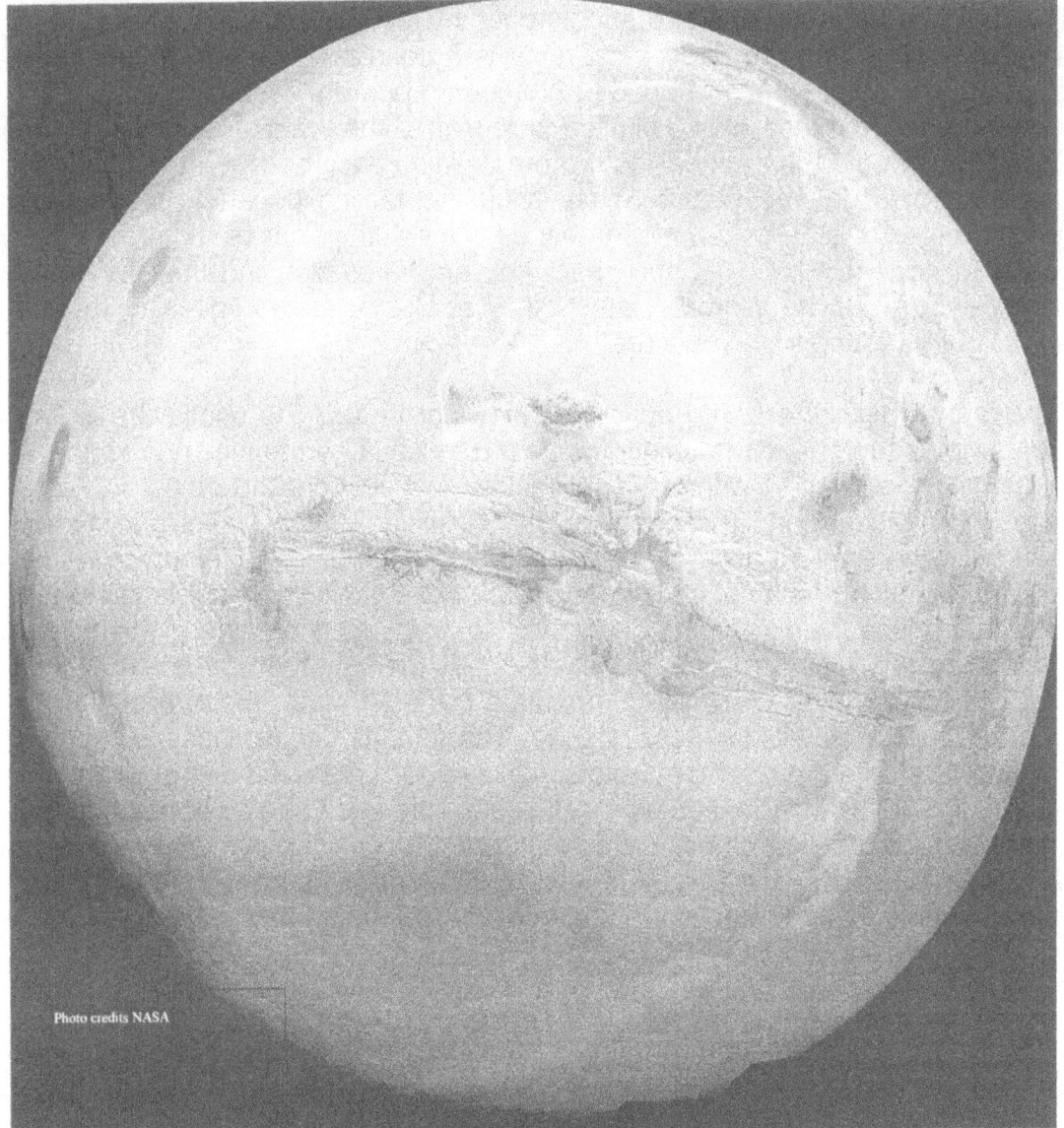

Photo credits NASA

Even the fact that the objects seem to be as near or far is defined by Kepler as a fact in $a^3 / T^2 = k$. This answers the question a friend of mine, Johan Boonzaier, asked me one night around a dinner table and which that night sparked life into my first Afrikaans book. His question about the blackness outside that could never end, was the question which finally got me inspired and set me off writing these books...It was his question about what space was and why space in outer space can never end. That was Johan's question and my trying to explain what time was and what space was according to my opinion was what started me off writing my work. Johan, that which you asked me what it was out there in the blackness of the night, that is not space, it is time, and therefore it can never end. If it was space, then there had to be limits, borders and measurements defining it but because it is limitless it is time eternal. It is the line of time being over grown and stretched limitless but still an eternal line. Borders will always define space being infinite but time is forever eternal.

That is the easy part to figure. By moving the object back in relation to what we view is not diminishing the space the object has because the object will hold the same space it had before. The object is as big at present as it was at the Big Bang event because what was there was there with no adding. What is present in the Universe is in the Universe and no adding or removing of what is in the Universe is possible. If the Universe grew the object had to grow in parallel with the Universe because the Universe got somewhat bigger than the size a neutron has but so does the object have much more space that what the neutron has. The size the Universe had contained what was inside the Universe at the time the Universe went bang. In that there is little to no change possible.

Therefore moving the object that is in our vision back does not diminish the space the object has. What does increase is the time differentiation between the position I hold and the relation of that to the new allocated position the object holds. Should I wish to reach the object at the speed of light, I then will have to diminish my space to that of a photon to be able to reach the speed the photon has. However, in the event I do not wish to surrender any space I hold for reasons of the survival of life, I then would have to travel taking much more time to reach the object at the location the object holds. Then we get to the question of what did it take to shift the object into the allocated slot it now has in the distance it  **now has. It took time to shift because it did not take space to move. The time it took the object to move to a new location was added between the moving object and me. It would take the same time but in a different ratio of moving to retain the time should I wish to move to the location where the object filled the same part of my view I have of space with the object in that space. That means since the Big Bang there was no space added but only time increased what parted the objects since then. Space is the retarding of time.**

As the cosmos present its evidence, we can see from such evidence how destructive overheating is. Forget pressure, because Newtonians over simplify everything with pressure and exploding. That might happen to a drum they fill with gunpowder but that is not applicable in the cosmos. In the cosmos, unlike in containers, there is no retaining wall that sets limits to pressure inside the container versus pressure levels outside the container. The cosmos has no pressure or pushing or pulling. It has a flow of space-time by concentrating time and duplicating space as it is driving space-time towards the centre. In any picture about any star there is no containing wall that keeps whatever is inside, inside. There is no limit to what the wall if the structure can contemplate before bursting. In the centre of a star is a point holding singularity and since such a point has no space and is immovable, space has to compromise by flowing towards such a location. We regard what we see at night as space and how wrong can we be?

Have you as you sit reading this part at this minute sat back and gave a thought about the light enabling you to read? Such a thought brings to mind the most simplistic answer one can imagine. The light hits the page bounces from the page and contact the lens of my eye where the lens conveys the photons becoming electricity to a part of the brain that translate the electricity to an understandable message and that makes one read. It is as simple as that! Ever gave a deeper thought about light streaming across the night sky, coming from ends of the Universe we do not even realise it is there? How does the photons manage to convey one complete picture coming from as far apart and as wide an area as it does? With a few photons connecting the eye or lens no one ever noticed the wonder of light. The photons reflect a view that seems as if coming from all the billions upon billions of stars. But most is coming from darkness covering an area no man can measure. Yet how many photons can actually connect to the lens of the camera or to the eye? Still a few photons coming from a single direction directly ahead eventually tell the entire storey. It is very simple to take the process of seeing by means of photon conducting very lightly and I have never heard one of the Brainy Bunch really in sincerity uncover the process to its utter and full potential. It is impossible that light from such an array of assorted sources can simply come together at the eye lens and show a picture of objects spanning across a Universe as wide as our mind can receive where the objects they reflect is beyond human measurement and the quantity is inconceivable many.

If the darkness was the representation of "nothing", then that should be exactly what we must see, nothing but the stars. Taken from the top picture some stars and leaving the rest to nothing is what we see in the picture below. A blind person sees nothing but when we look at space, we see something that we think nothing of as we see as space. One cannot have the ability of sight and see nothing except by closing your eyelids and then you see nothing. But in that case you do not see "nothing" in contrast of "something" you see "nothing" without it contrasting to "something".

Nothing is all about not being and not "not seeing".

By the ability to see the darkness renders the darkness something other than nothing and that changes the acquired value of the darkness from nothing to something. There is an eternal difference between something in infinity and nothing.

The arguments introduced up to this part of the introduction prologue only touches the most basic aspects of my work and by no means can such an introduction secure an opinion. Yet, not once through all my long investigation in the past thirty or more years have I found any other person claiming such views that I have brought about even in this skimpy way as I do in the prologue.

Light is much more than the medium science takes it to be. Light connects the Universe in a way we cannot contemplate. Light being far apart originating from regions not in the same time or Universal space connects in a way that present us with a picture holding the Universe in an understandable content. From the point we stand and we watch the Universe the significance of what we see surpasses the sense of understanding of what we are experiencing. How can the few photons that our lenses catch coming from such an area as the night sky cover transmit the complete picture of what we see. Take a few seconds and study the picture of the night sky then rethink the picture applying the full content in the picture to what the size of you eyes is. Think how big the picture is that your eyes take in and translate that area to the size of your eyeball in an effort to determine a ratio. One will be forgiven if one thinks of the ratio as eternal to nothing. Yet a few pages back I showed that according to mathematics there couldn't be anything as nothing. Consider the path the light followed from the source connecting to light from all other sources where all particles of the other light may come from and bringing a full picture to the lens one use to look through. In your mind connect a line from every atom producing light and connect the lines to your eyeball and see how you can manage to fit all the lines, as small as the lines may be.

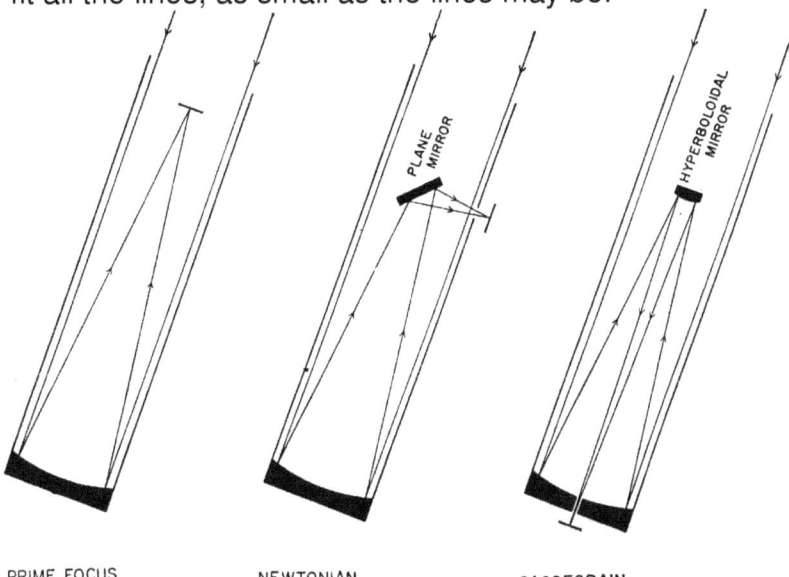

PRIME FOCUS   NEWTONIAN   CASSEGRAIN

If it is lenses that enable us to see what we can't see in outer space it also means we cannot see the light, which is outer space because we haven't got the lens to match the curb of outer space. Newtonians think of outer space as geodesic zero, with nothing in outer space but space. Geodesic zero means the light travels in a straight line from where it originates unhindered all across space to where the light connects the eye. Such an idea by itself is outrages because the stream of photons reduce in space to such a minute quantity that taken the area the photons travel and the space in vastness it covers, the chances of one photon coming across many hundreds of light years through billions upon trillions of cubic kilometres of space and selecting my eye to convey the electricity is less than infinite. Yet such conveying takes place every second of every minute. The position of the location of the second singularity, which is the precise duplication of the first singularity but in a diminished capacity, is obvious to miss when one is not applying a detective mentality, as one should in scrutinizing the cosmos. Culture will have us believe that when one sees a colour shining from an object the colour is associated with the object. Logic tells a different storey. A yellow dot is all the colours in the spectrum but yellow because it is disassociating with the yellow. That goes for red blue and all other colours we may visualise. I think the norm accepts this as scientific fact with very little argument or substantiating proof about that required.

If light came as individual streams of photon flurries, then our visage would translate that as such shown in the fragmented picture above. It would be a picture unconnected bringing across some photons in the manner where every object stands apart not being related in any way and that will be what we see, if it is anything that we see. That we know is not the case but that means geodesic zero is as much rubbish as anything Newtonians regard with simplicity and with careless thought. Geodesic zero means nothing and how can I see nothing as darkness because "nothing" is not darkness, nothing is "nothing" and the darkness I see is darkness showing the darkness as something.

What then about colours that are technically not colours as is the case with black and white? White is simple. By spinning all the colours in the spectrum the colour white shines through. Black is quite another matter. A friend of mine whom is one of the best painters I have ever come across told me that one couldn't paint black but have to make black a dark blue to show shade on the canvass. That apparently is his success in achieving the realism. He also went on to explain how many variations of dark blue form the shadows in one simple tree. This remark set my mind in motion. One cannot see black because black has no colour to show, but black is the colour most prevalent in the universe. One can see only by colour and since black is not a colour we should not see black, but we do.

If it was true about a yellow object not being yellow and a red object rejecting red and therefore not being red, the same must be true about dark and light. The bright object rejects all the light just because it is light from the outside. If it is light from the outside it has to be very dark inside. That too would count for what we believe it to be dark stars. Such dark stars must be most brilliantly lit because they keep all the light to their inside and well protected. They are dark because they keep the light on the inside where we can't see it and therefore out of our viewing range. The stars have gone so cold the stars have to conserve all heat to 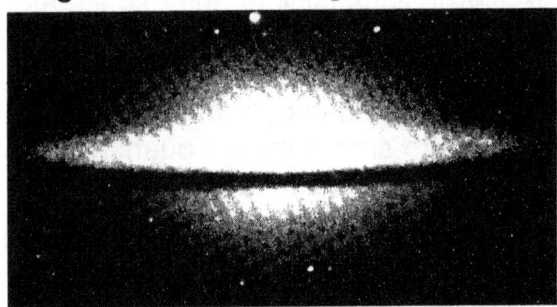 **remain in gravity. With light being the highest concentrated form of heat, it stands to apparent reason that the light would be the energy of prime choice to contain. The same must then apply to outer space in that outer space is conserving all light and by keeping all light, outer space is brilliantly lit. We just are unable to witness the light because our position is much concentrated where as the light being dark is expanded to the full. We are able to see the galactica because the galactica represents highly concentrated light in one reduced area. The darkness contrasting the light we see as darkness because the light is expanded to the ultimate. The fact that we can see the darkness makes the darkness light, which we are unable to see. However with the space stretched to the maximum the lens we see the light by has as far as our position goes, not even slightly curved because we are so small. Now you go and tell any mathematician in charge of theories this much and see how far you can get convincing him about your view. They wish to manufacture and design space whirls and not see reason.**

The fact that we see light means that the dark next to the light cannot be "nothing", If the darkness was the representation of "nothing", then that should be exactly what we must see, nothing but the stars. Taken from the top picture some stars and leaving the rest to nothing is what we see in the picture below. A blind person sees nothing but when we look at space, we see something that we think nothing of as we see as space. One cannot have the ability of sight and see nothing. It is light that we see and it is light that we use, which enable us to see. That proves the darkness that we see in outer space is light that we see without recognising it as such. If the darkness was the representation of "nothing", then that should be exactly what we must see, nothing but the stars. Taken from the top picture some stars and leaving the rest to nothing is what we see in the picture below. A blind person sees nothing but when we look at space, we see something that we think nothing of as we see as space. One cannot have the ability of sight and see nothing. It is light that we see and it is light that we use, which enable us to see. That proves the darkness that we see in outer space is light that we see without recognising it as such.

What puts us humans in a category one higher than animals (or so we like to think) is our ability to think about that what we can see. The less develop an animal is the more it has the attitude of eat or be eaten. The higher developed animals are the more the animal find reason to argue. One may teach a crocodile not to eat you if you start feeding the animal. That is a mindless reptile and yet it can think above eat or be eaten. What we see is not merely the truth and it requires reasoning to see the truth and substantiate between culture motivated observations and thought through decisions.

When the star is encountering adverse conditions, the flow will interrupt the even-handedness anytime will set a new standard. It happens all the time and every time the centre fails to set a standard that the flow of space-time in the star can meet. The relevancy of time sets in place a new standard and this comes about by the use of a principle we think mostly of in aviation. It is the principle we refer to as the Coanda principle. It is where motion creates a flow of space-time, which establish a centre and where that centre performs demands that the flow has to initiate. The containing of the space is as much set by the time of the flow as the retaining of the centre. It is a proven dimension implicating Kepler's vision of gravity.

There is a mad hype about reversing time and not only amongst us the uneducated mindless, but even more amongst the better Educated Brainy Bunch. It is remarkable that every one wishes to reverse time by going back into history but never to speed time up going onto the future and haste their oncoming date to die. We know all about the spectacular machines where some even wish to employ a multitude of Black Holes and reduce space to squash the daylights out of time. To return to the past, one has to reduce the **k** factor of every atom in every star and every star incubating galactica through out the entire cosmos. It involves a far more complicated process than just move down some imaginary double Black Hole that creates an artificial space whirl everywhere they so wish. As **k** develops so does time expand, so does space explode and so does time within the Universe progress. $k = a^3 / T^2$ Kepler said space a^3 will grow by dimension of **k** while time T^2 will decline $k^{-1} = T^2 / a^3$. To reverse time every point holding **k** has to reduce k^{-1} back to k^0.

The point of **k** connecting k^0 is to differentiate space-time from its origin since singularity throughout and everywhere is equal. The fact is that $k^0 = 1^0$ no matter what is involved. Getting the growth growing away from the point of origin is creating time distinction away from where all sides meet in singularity and singularity is the place where there is no space dividing the Universe without showing division but is still dividing. It is where there are no boundaries and yet every aspect of k^0 within every individual particle commits a boundary in relation to the singularity that is maintained. As we are part of the 3D and on top of that locked in time motion it will therefore be very wise for us not to try and understand the fist dimension or singularity at this point. In the second dimension there are sides committing motion to the third dimension but in the first dimension all is alike with separation coming about merely from being a unit. With every singularity progressing by **k** from the centre of the galactica since the Big Bang, **k** became the measure of progress. It progressed from a Universal governing singularity charging an accumulative control that responds to the growth of individual singularity forming and from that more individual atom singularity comes about. As singularity finds heat through gravity it progresses by establishing less support in the atom's individuality and by passing support onto the governing singularity.

The atoms are at task to remove heat from space thereby eventually they remove the atom's individual independent singularity to favour the stars governing centre singularity. The factor time T^2 provides the duration of time but **k** indicates time development and if our time travellers wish to return back to the past (to do what only they would know what they wish to do in the past) then they have to reduce **k** by taking time and **k** back as far as they need taking **k** back. Time is not just a direction but time is progress of stars, galactica and atoms forming countless Universes enrolled into countless Universes. Remember the least singularity is equal to the ultimate original singularity by virtue of value being 1^0 and that makes whatever point there is holding singularity as much a Universe as the entire Universe forms a Universe. Those whishing to travel through time must accept the reducing of space (of all space known and visible as much as unknown and invisible) and the increase of heat that accompanies the journey. It will be a lot easier to shoot into Jupiter and by such reduction of their personal **k** establish a duration extension of time. There was a time when the Sun had **k** as relevance being the very same as Jupiter now holds and Jupiter then was presumably even more frozen and much less than Pluto is at the present. Being frozen back then adhered to a much different meaning than being frozen now. Coming to this conclusion is simple. All one needs to do is study Galactica and see what the galactica lectures when they say in light what they say at night.

It is not only the relevancies of heat and time putting the Big Bang Universe much different from our perspective but the relevancies produced a completely different Universe altogether. It is not shocking that during the time and while experiencing the Big Bang the Universe was a nice average temperature of 10^{34} K because back then with conditions applying, it was just a normal day during the Big Bang. It was another Universe, one being in the same space that a neutron holds today. That fraction of space in time had to accommodate an entire Universe potentially filled with everything as we now see.

The Universe then had completely other rules than we have today. It is all coupled to the relevancy we find that singularity holds whereby gravity developed space-time. Everything changed as **k** extended and that proves that the factor **k** is the determining factor of space a^3 and time T^2. The progress of **k** unlocked all other factors including layers in galactica, which in turn unlocked stars from the galactica cradle. The stars in turn had layers developing by measure of **k** and with that the Sun also developed because all other stars develop in this manner and so does the cosmos develop using this manner.

Then again there was a time when the sun was the same frozen state as Pluto now is but freezing then meant a temperature of what we now consider to be 6500^0 C throughout the entire outer space. Outer space was a pretty cool 6500^0 K keeping the surface of the sun nice and icy. It is all in relevancy because what is freezing to one is glowing to the other. At a point 6500^0 K was many times colder than was outer space, but outer space represented one end and matter indicated another end of the relevancy there was and lying between what we measure as hot and cold. It is apparent that in the sun hydrogen freezes heat to a liquid state at 6500K and with oxygen further down towards the colder centre parts of the sun it is any person's guess at what temperature oxygen will freeze to a solid state in the sun. We also know that deep inside the sun where things really get cold at 18×10^6 K it gets so cold a hydrogen proton freezes to anything including an iron cluster. But with oxygen freezing to point of fusing the condition are considerably different with the governing singularity setting other rules as to what we with life can understand. The variations in conditions are in relevancy with the sun's governing singularity and is setting grounds, which apply different conditions to that what we find to suit us living on Earth. I do not wish to make presumptuous statements, but such rules apply very similar factors as the rules that bond compounds where singularity locks space-time of different elements in a relation and only by reaching specific counter conditions can the compound unlock and set free the elements. In that manner stars will regulate layer condition but not by creating compounds...no, only by producing similar type of rules. This means not only does the Big Bang look different but also, everything about the Universe was different then from what is now applying.

As sure as the Sun shines today so will Jupiter one day also shine as another Sun and then Jupiter's moons will be planets as large as the Sun's planets presently are, but by that time the Sun will be something awesome and awful. With the ratio fitting this tidally, the time duration and NOT TIME AS SUCH but the time it takes time to tick becomes infinitely shorter as it is coming from the eternally longer. In that is found also just a ratio. This goes totally against Newtonian religion and I am about to explain why that is in a minute. At what temperature will water boil on Jupiter and the answer will most definitely not be equal to a temperature it boils on Earth. Even on Earth the temperature of boiling water runs along a spectrum covering many scenarios, but the temperature of water boiling on Mount Everest is not equal to the boiling point of water at the Dead Sea level. Ignoring such indicators only brought scientific miscalculations and scientific mistakes. If water does not boil evenly at all levels on Earth how can science establish constants in outer space? It is known that even on Jupiter the freezing point connected to hydrogen varies as one continue down the atmospheric layers of the Micro star. At one point the equilibrium between the temperature of the Sun core and that of outer space matched setting the singularity within the core free and allowing the Sun to have a free inner core that then began applying gravity individually. We must presume that the outer regions of the Sun then were as frozen as Pluto is at present. We may think it was hotter but it was just less spacious on the outside. The cosmos was a lot smaller but that only made it more concentrated. This statement I shall retract but in order not to get ahead of myself we will keep using this statement for the time being. As space grew and reduced at the same time the growing in space by reducing in heat intensity came as a result of a changing **k** factor that represented both the demise of time as well as the increase of space. This Kepler introduced while no one took any notice until Hubble came. By that time every one forgot to look at Kepler again. Claimed space increases their space while time demise as the relevancy reapply in favour of materials. Materials holding space a^3 is inseparably linked to time in the square $k = a^3 / T^2$ as structures in orbit apply duplicating motion and $k^{-1} = T^2 / a^3$ as contraction recoups time that brings about motion. As space increases time demises. This I say full willingly knowing such a view does not represents the Newtonian and therefore the overall human outlook because they focus on what is present in what they can calculate by using constants.

> **In measure 1 cubic meter per second of motion is one kilogram of mass per ton of gravity**

We measure a distance by the length of a meter, which is a distance we take from where a line begins to where a line stops during specific time duration. That we then convert to a cube, which we attach to a mass by the thousand that we connect to the distance the Earth travels while rotating around the axis when turning yearly around the sun as well as applying the gravity. Our measure of $\Pi^3=\Pi^2\Pi$ is completely different to the same measure we find in Jupiter or in the case of the sun. Every cosmic structure will have gravity at Π^2 but the gravity will not be 9.81 Nm/sec as we have on Earth. Gravity at Π^2 but the gravity will be gravity aligning with Π in relation to the governing Π^0 we find applying in that Universe. That is why the sun can freeze hydrogen at 6500 K and Jupiter can turn hydrogen into liquid at -150^0C, Jupiter takes less cooling to also have hydrogen in a semi liquid state. Gravity on Jupiter will be Π^2 no matter what because the Jupiter space Π^3 has a built-in seclusion from the rest of the Universe. Again the space Π^3 is directly in relation to the relevancy Π that we find that Jupiter has, which stands in relation with Π that is the extension of Jupiter in singularity Π^0. That makes Jupiter a nice other world and our measure by meter per second and weighing in kilograms will be very different on Jupiter than we have on Earth. A six-foot man will most probably be some four-foot down at Jupiter's sea level where hydrogen becomes a sticky metallic thick liquid, but the man would not be an inch smaller. The man cannot be shorter because the man remains the same man as the man was on Earth. The man did not change for he only came into respecting different relevancies. The foot in distance has changed and the six in number could have changed but the man stayed the same.

Our tunnel vision comes from our stance where we see the cosmos we wish to see because we also only fit into a small slot. Our slot is blocking our view but don't tell the Newtonians that! As **k** develops **k** has to develop in all aspects of the Universe, that is if we are to believe in the Hubble constant (another nice little constant with no applicable use anywhere). That means the layers in a star and in a galactica is the prone ones to changes brought about by the rules that the Hubble shift adheres to and the layers brings growth to singularity as singularity in governing stars come to life. They take charge as the atom singularity progress and as the atomic singularity shifts the dominance to support the governing singularity in the star centre. Then **k** progressed further and with that the **k** within the Sun and indeed all the solar structures progressed in equal terms but not alikeness since every one has in place the charging of its space-time its coming as a result of another and most different singularity. It is this relation we have with the Earth and with Earthly relevancy we hold so dear in our occupying space on Earth that the Earth **k** holds every aspect including our minds and our thinking under control. The Earth is our Universe and that we can see from all the constants and ultimate limits we place on the Universe in order to bring the Universe in line with our Universe which is the Earth and all standards going according to the Earth.

We tend to see the Universe from our perspective we have where we are filling the **centre of the Universe**. With all the absolute phenomenal achievements science accomplished and more so the past sixty years, by going to the moon and splitting atoms and visiting planets and... was there ever one that took the time to find the **centre of the Universe**? How can anyone tell how much mass gravity attracts in the entire Universe in relation to the critical density of the entire Universe when you are incapable of knowing where to look for the **centre of the Universe**? The Universe with gravity's attracting action must be pulling us to the **centre of the Universe** and because you can't judge the direction we are going, you don't even know where to find the **centre of the Universe**. When any person is standing on any place anywhere, while viewing the Universe, that person is filling the **centre of the Universe**. That is not only applying to Americans in particular, but to all persons that were born through childbirth. This however does no apply to animals but that is in another book reserved for another day.

Let's get more personal. We tend to think of our position as having the position only the most important person in the Universe can have because every one thinks of himself or herself as the most important individual that is holding the **centre of the Universe** in the entire Universe. Einstein said gravity is where the Universe draws flat, and the Universe can only draw flat where the **centre of the Universe** is because only gravity can draw the Universe flat. Consider this while looking at the night sky outside where light pollution has not destroyed the view... All the light that come across and travelled all of the vacant space from any and all possible positions in space runs directly towards your position using a straight line towards you where you are filling the **centre of the Universe**. With you being able to draw the entire Universe flat so that all the light through out the entire Universe come together to meet you in person in the position you hold, you must therefore have the most intense gravity by your effort of drawing the Universe so flat, in order to have all light running directly to you. Not allowing even excluding the effort of one photon, all light is heading to meet you where you are in that centre spot and not one photon will pass you by. Not one photon dare miss you because if they do they miss the effort that all light has to accomplish and that is to locate you as the person filling the **centre of the Universe**. Should you decide to shift your position to any other place in the Universe, you will shift the **centre of the Universe** to that location as well because the light will track you down in your new position. If you install a camera on Mars, the light is obliged to acknowledge your relocating the **centre of the Universe** at your will to reposition you're taking control of that **centre of the Universe**.

All the light that ever left its destination crossing the vast spaces of the Universe, excluding no particular light, travelled all the way just to find you filling the **centre of the Universe**, right where you are. By you're standing anywhere, you fill the **centre of the Universe**, and the entire Universe admits to that because all the light comes to meet you there. If you shift from the North Pole to the South Pole you will shift the **centre of the Universe** because all the light travelling throughout the Universe will find you where you then moved the **centre of the Universe**. The light left its destination billion years ago as it travelled through space at the speed of light so anxious it is to acknowledge you're being in the very **centre of the Universe**. No photon will be able to pass you by where you are in the **centre of the Universe** because all light is heading your way from their starting positions. No wonder every person born has the idea they were born to fill the **centre of the Universe**, which we do fill. The Universe is spinning around you or I, which is filling a centre where all motion is connected. It implicates gravity as wide as can be... Some things mathematics is able to explain but other explaining goes beyond mathematics.

Some aspects of the Universe go beyond mathematics and some even go beyond words. It is our task to find space, to find time and moreover it is our optimal task to find the Universe. This line of thought is in concept a joke but as much as a joke it may be it is the truth as no other. It also is the scientific Newtonian inspired approach to science that brings the thought pattern of truth in all people to mind. Because we use the Earth and on the Earth a meter is a meter in a second, we believe that meter per second will also be a second in measure of all meters everywhere and the distance duration will be the duration of all seconds per meter everywhere. Let's see how much this joke is a reality in the minds of Newtonians.

Being in the **centre of the Universe** is frightfully Newtonian. It is very clear how Einstein and his compatriots followed their genius in their arguments when they argued about the critical density and the manner in justifying an attempt that is as much a cover up for a scandal that has no political proportional rival to date by correcting Newton's obvious misconception. However fortune favours the brave. They went about shouting about not enough mass being the culprit for the expanding that Hubble uncovered, which was contradicting the compressing theory Newtonians so stubbornly cling to. Nevertheless the occasion presented the man and the location supported the deed in rectifying the error of directional flow of what the Universe was up to. To this day no one came up with a reason what besotted the Universe to go in contradicting the direction of development to what Newton declared. To count the mass they were fortunate enough to be in the **centre of the Universe** because only from such a vantage point could they see and measure the entire Universe. From where Newtonians stand even to this day the Newtonians have the fortune of seeing the entire Universe from edge to edge to edge and so forth...and that too applied to Einstein. Einstein could see all the edges and Einstein also new that beyond the edges was nothing more than just the edge of the Universe. Although we know the Universe is without an edge, Newtonians being in the **centre of the Universe** can see the edge that cannot be there but to Newtonians it is being there because by magic Newtonians create gravity. They see the edge of the Universe every day from every possible telescope and to hell with those saying the Universe are eternal without borders and edges! So there is an edge where no edge can be because they are Newtonian and Newtonians fill the **centre of the Universe.**

However it is not for us to criticize so therefore let's journey back and see what Einstein saw...and remember we are back in time so tenses becomes an issue. When doctor Einstein and his fellow doctors look outside at night they see the edge of the Universe to the left of them. They can see where the Universe ends in that direction. Looking to the right they see the end of the Universe to the right of where they are looking because there to the right where they look, they see the Universe ending at the edge. Looking to the front the same happens and to the back the same happens with edges in all directions. Even when looking up into the night sky they can clearly see where the edge of the Universe defines the end of the Universe in that direction. It is to their fortune that they are where they are because by being where they are they are filling the location holding the **centre of the Universe** in the most magnificent place they could ever choose to be. They are in America and we all know that Americas is very much the **centre of the Universe**. More so is the fact that they are in an institution called Harvard, which puts them in the Academic **centre of the Universe**. By them being part of the physics department of Harvard brings them in line with the astro physics Academic **centre of the Universe**.

Things are going from good to better to best to excellent, because with then being in doctor Einstein's office they are in the brains **centre of the Universe** and with doctor Einstein being in their midst they are placed by his presence and intellect smack in the **centre of the Universe**. Their place in Harvard's physics department right inside doctor Einstein's office standing next to doctor Einstein puts them smack in the centre of the one half of the Universe. Now they do not have to worry about finding the bottom half of the Universe because America having Harvard with a physics department having an office where doctor Einstein presides takes cover of the bottom half of the entire Universe and puts their bottom half smack in the **centre of the Universe**. Doctor Einstein is the living presentation of everything that is not stupid as doctor Einstein just has to walk outside and look at the light coming directly to him. One spin of 360^0 would ensure him that being doctor Einstein and all... he then must be in the **centre of the Universe** because he can see the edge of the Universe in every direction possible! If he wasn't the **centre of the Universe** there was no way he could measure all the mass in the entire Universe because his allocated position of not being the **centre of the Universe** would bring obstruction to part of the view required for the measuring task in hand.

That means being in America and more so within America's Harvard Physics office that takes part of the bottom **centre of the Universe** and doctor Einstein in person is then taking care of the top half of the Universe finding the location that presents the **centre of the Universe** where the entire Universe aligns at that point. I know every one has sleepless nights wondering why they gave the problem of measuring the entire Universe form edge to edge to edge to edge to determine the critical density calculations to a person such as doctor Einstein. Wonder no more! There is a possibility that it has something to do with his mathematical abilities but that would not count for much if he was not able to see the Universe from edge to edge to edge to edge to… But with doctor Einstein in the place where he is, he can see all the stars sending light directly to him and telling him how big and how far they are. If he were in the incorrect place in the Universe his measurement from not being in the **centre of the Universe** would no have been trustworthy at all.

With doctor Einstein being doctor Einstein, he knew he was the most important set of brains America could present to the Universe. By the Universe realizing this fact and acknowledging the fact while sending all the light to them at the **centre of the Universe** and without causing delays, the Universe responded by sending all the light at the speed of light to doctor Einstein. The light came from near as it came from far. It came as much from the very edge of the Universe to the right hand side of doctor Einstein as much as it came from the very edge of the Universe to the left hand side of doctor Einstein. Then the light came from the front as far away as it came from the back of doctor Einstein. From the top as well, all the light travelled as far as light can travel and travelled as fast as only light can travel just to acknowledge and support doctor Einstein in his task to calculate all the mass in the entire Universe because he has to prove Newtonian incorrectness automatically correcting by self-preserving determination.

Of course with him being in America and at Harvard's physics department and moreover in doctor Einstein's office placed the bottom half just as accurately in the centre of the Universe as doctor Einstein found the top half to be aligned. If it was not for America and if it was not for Harvard's physics department and doctor Einstein's personal office the bottom half of the Universe might have mismatched its effort to align with the top centre and then the lot was not in the **centre of the Universe** from where they could see every possible edge of the Universe. But with the fortune of things being as they are the top and the bottom halves of the Universe matched and in that doctor Einstein could now fill the entire **centre of the Universe** on top, at the left, at the right, to his back and to his front as well as the bottom half of the Universe. If that was not the case then what a tragedy that would have been because only from being in the **centre of the Universe** could doctor Einstein view the entire Universe and see what there is in the form of mass to calculate and measure every star there is in the entire Universe.

Fortunately for mankind the world had a person such as doctor Einstein to fill such an important position from where he was then able to see and measure the entire Universe. How gratefully we should be to the Academics of America for allowing us to share America's **centre** position **in the Universe**. More than thankful we must be for the Academics that allowed us in sharing the physics department of Harvard's central position in the entire Universe because from there it was a hop to get into the office of doctor Einstein and share his position of seeing all the light from every corner the Universe has to share with him and use such a marvellous position in the entire Universe to gauge and measure all the mass we can see…and if you say Newtonians Academics don't put them and their position they have from where they see them being and filling in the **centre of the Universe** ...then please do a rethink! The theory they present puts conditions we find on Earth used by the entire Universe in the entire Universe. Being on Earth we can see that water freezes at zero Celsius and we can see that it is one bar of air pressure that we find at sea level. We have the element table with solids and gasses nicely arranged for us by nature in the table in column order and we know that absolute zero is absolute zero because absolute zero is what we measure when we measure absolute zero. How can the Universe have the tenacity to have absolute zero anything else than what absolute zero should be where we measure absolute zero.

The Universe has not the capability to change anywhere because if it did dare to change we will see such change as we fill the **centre of the Universe**! From the Newtonian stance the Universe grew from the size of a Neutron, however its official Newtonian policy that the Sun was the same since time began and the atom was always what we measure the atom to be. With Newtonians filling the **centre of the Universe** we know the Universe can grow and expand but that feat is quite impossible for the Sun and the planets to achieve. It is completely anti Newtonian to think that the solar system is getting bigger just because the Universe is growing bigger. All Newtonians would recognise change the instant change occur because from filling the **centre of the Universe** Newtonians will notice change immediately and after all one has to consider that Newton said the lot is contracting. Saying anything to the opposite will be quite sacrilegious to Newtonian religiosity. The reality is that the Universe grew and the Universe still grows even in our part of the Universe. Amidst all this evidence Newtonians have a constant speed of light, a constant time since time began, a constant gravitational force, a constant expansion and every other aspect that will bring along nice and easy calculations, so the Universe will have its constants just to keep life a little simpler for the Brainy mathematicians.

Go ask every Newtonian and that Newtonian will tell you the Universe has no edge because it is limitless, but being as important as only Newtonians may be they fill the position where they are able to see all the edges the Universe cannot have. However I should warn any one that listens, don't tell Newtonians of their double standards. When I confronted professors in the past and accuse the Academics of limiting the Universe to benefit their views about claims they make or to support Newtonian claims, I am compromised being the one referred to as the incoherent, as the raving idiot because they say Newtonian science will never do such a thing.
Then the next minute, I see Academics limit the Universe to having an edge, which they find in very clear telescopes. They see a boundary where the Universe ends. To prove my case I present I challenge anyone to visit the web site and go and visit the many such web pages carrying this very claim. A red giant was found on the edge of the Universe or some galactica conjunct with some other galactica at the very edge of the Universe. It is such normal every day Newtonian practice to use a double forked tongue. It states clearly that science caught big bright stars on the edge of the Universe. Any one can glance at this on the condition they have excess to the web page and can use

the web page. This clearly shows the lack of understanding on affairs Newtonians claim to be knowledgeable of. Those Academics (the lot of them do advocate that the Universe does not end) are those giving the Universe an edge…to do what with. What ends at the edge or the border and what bring the edge about, forms the wall that allow no more of the Universe to continue. However I am sure that there must be some persons in education that will share my view that Newtonian science did not yet leave the shores as Columbus' sailors did when they found no edge of the world. But Newtonians science did manage to take science much further than science was before. Newtonians managed to shift the edge of the world so far it became the edge of the Universe and gave the Universe with no end, an end. It is about time that the entire philosophy of cosmology is overhauled and is revised from the backwardness of five hundred years ago to a more fitting approach in the five hundred years that time went on.

Then they claim the Universe is expanding…expanding where too. Where can anything go that covers the lot and has nowhere to go? How can anything get bigger when such a thing is as big as anything will ever get. How can the Universe get bigger when it is limitless? Notwithstanding what logic needed they will tell you the Hubble constant is about the Universe expanding that which is not expandable and going where there is no going too because (I guess) the edges that are not there is shifting further out to nowhere. The Universe is getting bigger but where is the Universe getting bigger too because wherever it is going the Universe is surely already there! There can be no place without the Universe already being there so where is the new claimed territories that is gained by a growth that it is gaining from what because it clearly already claims all there is. I realise that Newtonians being in the **centre of the Universe** can claim to see what we others with less intellectual means are unable to see for we do not have the grant in privilege to see the Universe from the **centre of the Universe.**

It is rather comical to think that Vasco Da Gama and his sailors would not set sail on a voyage of discovery in fear of confronting the edge of the world while currently our cosmic sailors in waiting desires to get going on such a ridiculous voyage and reach the edge of the Universe because they can see the edge of the Universe. How much did things change just to remain the same will you not think? Today the schooled Newtonian opinion about cosmic science is that we just have to hop into some craft, blast off to the unknown into the unknown at the speed of light and send a post card back home when we get to the edge of the Universe. …And all the while Newtonians have no clue how light travels. They show some mat-like surface with graded blocks that should represent space and time but it puts space and time in some single dimension holding a square of some sorts and present that as the travelling road light supposedly takes as it journey all the way to them where they are in **centre of the Universe.**

Whenever I am presented with this explanation I so dearly wish to ask the person presenting the argument what he did with the rest of the Universe? Where did that person find a place to go and hide the other five sides? Only nothing can vanish and the cosmos is definitely not nothing. He took the three dimensional six sided Universe away, put it somewhere I don't know and then 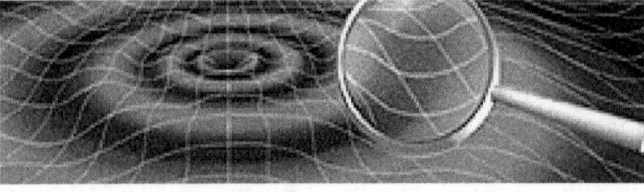 left only one of the six sides where I am suppose to see only one of the six sides…and how am I suppose to know where he put the rest of the Universe! Here he is so smartly showing a stupid bloke such as me what happens when a Black Hole comes about. That is fine and that much I truly understand even with me being as stupid as I am. What I do not understand is what did he do with the other five sides of the Universe. He is showing one side of the Universe that has gone flat, but what did he do with the rest of the six-sided Universe. How did it disappear and where is it gone to…how long will it stay there and how is it coming back…who is strong enough to bring it back…it is all viable questions asked in sincerity…therefore even in my stupidity I deserve an answer, after all it is my Universe too. Just look what did they do to my Universe…and who is going to repair it! How did the Universe get that flat because gravity is not in outer space…gravity is in the atom no, moreover it is in the proton. It is the proton pulling space-time flat and not outer space pulling flat. They show a centre and they show a flat Universe with light of all things travelling about a flat Universe. How light which is the very focus of the three dimensional purpose of the 3D Universe can go flat and travel in a flat Universe along where no travelling space is available, but the light travelling is just the thing that is establishing the three dimensional space we see. How light under those circumstances can travel flat through no space is beyond that which I shall ever be able to understand. Fortunately, I am the stupid one around and they are highly educated. If this is true then the electron should pull the atom flat because the electron is the epitome of light and what density light can ultimately achieve.

It is 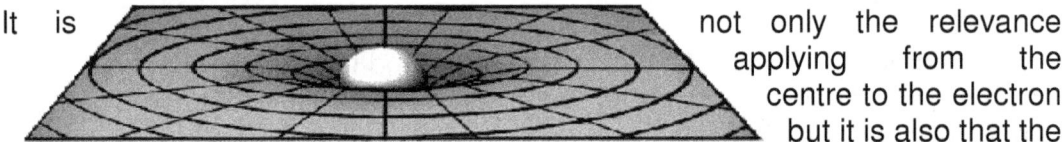 not only the relevance applying from the centre to the electron but it is also that the electron is holding an allegiance with the centre in the precise manner as planets do in relation with the Sun. It is because of the dual relevancy that the electron stubbornly clings to the newly elected centre before being overpowered by the Earth providing the controlling centre. It is all about relevancies attaching centres that formed when creation came about. To break those relevancies we have to take time back to before those relevancies because of the gravity gluing the relevancies in a unit.

So we have heard all the hype and the brilliant mathematical expression of the impossible where the star goes bang and the gravity goes mad and it implodes (how ever that may be achieved) and the whole cosmos goes bananas because a star has gone lost for one eternity and now has died a tragic death. That is so complicated that there is little available to explain. That is something about gravity going nuts with one precondition; how does gravity go loony and form a Black Hole when all the mass that is supposedly producing the gravity that is supposedly forming the Black hole just went on a trip to the outside out of that which forms the star as a unit? Then the Academics call me incoherent...

The answer to my explaining is so simple it is laughable. A star is about fusing atoms together and on that am I correct or incorrect? Then what happens when all the atoms fused together in the space of one atom and what gravity will such an atom produce that is consisting of the centre where all the atoms that the star had through out its entire existence accumulated into one point holding singularity? If only a handful or maybe just one atom remains which is the final result in fusion of what ever contraction finalized all the possible fusion between very available atom of all the atoms in the star, and one atom ends up with all the gravity the star had which was initially delivered by all the atoms in the star that produces gravity and such gravity is now within the space that one atom holds, then the gravity will be devastating. The fusion has to end somewhere because by fusion the star is heading in a direction that will eventually combine all the atoms in the star.

Curved Space-time

The figure below next to the spiral represents a two-dimensional slice through three-dimensional space showing the curvature of space produced by a spherical object, which is perhaps the Sun. Einstein's view is that the planets follow the curvature of space around the Sun (and produce a tiny amount of curvature themselves). That again is crossing monkeys with watermelons to bread marble. They picture a two dimensional (flat bedded) mat like surface that light uses to travel by. According to the Educated Wise, gravity will pull the Universe as flat as the topside of a mat. However, some of the content within the Universe escape the fait the Universe has because some things in the Universe does not draw flat although the entire Universe supposedly draws flat.

I have very disturbing news for Doctor Einstein and his followers, those I call the Brainy Bunch. It is the fact that space goes flat. However, they are looking at time and the presumed space they have in mind is actually time that

Our Super-Educated first puts on the table a flat surface. No one ever try to bring across any explanation or reason when the surface went flat or is flat except that gravity is making it flat. They left the guessing to us to fill in what produces gravity in outer space where bodies float around centre objects because such explaining might just mesmerize their theories while they wish to mesmerise our brains. The Universe does draw flat. I admit, but what is the Universe that goes flat?

goes flat. For space to be space the space has to have three dimensions where three are opposing three more sides. That is space a^3. Where space becomes flat the space is moving with time and such space then has motion in relation to an ever-changing position and location endorsed by the square of time T^2.

However, it is the motion of the three-dimensional space that holds the square value and not the three dimensional space going square by losing one dimension. Time or motion of space has the square which gravity also has being Π^2 The square apply to the moving of space Π^3 in the third dimension but it is very unrealistic to place space Π^3 by measure of losing one dimension. Moreover it is confusing and more so to the Academic preaching that as a fact in physics. It confuses the rest in believing what the rest of what the person has to say. If such proposal is made, one require a reason why five dimensions went away because gravity is not space therefore gravity is the motion of space. One cannot say gravity pulled space flat. How flat is flat and what is flat? Singularity is flat but singularity has no dimensions. That I can prove because I can show where singularity is because of singularity not being there.

Then in this carpet that represents the flat two dimensions there appears a hole in the centre. The presence of this hole returns the whole picture back to being a three-dimensional surface because the whole produces a third dimension to the point where gravity is pulling everything into a flat two dimensional state, but the hole is adding one dimension to the bargain of the flat square. How did gravity produce a flat two dimensions while equally at the most intense point, gravity produces a third dimension in the location where gravity is supposedly lurking to give is our flat Universe. Then comes injury to the insult already inflicted on our weak minds; they place a three dimensional ball into a flat Universe to show us a three-dimensional hole in the flat Universe. If that is not done by the magic of Newtonian gravity, then there can be no other acceptable explanation about the whole affair. They have no valid reason for the Universe to go flat except blame it on gravity. However, gravity is the strongest where space is the least. That would then allow gravity to curve the space-time surrounding the sphere from the centre, evenly, in all directions equally. It would put in place the seven degrees the circle that the sphere in form insists to have. It would use Π by many dimension placing Π in relation to the centre where gravity is the strongest. With this in mind there is no need for all the hype about the curvature of space-time, we may just refer to the sphere being in place.

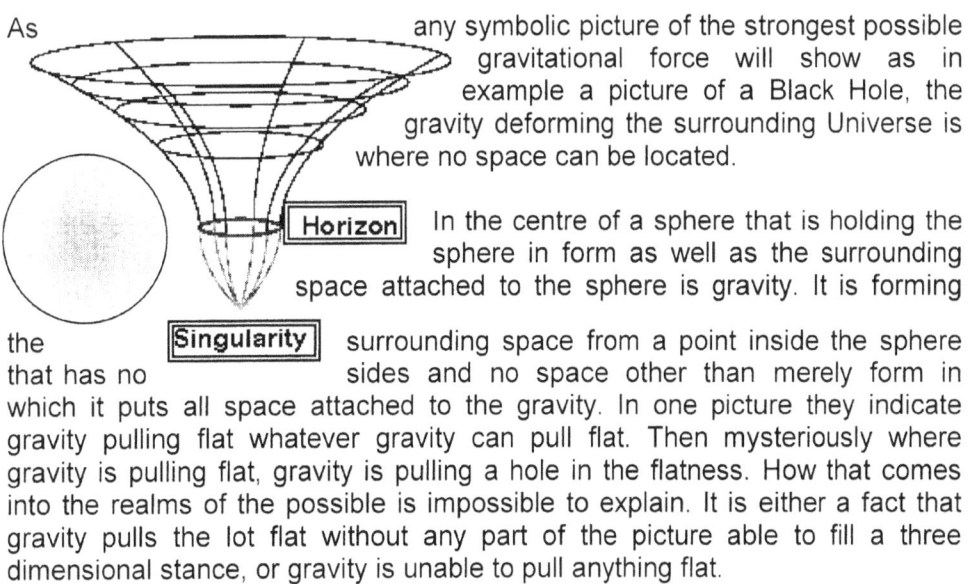

As any symbolic picture of the strongest possible gravitational force will show as in example a picture of a Black Hole, the gravity deforming the surrounding Universe is where no space can be located.

In the centre of a sphere that is holding the sphere in form as well as the surrounding space attached to the sphere is gravity. It is forming the Singularity surrounding space from a point inside the sphere that has no sides and no space other than merely form in which it puts all space attached to the gravity. In one picture they indicate gravity pulling flat whatever gravity can pull flat. Then mysteriously where gravity is pulling flat, gravity is pulling a hole in the flatness. How that comes into the realms of the possible is impossible to explain. It is either a fact that gravity pulls the lot flat without any part of the picture able to fill a three dimensional stance, or gravity is unable to pull anything flat.

One cannot depict gravity as being selective to prove the thinkers thoughts. If the ball in the hole is three-dimensional the hole is three-dimensional making the picture with the light portraying the picture three-dimensional. Or otherwise everything is flat without having a hole to fit a ball. If gravity pulls the Universe flat then gravity is quite unable to curve the Universe by the same margin. Again we arrive at a Newtonian fork tongue. This can only happen when one put mathematicians in charge of theory. It is still the same double standards to give the Newtonian view validity. Gravity is having the Universe go flat or so did Einstein portray our gravity stricken Universe. Gravity also produces gravitational lenses…but they never mentioning whether it is while or before it is going flat because there is a choice to be made. It is either going flat or it is having a lenses but it cannot have a flat lens.

That is space a^3. Where space becomes flat the space is moving with time and such space then has motion in relation to an ever-changing position and location endorsed by the square of time T^2. However, it is the motion of the three-dimensional space that holds the square value and not the three dimensional space going square by losing one dimension. Time or motion of space has the square which gravity also has being Π^2 The square apply to the moving of space Π^3 in the third dimension but it is very unrealistic to place space Π^3 by measure of losing one dimension. Moreover it is confusing and more so to the Academic preaching that as a fact in physics. It confuses the rest in believing what the rest of what the person has to say. If such proposal is made, one require a reason why five dimensions went away because gravity is not space therefore gravity is the motion of space. One cannot say gravity pulled space flat. How flat is flat and what is flat? Singularity is flat but singularity has no dimensions. That I can prove because I can show where singularity is because of singularity not being there.

Then in this carpet that represents the flat two dimensions there appears a hole in the centre. The presence of this hole returns the whole picture back to

$$g = \frac{GM_\oplus}{R_\oplus^2} = 9.8 m/s^2$$

being a three-dimensional surface because the whole produces a third dimension to the point where gravity is pulling everything into a flat two dimensional state, but the hole is adding one dimension to the bargain of the flat square. How did gravity produce a flat two dimensions while equally at the most intense point, gravity produces a third dimension in the location where gravity is supposedly lurking to give is our flat Universe. Then comes injury to the insult already inflicted on our weak minds; they place a three dimensional ball into a flat Universe to show us a three-dimensional hole in the flat Universe. If that is not done by the magic of Newtonian gravity, then there can be no other acceptable explanation about the whole affair. They have no valid reason for the Universe to go flat except blame it on gravity. However, gravity is the strongest where space is the least. That would then allow gravity to curve the space-time surrounding the sphere from the centre, evenly, in all directions equally. It would put in place the seven degrees the circle that the sphere in form insists to have. It would use Π by many dimension placing Π in relation to the centre where gravity is the strongest. With this in mind there is no need for all the hype about the curvature of space-time, we may just refer to the sphere being in place. If the ball in the hole is three-dimensional the hole is three-dimensional making the picture with the light portraying the picture three-dimensional. Or otherwise everything is flat without having a hole to fit a ball. If gravity pulls the Universe flat then gravity is quite unable to curve the Universe by the same margin. Again we arrive at a Newtonian fork tongue. This can only happen when one put mathematicians in charge of theory. It is still the same double standards to give the Newtonian view validity. Gravity is having the Universe go flat or so did Einstein portray our gravity stricken Universe. Gravity also produces gravitational lenses...but they never mentioning whether it is while or before it is going flat because there is a choice to be made. It is either going flat or it is having a lenses but it cannot have a flat lens.

In a picture of outer space Newtonians place a gravitational constant. That means there is a form of gravity keeping order to the vastness of outer space. In such vastness light is known to travel in a straight line between points. Gravity is pulling objects to a centre where mass is concentrated. In this vast region without any end they claim a specific gravity and that gravity has a specific value with a specific name being the gravitational constant. If it is gravity being out there in the blackness, then it is pulling to a centre. The biggest Newtonian question then is: Where is that centre and what produces the centre whereto the pulling is going. Where is the centre of the region thought of as outer space? If outer space had gravity there has to be a centre to which the gravity is pulling or moving objects. The only centre there is, is the centre material form. It is the centre of the Sun, or it is the centre of a galactica. That is material and can hardly qualify as outer space because such space is openly controlled by material moving about in such a space. Other than that there is complete lack of gravitational evidence.

In the use of the formula there is a deliberate admitting to space-time being used. In the formula they put 9.8, which are the square of pi to a unit of meters (space) divided by seconds (time) to the square. However that is applicable to the Sun as much as all other objects. Then the formula is taken to apply to Black Holes where the concept science

$$R_{grav} = 2GM/c^2$$

then put foreword changes totally. It becomes Using the formula in the Black Hole suddenly calls for the square of the speed of light to come into the formula as a factor. The square of light comes in to support the diameter of the star. How can they get the diameter of the star married off to the speed of light? The one concept is a length a distance between points of compacted material and the other is the flow of liquid space through time. The one concept is measuring distance and putting that into and as an additional factor, which by any standard, is inexplicably unrelated to the distance of the diameter, but still comes to support the square of the speed of light. Motion

supports the diameter of material by the square.   Light cannot go into a square because the speed of light is the epitome of motion in space through space. There cannot be a speed of 1.005 times the speed of light or be twice times the speed of light. The speed of light is the ultimate and the optimal. It is where velocity ends and no increase can come into furthering the calculation. Yet the Brainy Bunch has this unexplained novelty of squaring off the speed of light where we locate a diameter to the measure of the third dimension as r^3. I do not say that doing that is incorrect, but the motive and the explaining why in could be corrects totally lacks in science and even their realising the incorrectness passed all of the Brainy Bunch by. I also do not say going square by motion in conjunction of the diameter is incorrect, but they do it by using all the wrong reasons which they never mention. If that is true and they can use C^2 in place of a diameter to perform as gravity, then they have to admit to my statement that gravity is no force but pure motion. I say gravity is the difference there is in motion between particles having unequal relevancies to a specific centre controlled by singularity and the gravity is the extending thereof. Still gravity is the motion of space in time through time.

Just below is a sketch. The sketch represents a ball thrown through space and through time. The ball is representing the movement of time. See how equal space and time is? See how space and time form a mathematical vector. See how easy it is to draw a line that represents space and represents time in precise proportions. Does that not make the life of all mathematicians much easier? Can any one argue that space and time is represented even handed by equal proportions just to make life much simpler to calculate the Universe? This (I suppose) must be where gravity draws space flat and time flat and space multiplying by time gives some square from which a vector then forms. How flat can time get and what happens when space goes flat but more still, how does the flat then multiply the flat to form a flat square. There is no mention of dimensions although we all realise dimensions is the issue concerning the Universe.

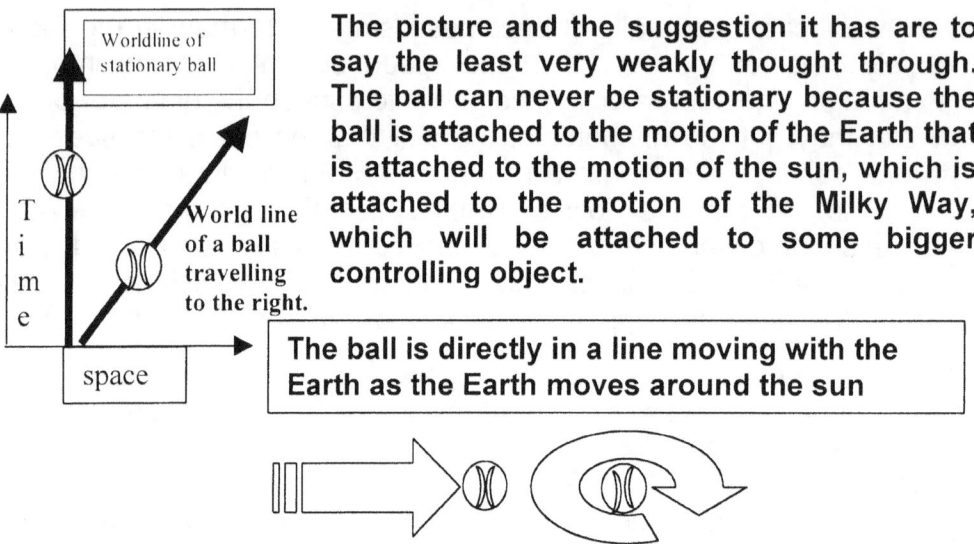

The picture and the suggestion it has are to say the least very weakly thought through. The ball can never be stationary because the ball is attached to the motion of the Earth that is attached to the motion of the sun, which is attached to the motion of the Milky Way, which will be attached to some bigger controlling object.

At the same time the ball is spinning with the Earth around the axis of the Earth. At no time can the ball ever be stationary in relation to the rest of the Universe and that relation in motion is time.

Please note how the Newtonians suggest that time and space is both the same value and is equal in partners being one and the same of value although being far apart in differentiation. If that was the case then the Universe was one because the one in space would cancel the one in time making space eternal and time infinite, and then they claim to be the highly taught. How they can bring space in as a flat and the same as time although in division of time (space dividing or per time or space-time) notwithstanding Einstein declarations goes beyond any normal logic. How they can put time and space as equal partners and still find that space can move through time also is beyond what silly old me can explain. Fortunately for every one concerned I am the incoherent one. Time is always in control of space by motion of space. Time positions the location of space in the relation time where time places all in relation of all space to the rest of the Universal spaces. If science wishes to put space in a motionless stance science should produce evidence where space is motionless or bring proof where time can find the ability to stand still.

The time that Newton froze on paper in his establishing of a single t that is representative of time is effective in remembering the viewer of an event but that cannot be the event that is part of the present any longer. If we reduce the moment to a snapshot, the picture we focus on can only be what we see the image to be. The image remains in ink on paper, which froze not time but an image of time. It holds an image that was part of time for a very short instant and then forms that which was how the event occurred during the time from where the camera shutter opened T_1 to where the camera shutter closed T_2 and the time frame T^2 was then during the open period of the camera shutter. As soon as the shutter shuts, time moved on and another T^2 formed leaving the image taken as time serving an image never to repeat again.

Afterwards the image we see is not time. It is an event that occurred during the flow of time at a specific stage in the flow of time. It did not freeze time but took an image of time distorted forming space as the picture represented **t** at that moment of T^2. When looking at the picture, the looking at the picture also became an event. That event happened during a specific T^2 that went from where one is taking the first look to where one is looking away from the paper. The event lasted while carrying the first dimensional image of an event gone by. That is at that stage a representation of **t** in another milieu of $a^3 = T^2 k$.

It is not time standing still. If you show the picture to a horse, the horse will try to eat the picture because the horse will be unable to recognise the image in ink on paper. The last thing the horse will experience is a freezing of the moment. The **t** in the single is when mathematically presented as only **t** indicating a mathematical single flat dimensional view of time being part of paper by means of ink. The image we recollect is in our minds and not in time. It is an image that is then correctly applied because it represents a reminder of a four-dimensional event $a^3 = T^2 k$ that went single dimensional because the moment in the fourth dimension which was then frozen in a single dimension on paper. With the paper being part of space-time while the fourth dimension $a^3 = T^2 k$ soldiered on and time will always be representing T^2 as Kepler stated.

In $4\pi^2 a^3 / T^2 = G(m+m_p)$
$a^3 = T^2 k$
$a^3 / k = T^2$ but
$k / a^3 = 1 / T^2$
$k = a^3 / T^2$ = singularity
$a^3 / T^2 = G (m+m_p)/4\pi^2$
and $a^3 / T^2 = k$
then $k = G(m+m_p)/4 \pi^2$

All Newton's changing was possibly done with good intensions but even that I doubt. The end result however was in some cases far from good, as it does not do such great credit to Newtonian insight into cosmic affairs. Only Kepler and only Kepler unaided without the intimidation and interfering of Newton can explain the Coanda effect. I grant the fact that the Coanda effect was discovered before Newton saw himself fit to change Kepler, but only Kepler can explain the Coanda gravity effect when Kepler is without the attentions of Newton.

But I showed that $k = a^3 / T^2$ and Newton's claim is that $a^3 / T^2 = G (m + m_p) / 4\pi^2$

Time is in the square, and that is allocated to space having a cube. Kepler said gravity is $a^3 = T^2 k$ at a time even before gravity got a name. But reducing the dimension of time to a single **t** one will find the ability to mathematically design the paper on which the photo image will be printed in time T^2 using space a^3 in the third dimension to apply the ink in the third dimension. Printed on the paper is an image that is not part of space-time while the ink used is space-time and the paper is space-time. The ingredient of ink on paper all hold different values since the image has a value we as humans grant the image to carry such a value. The image, the ink and the paper all hold different relations but the image only relate to thought in our mind. The image has no **k** indicating only references with and to what forms part of a realistic different singularity coming from the Earth centre and connecting to individual atom groups, forming individual as well as group space-time. We are all so painfully aware of Newton's claims about a spinning wheel standing still because of the fact that Newton could get time to stand still and then when time was standing still he would divide the one in the other and best of all he would get zero.

If Newton's vision about the wheel not producing work where does the work that electricity do come from. The generating is done by a "wheel" turning and the "wheel" contracts space-time by measure of heat (yes electricity is just heat and because it is just heat it give burn wounds) If Newton is correct, no driveline can form because by driving a line in infinity (1^0) takes charge of the motion and drives what is in need of driving. I am aware of this hornets nest I (again) disturb and was marched from campus about this on previous occasions but I cannot agree with what is disagreeable to me. The whole idea of time standing still is wrong.

$$\Omega = \frac{d\theta}{dt} \quad \tau = \frac{d\mathbf{J}}{dt} \quad d\theta = \frac{dJ}{J}$$

$$\Omega = \frac{1}{J}\frac{dJ}{dT}$$

$$= \frac{\tau}{J}$$

$$= \frac{rmg}{J}$$

$$= \frac{rmg}{I\omega}$$

$$\mathbf{J} = \mathbf{r} \times \mathbf{p}$$

$$\frac{d}{dt}(\mathbf{r}\times\mathbf{p}) = \frac{d\mathbf{J}}{dt}$$

$$\left(\frac{d\mathbf{r}}{dt}\times\mathbf{p}\right) + \left(\mathbf{r}\times\frac{d\mathbf{p}}{dt}\right) = \frac{d\mathbf{J}}{dt}$$

$$\frac{d\mathbf{J}}{dt} = 0$$

What Newton suggests is having a wheel that has one side no top or no side at the bottom because the two sides are not relevant. The one is indifferent to another as where one part divides into the other it results by forming zero. While the wheel is spinning one may not remove the one side and then claim there is no attachment between the top and the bottom. That would mean in a graph the top is not connected to the bottom because a wheel spinning is a graph moving against time. It is the principle all driving is done and not the least electricity. If it is that dJ / dt = 1 then that is exactly what Kepler said when he said $k^0 = a^3 / T^2k$. Kepler also said the motion brings about the filling of singularity $k^0 = 1^0 = 1$ when he said that the space is filled by the matter in the motion through the time period. All evidence we find in Kepler proving the Coanda principle point to the oppesite being true. The spinning forms space-time by forming the atom.

In spite of Newtonian Culture and in spite of Newton's declaring of such a principle there is evidence of zero being a part of Mathematics. If it was space-time could not be a facto one can find in the Universe. Space-time can only result from singularity being the pivotal piece in the Universe. From the centre must flow space-time and zero would absorb all space-time.

The definition of space-time is as follows:

Space-time is a four dimensional position of the Universe where the position of an object is specified by three coordinates in space and one position in time. According to the theory of special relativity there is no absolute time, which can be measured independently of the observer, so events that are simultaneous as seen from one observer occur at different times when seen from a different place. Time must therefore be measured in a relative manner as are positions in three-dimensional Euclidean space, and this is achieved through the concept of space-time. The trajectory of an object in space-time is called world line. General relativity relates to curvature of space-time to the positions and motions of particles of matter.

In view of the definition of space-time I wish to elaborate on my view of singularity and my deriving of space-time from the likeliness that singularity may produce space-time. In the past singularity was mentioned in the manner one would speak of a ghost hiding in a haunted Black Hole. Let's put singularity in the clear. Singularity is within every sphere due to the natural shape or form the sphere is committed to. According to Einstein singularity is a mathematical reality within the Black Hole but much more so in every sphere. Einstein may be the first to name it and Galileo (unwittingly) may have been the first to define it as Kepler was the first to formulate singularity, but in mathematical terms singularity is the most basic principle. At this point I wish to establish a fact that seems lost in all other grandeurs of cosmology. When tracing the radius down into the sphere the radius stars where all lines start and a straight line cannot begin at zero or nil it can only start at infinity. Such a statement will hardly seem appropriate but the relevancy of this fact has no limits.

POINT OF INFINITY ▶

If the line started at zero there was no line to start because zero multiplied by whatever results in zero as the answer. That must also be the cosmic starting point. Einstein introduced such a point and named that point singularity. When looking at the cosmos from whichever angle the indications seem to be that the fact that the cosmos is entirely in motion. It is forever spinning and it is going too as much as it is coming from. Everything is on the move and always encircling something of greater importance. A top can spin but the parameters of its spin are limiting the motion it can apply. By not spinning the top is still spinning as the Earth are doing the spinning on its behalf.

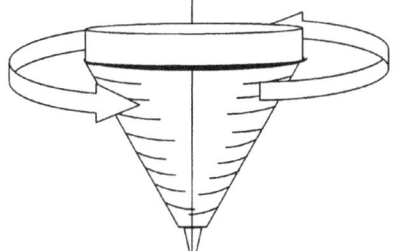

When spinning too fast the top fights something because the alignment keeping it upright starts to tarnish. The same apply when spinning too slowly but that makes sense. It is the fact that the same affect comes about when spinning too slow that triggers the questions. Why would the top stand upright by spinning. It must be because singularity charge the top into a cosmic independent reality.

Newton made a brief calculation as a young man that saw an apple fall from a tree. Seeing this he jotted down a formula and chucked it away. His piers and elders picked up the trashed paper with the calculation, and got all excited by the logic implications it had. $F = r^2 / (M_1 M_2)$. The mass of the two objects destroys the radius between the objects. Everyone went ballistic, proclaiming him as an instant genius, the one the world was waiting for after the crucifixion event.

I do not, for one second, deny or dispute the revelation. What I do encourage, is to place the event into its correct context. It was merely, and simply an apple that fell from its branch to its roots. The apple did not pretend to be a meteorite that fell from the heavens. If it were a meteorite, I am sure, with the man's genius, science would be somewhat different at this stage. However, as a young man, being very impressionable, as all young men are, and with the attention this brought about in the world of science, the matter overshadowed the fact.

$$\Omega = \frac{d\theta}{dt} \quad \tau = \frac{d\mathbf{J}}{dt} \quad d\theta = \frac{dJ}{J}$$

$$\Omega = \frac{1}{J}\frac{dJ}{dT}$$

$$= \frac{\tau}{J}$$

$$= \frac{rmg}{J}$$

$$= \frac{rmg}{I\omega}$$

$$\frac{d}{dt}(\mathbf{r} \times \mathbf{p}) = \frac{d\mathbf{J}}{dt}$$

$$\left(\frac{d\mathbf{r}}{dt} \times \mathbf{p}\right) + \left(\mathbf{r} \times \frac{d\mathbf{p}}{dt}\right) = \frac{d\mathbf{J}}{dt}$$

$$\frac{d\mathbf{J}}{dt} = 0$$

I am not disputing Newton; I am disputing the relevance of Newton's scientific breakthrough. It was not two objects of cosmic proportions, colliding in a show of the spectacular. It was, after all, only an apple falling from a tree and not that big an event. With this miracle he revealed, Newton found he was competent to improve on the work of Kepler and what Newton saw about what Kepler found was to Newton's mind the proof of total mathematical incompetence. He (Newton) saw a circle and without Π there can be no circle. Further more, since he was the founder of the invert four square principle, the principle also had to be included the make the picture a smart Newtonian picture and with that remove Kepler as such.

$\frac{dJ}{dt} = 0$ Newton, and science, made one enormous blunder, from this stance. They took the radius of a wheel not to have any influence on the wheel. In doing that, they removed the very fact that keeps the universal attachment together. They put two objects in an attaching relevancy and then announced no relevancy. Doing that is breaking the most fundamental mathematical principle.

$\frac{dJ}{0} = dt$ or $\frac{0}{dt} = dJ$ This disputes mathematics. DJ / dt can have any number from eternity to infinity, only excluding one; it cannot be 0. By placing the one in division of the other, you bring in relevance. You cannot then say there is no relevance. By doing such, you proclaim that one of the factors is non-existent. In both cases, one of the factors then does not exist. Such a claim is incoherent, because you proclaim that a circle has no radius, or a radius has no circle. When calculating a circle, you multiply either the square of the radius by Π, or the quarter of the diameter at a square by Π.

$\frac{dJ}{dt} = 0$ constitutes a circle and is also therefore $\Pi \times r^2$ = CIRCLE

If you remove r it then is $\Pi \times r^2 / r^2$ = CIRCLE.

You cannot then say $r^2/r^2 = 0$ and therefore $\Pi \times 0 = 0$. That is nonsense. $\Pi r^2/r^2$ will always be $\Pi \times 1$, and that is the eternal circle.

When looking at any rotating object, there has to be a point of no rotation and no rotation means "no rotation", not no existence. No rotation means a factor of 1, not zero. That then is singularity. The eternal Π, the Π that may not have significance but still it is a Π of value.

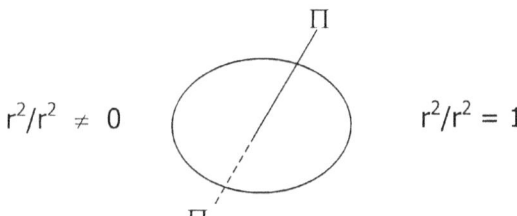

The relativity remains one, eternally one, but it cannot be zero. Therefore, dJ/dt cannot be zero.

dJ/dt can be eternal or infinitive or at the worst it can be dJ/dt =1 but dJ/dt ≠ 0

When explaining this to any child, they can immediately see that. Explain this to any Newtonian High Priest and he may have you removed forcefully from campus. I cannot find one Newtonian, large or small to accept that.

What is it the Newtonians fail to see? If an electron is orbiting around an atom, 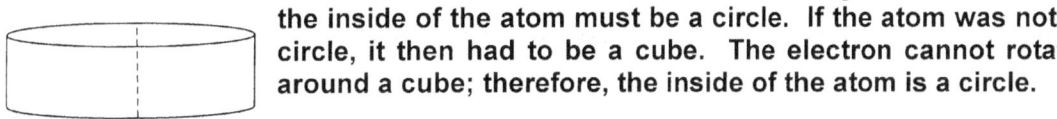 the inside of the atom must be a circle. If the atom was not a circle, it then had to be a cube. The electron cannot rotate around a cube; therefore, the inside of the atom is a circle.

In a circle, there is a radius that initiates the circle. The calculation of such a circle is $\Pi \times r^2$.

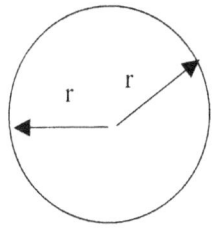 The radius r runs from the circle outwards, from a circle centre point towards Π, the value of the circle. In the centre of the circle, there is a point where the radius starts. It runs outwards from that point in all directions towards the circle Π. Technically, there then has to be a point where r is infinite and not zero, an absolute infinite. However, the circle therefore remains Π. The circle does not disappear; it remains there for all to see. It is only the radius that almost disappears into the infinite, but it does never become zero!

$$\frac{\Pi r^2}{r^2} = \Pi$$

If one removes the radius from the circle, the circle remains, only holding the value of Π. By removing the value of r, Π becomes singularity with no place to be. Singularity is the place where there is no space to be in place. However, Π remains because once r receives the slightest of space Π will find space. Then the circle will grow to Πr^2 and r would determine the space. Without space, there is no r but there is a circle with the value of Π.

Singularity is in every single rotating object, be it the proton or the combining effort of all particles in the universe. That is what light and the photon is. It is concentrated heat that the sun (or any other generator of electricity) connects heat to singularity where the heat receives either temporary connection to singularity or a small piece of individual singularity.

All spinning matter has the point where the spin is still there but the radius is to small to measure by any means. That point is standing still in relation to the rest of the spin. In relation to that logic I do not except Newtonian science holding the radius of s spinning object unaccountable in the spin, whether the spin is applying or not.

Applying Newton's second law F=ma

One arrive at the formula
$GMm / r^2 = m(\omega^2 r)$

By replacing $(\omega^2 r)$ with $2\Pi / T$ we obtain Kepler's third law
This law predicts that $T^2 = a^3$

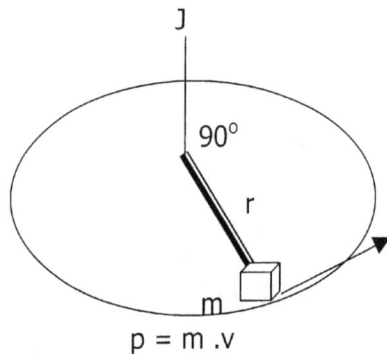

The mass (m) multiplying the speed (v) forms a new value J AND THEREFORE j CONTINUOUS TO IMPLY $J = I \omega$

$J = r \times p$ where $p = (v = r \times \omega)$

$J = r.m.v = m.r^2.\omega = I.\omega$ and becomes interpreted as $J = I\omega$

This establishes that $r = dJ / dt$

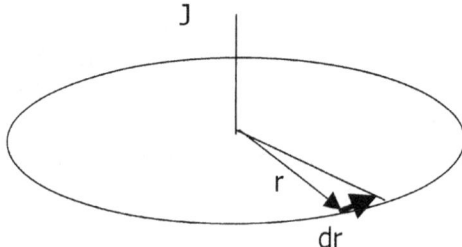

$r = dJ / dt$ In the case of planets in orbit around the sun r forms a value of zero because $dJ / dt = 0$.

What this statement implies is that r does not exist. When anything has a value of zero it is for all purposes non-existent. Only when an object is following s straight line can the radius be non-existent because the radius alters value through time development.

Taking the argument back to Kepler's law,

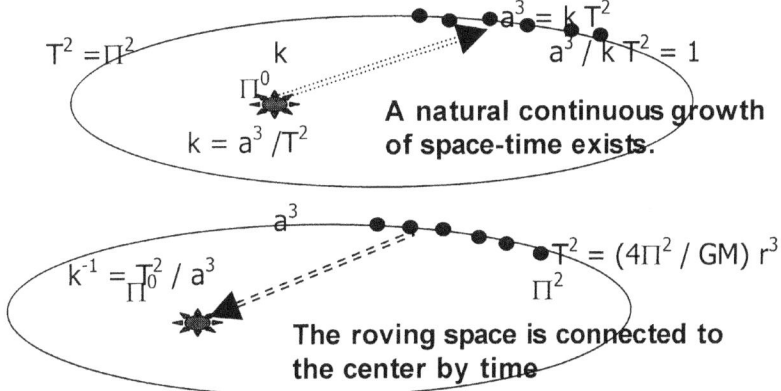

The spinning or not spinning is not part of the issue because at the point of absolute singularity the object never spins. Therefore spinning or not spinning does not apply to the point of singularity because singularity never spins in any event.

<u>Since Newton became an institution forming the King bee of the academic cartel world wide The Brainy Bunch had Newton's vision written in the minds of the future generations almost at gunpoint...well definitely at an academic gunpoint.</u>

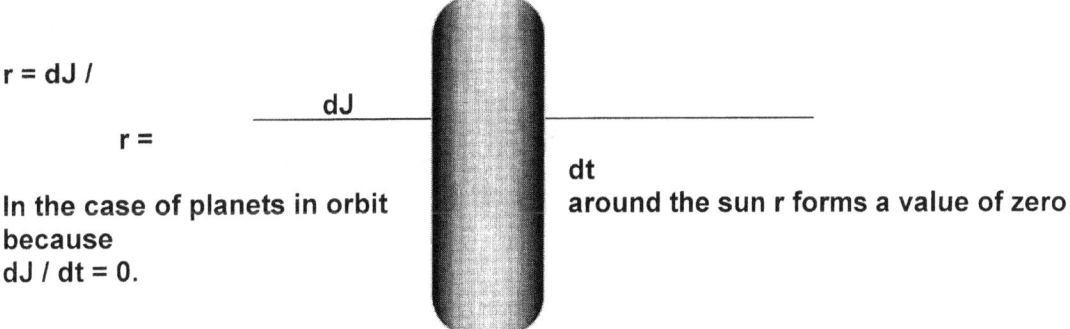

$r = dJ /$

$r =$

In the case of planets in orbit because
$dJ / dt = 0$.

dJ

dt

around the sun r forms a value of zero

<u>I am not the brightest in the world that I admit, but one thing no one can do, not even if you are the one and only Isaac Newton, is that you cannot place any relevancy in a relevancy and then claim it not to be in a relevancy because such a relevancy does not suit your taste.</u>

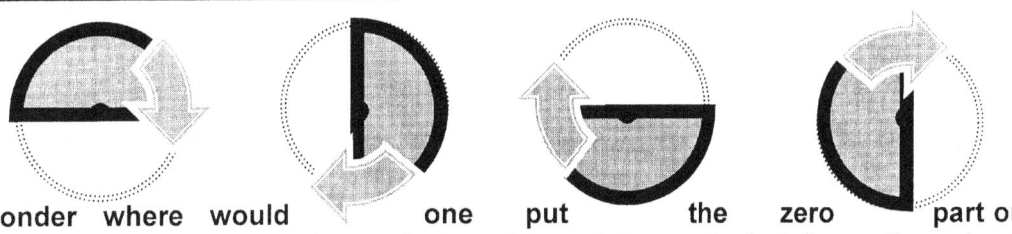

I wonder where would one put the zero part on the spinning wheel and what part must be excluded from the wheel. What Newton suggests, is a wheel has one side on top and no side at the bottom. While the wheel is spinning one may not remove the one side and then claim there is no attachment between the top and the bottom. That would mean in a graph the top is not connected to the bottom because a wheel spinning is a graph moving against time. It is the principle all driving is done and not the least electricity.

You cannot put something in relation to another object and then decide there is no relevancy in the relevancy

r = dJ /

r =

dJ / dt ≠ 0.

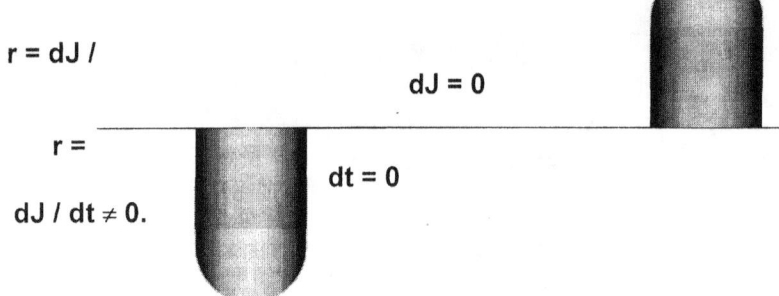

dJ = 0

dt = 0

If dJ = 0 then dt = 0 That is a mathematical principle, much larger than even Newton

Newton, and science, made one enormous blunder, from this stance. They took the radius of a wheel not to have any influence on the wheel. In doing that, they removed the very fact that keeps the universal attachment together.

$\frac{dJ}{dt} = 0$ This disputes mathematics. DJ / dt can have any number from eternity to infinity, only excluding one; it cannot be 0. By placing the one in division of the other, you bring in relevance. You cannot then say there is no relevance. By doing such, you proclaim that one of the factors is non-existent.

$\frac{dJ}{0} = dt$ or $\frac{0}{dt} = dJ$ In both cases, one of the factors then does not exist. Such a claim is incoherent, because you proclaim that a circle has no radius, or a radius has no circle. When calculating a circle, you multiply either the square of the radius by Π, or the quarter of the diameter at a square by Π.

Quite the very opposite is true and the rotating wheel in fact is the moving wave.

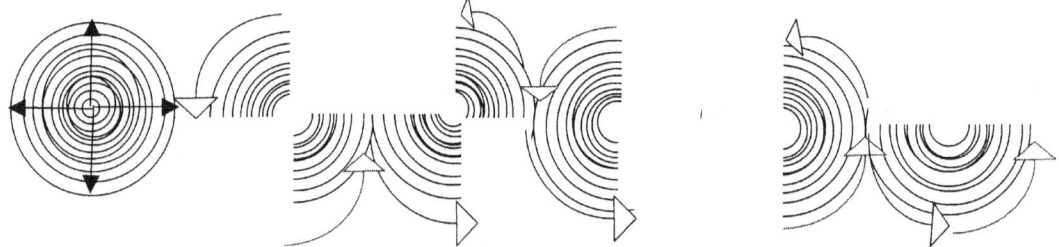

The graph which is a cornerstone of mathematics and which is used extensively in various calculating procedures would not function because at the end of a cycle there will be no cycle.

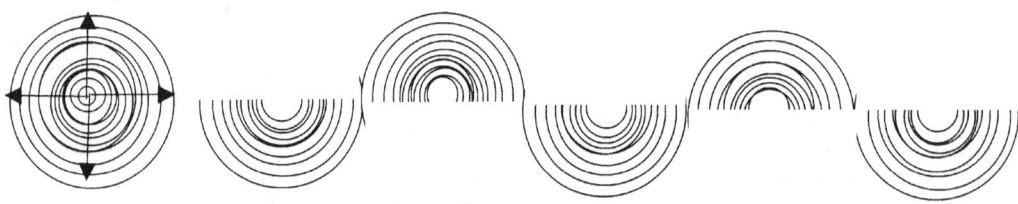

Every quarter of a rotating body is opposing the opposite sector directly and completely.

That proves that every rotating object holds the form of a wave and mostly it etiolates Newton's claim totally that no relevancy exist between the rotation and the axle which is pointing towards motion. That proves that while spinning, an object holds an absolute relevancy of one in order to maintain equilibrium as to ensure rotating motion.

$\Pi \times r^2 = $ CIRCLE

If you remove r it then is $\Pi \times r^2 / r^2 = $ CIRCLE.
You cannot then say $r^2/r^2 = 0$ and therefore $\Pi \times 0 = 0$. That is nonsense. $\Pi r^2/r^2$ will always be $\Pi \times 1$, and that is the eternal circle.

When looking at any rotating object, there has to be a point in the infinite middle where the one side rotates in one way and the other rotates in the other direction opposing the opposing direction. That point in infinity is the point of no rotation and no rotation means "no rotation", not no existence. No rotation means a factor of 1, not zero.

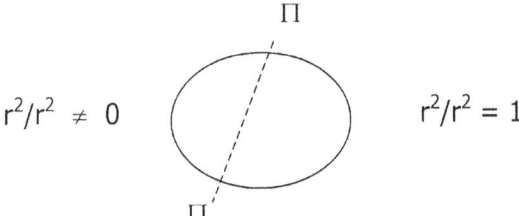

That then is singularity. The eternal Π, the Π that may not have significance but still it is a Π of value. The relativity remains one, eternally one, but it cannot be zero. Therefore, dJ/dt cannot be zero.

> dJ/dt can become eternal or infinitive or at the worst it can become one
> dJ/dt = 1

When explaining this to any child, they can immediately see that. Explain this to any Newtonian High Priest and he may have you removed forcefully from campus. I cannot find one Newtonian, of any significance being large or small to accept that. By not having a wheel rotate, the wheel becomes the factor of one, and the rotation becomes zero. The wheel does not disappear. In the cosmos, everything is rotating because nothing ever stands still. Therefore the mean equilibrium, the common factor there is to share, has to be one, eternity, the eternal Π, because all rotating objects has Π in singularity, and sharing singularity, gives every object in space a relation with all other objects in space. After trying for many years to bring them the candle, I concluded that Newtonians are incapable of realizing that mathematical principle as reality.

If Newton had said that dJ / dt = 1 then that is exactly what Kepler said when he said that in the centre of space–time singularity is allocated a position of control $k^0 = a^3 / T^2 k$. Kepler also said the motion brings about the filling of singularity $k^0 = 1^0 = 1$ when he said that the space is filled by the matter in the motion through the time period. Motion establishes space in time and cannot be zero because THAT is what gravity is. It is the motion of space-time and that can't be zero. If gravity were equal to zero the entire Universe would stop existing.

By using this idea and tracing it to Kepler we can establish the atom from what forms the Universe in the Universe fundamentally being the atom $a^3 = T^2 k$.

What Newtonian mathematicians should conclude is that space is motion and motion is time in space $T^2 = a^3 / k$ Kepler's said space –time is $k^0 = a^3 / T^2 k$.

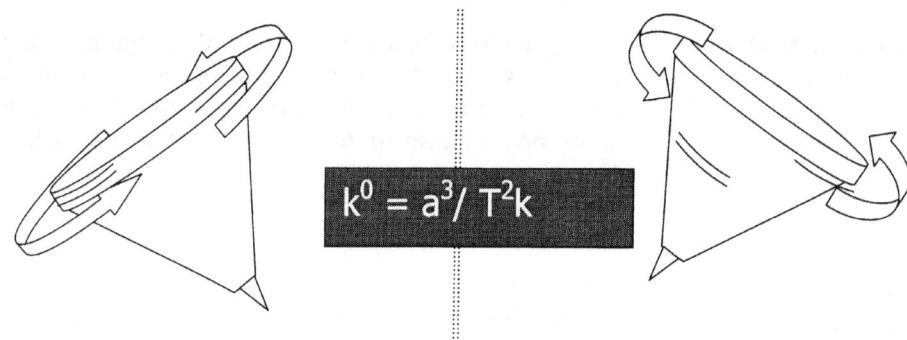

The spinning top is all the evidence any one needs to come to such a conclusion. I know probably as much as any graduate about cosmology but lack certificates to prove my knowledge. I am not part of established science. In my developing of knowledge accumulation I came to some conclusions about cosmology that are unique and divert somewhat to drastic form the accepted norm. Most of the work I see the same way as the norm does but in a reverse. Allow me a short explanation

We have to be clear about what we think of when we think of the Universe. Most people think of a picture recalling the black night sky when thinking of the Universe and that thought is most incorrect. Einstein was most correct when he declared the Universe was going flat where gravity is at its utmost, but the concern we should have is not with the mathematics being valid or not but with the vision about the Universe being what we think of and where we place the Universe. The Universe is in the centre of what is spinning and the biggest single particle that is spinning in total independence of the rest of what forms a total Universe is the atom. The atom spins and by the motion the atom evokes the universe forming what must be the group effort of all the atoms then spin by the motion the atom renders the rest of the larger Universe. The Universe is the part that allow the rest of what the Universe establish to spin. What spin you may ask. Kepler said it without saying it: $k^0 = a^3 / T^2 k$ and not even Einstein with his super human mathematical skills could say it better or more accurately.

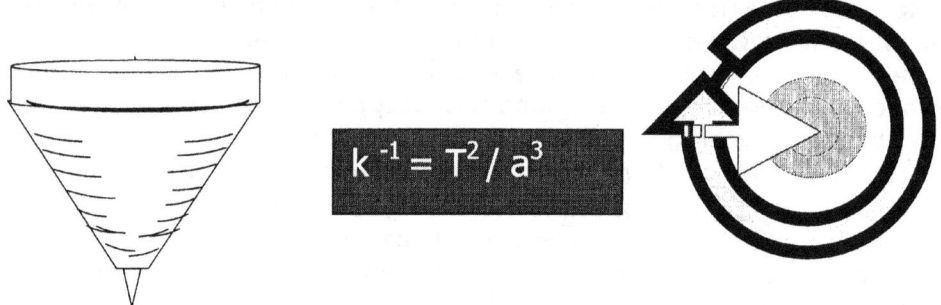

The motion established by singularity results in the implicating of the Coanda effect as much as the motion establish the Coanda effect. The spin realises the space limit while the space limits attaches the motion onto the space in the time within the time.

Let's get back to Kepler and that which Newton discarded. Newton proved that one factor dividing another factor could result in the division bringing about zero. The revolution ends where the revolution started with nothing accomplished. Newton was one man and all men are fallible and prone to errors. Newton saw what Newtonians whished to see and that might be inexcusable but to Newton it served a purpose. He could allow his way of thinking about gravity work and for three and a half hundred years it did. What is truly amazing is that not one person ever got wise about this matter. Remember, to this day we are talking about the brightest minds the human intellect can offer was dealing and working on this issue and not once, even doctor Einstein in person, questioned this madness. When a line within a sphere divides and divides as many times as one might divides achieve then one may get to a appoint where Πr^2 the r would go eternal flat as much as becoming r^0 it then is still not zero r=0. At best it could be r^0 **$k^0 = 1$**

With the top spinning the Coanda effect steps in and do justice to Kepler's formula.

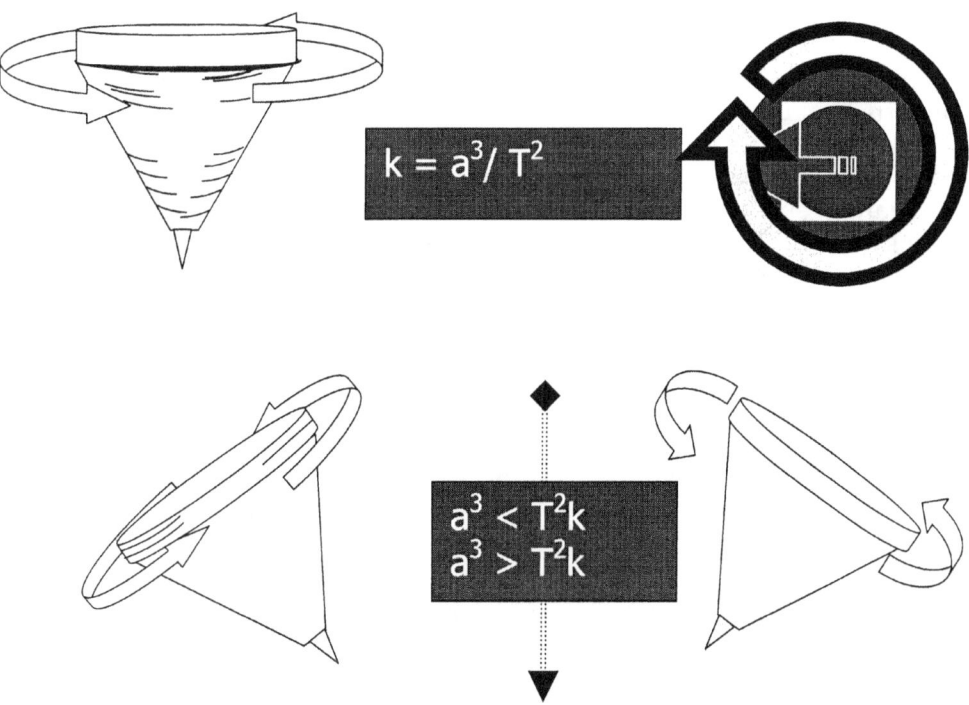

Time is always a displacement of space in relation to the implication of singularity, and comes about between two points in space relating to the centre of singularity as positioned by k, either too the value of k or too k^0.

I have asked as many persons as I do not care to remember why the top sinning will remain spinning around one point while turning. The answer I receive from the most educated to the schoolboy is always about momentum. That is a very simple answer and to say the least a little too simplistic by further analysis. Why would the spinning top go of centre when spinning higher than a specific velocity and lowering the velocity it would stabilize and run square to the Earth only after that it will go oblong and then fall. I could go on about different positions bringing across different momentum of thrust but I do not wish to insult your intelligence because I am aware that you are familiar with all the law. When the top is spinning it is spinning about its own axis and when it is not spinning it still remains spinning about the Earth's axis therefore when it is spinning it is also spinning about the Earth's axis. Therefore the limitations applying can only result as an influence coming from the Earth's axis.

The second question now comes screaming across and that is in what manner could the Earths axis ever affect a spinning top since the spin and he spinning top is a gross mismatch to what ever standard the Earth may introduce. It is clear that spinning objects do influence each other in contrast to Newtonian opinion.

If you remove r it then is $\Pi \times r^2 / r^2$ = CIRCLE. You cannot then say $r^2/r^2 = 0$ and therefore $\Pi \times 0 = 0$. The smallest one can go is get a factor of Π leaving form.

Every round object has a point establishing a very centre, a middle dividing one side from the other. That division determines the space from one side away from the other side. At one point there must be a point that does not fall on either side of the divide. Such a point will still be a circle, because from that side the circle divides into two sectors.

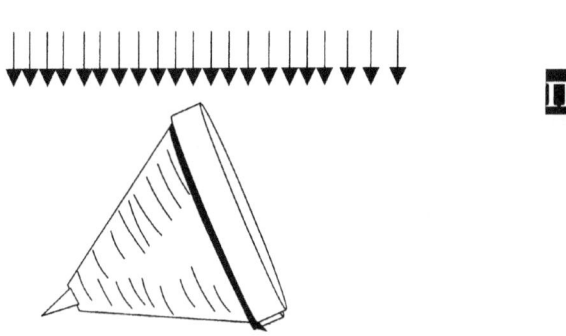

In every spinning object there is a point of infinity, a point that does not turn because it holds the dividing spin. However when such a point becomes a line that cannot spin a new Universe is born in the midst of many others. At the birth that point diverts space outwards and from that point the spin is either clockwise or anti clockwise in all directions. As I pointed out no line can start at zero because then there is no line and no rotating point can start at zero because then there is no rotation. Calculating a square involves two aspects that we think of as sides.

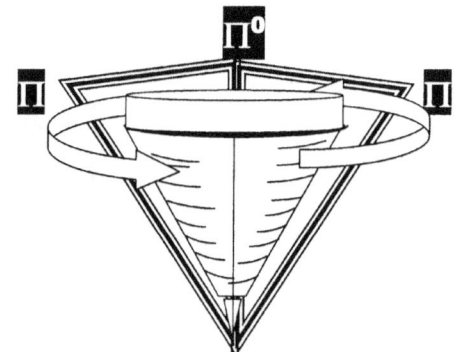

There is a Universe in differences between the top lying down without any individual motion and ostentatiously independent, self assured spinning top that even produce a sound to match the occasion. While without motion the top submits to the contraction lines running as the straight line holding half the value of the square being 180°. The top seems dead as it surrendered its long-term position and would eventually succumb to the Earth's gravity by relinquishing the structural independence it has. Then the motion brings life into the top and gives the top reasons to fight the Earth by fighting for independence. The top just became independent by the motion it received from the combined efforts of all the independent atoms forming the structure of the top.

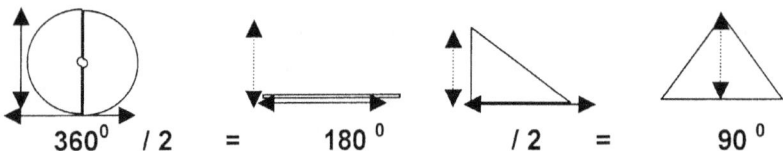

$360°\ /\ 2\ =\ 180°\ /\ 2\ =\ 90°$

The circle is a square holding a round shape, as the straight line is a square holding one side to infinity. Calculating a circle involves two aspects where the one is either the radius or the diameter that is double the radius. The other is the factor Π

However to say $\dfrac{dJ}{dt} = 0$ That is nonsense. $\Pi r^2/r^2$ will always be $\Pi \times 1$, and that is the eternal circle. No rotation means a factor of 1, not zero. That then is singularity. The eternal Π, the Π that may not have significance but still it is a Π of value.

$\dfrac{\Pi r^2}{r^2} = \Pi$ That is the smallest any circular dividing might go. That leaves Π and as such that indicate a round shape. Going that small leaves us with the curvature of space-time, which must then represent space-time.

From the graph one can establish the link in the circle's rotation around a conforming unit being singularity.

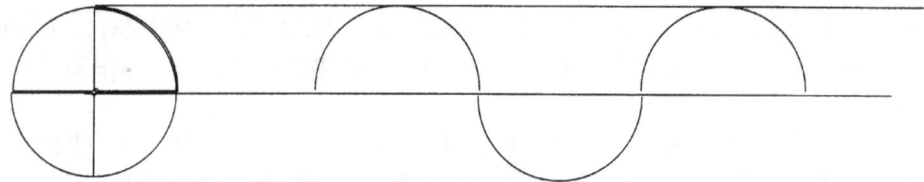

Saying that one therefore has to admit that the smallest spot has to hold space because the most insignificant dot can transmit light and being able to accomplish that, one must accept it to carry a value of something. If that spot had the value of nothing, it means that spot was not there to begin with. Holding space-time one should return to the original formula indicating space-time in as much as $a^3 = T^2 k$ where $a = R$ and $T = T$. Being time it has to alternate positions and that can therefore only apply to k where k will indicate a relation to the space-time in question or the relevancy to singularity being $k^0 = 1$. **By receiving k on top of the singularity from $k^0 = 1$ to $k^0 = a^3 / T^2 k$**

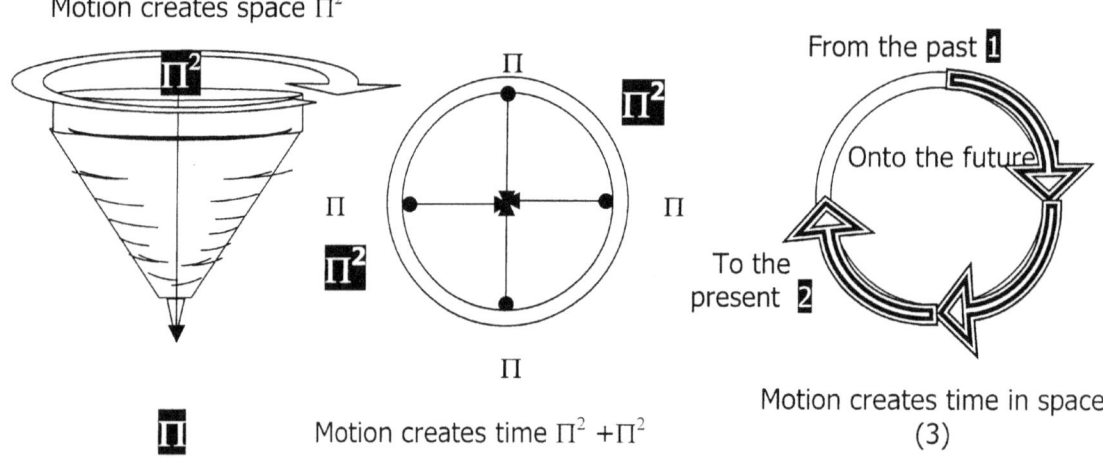

$R^3 / T^2 = k$ or $R^3 / T^2 = k^0$. With this fact established we then must return to the value as indicated by singularity being Π. In this we find that $\Pi^3 / \Pi^2 = \Pi$, weather k is Π or Π^0. This brings about the value relating to space-time relevancies as a formula consisting of $\Pi^3 / \Pi^2 = \Pi$ in various forms and relations.

One also must keep in mind that there are ALWAYS four sides relating to the Universe from any point holding singularity, and since every point in the Universe contains singularity in what ever form, very spot in the Universe comprises of four points initially extending to the next spot by means of $\Pi^2/4$, which we know as the Roche factor.

What I am talking about is not rocket science or re calculating the Universe to find the critical density or establishing time bypassing by space whirls and all the other magnificent things our most brilliant mathematicians keep their minds busy with. No, I am talking about the most basic mathematics primary school children work with. If a cycle did not bring work how do science calculate electricity generating?

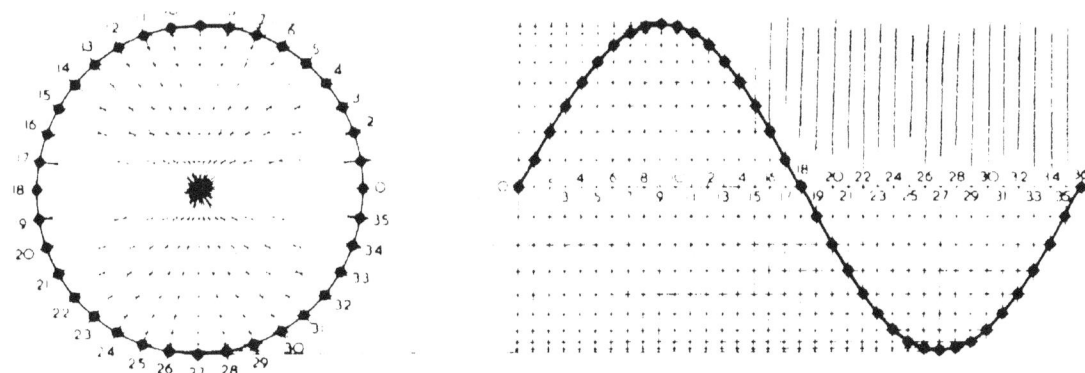

The motion the top has is directly a mirror image of what we find in the sine wave, which is used to calculate electricity. There is the line or Earth or singularity from which space-time diverts.

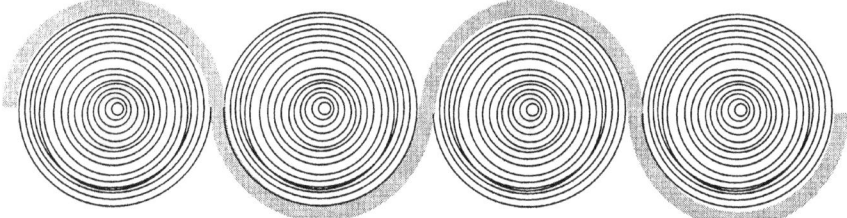

From such a relevancy there then must be four different values relating to singularity and since the atom has a proven relevancy of $(\Pi^2 + \Pi^2) \Pi^2 \times \Pi \times 3$. In the motion that replace the motionless, the motion made the motion less an atom by putting the object through the commitment of motion into an independent Universe where every aspect of the Universe becomes an Individual atom that maintains its independence as all atoms do.

It started with a dot, because that is the only form, size and dimension mathematical logic will allow our brain to accept. From the one dot had to come a second dot and a third dot. The dynamics of such a dot is smaller than we can understand because such a dot is in negative relation to what we see Π to be, and the deeper we delve in finding the smallest fragment where space started, in the spot where time is still eternal as much as we can accept eternity to be. The reason why we should first locate the spot is because we can only work from that point forward. By working forward we have to work backwards to locate where we are heading. The cosmos started at a point and where such a point is, we will find the Universe. Every one knows where the Universe is, because we can see where the Universe is, but if we can see where the Universe is, then we should find the centre of the Universe in that spot. Einstein theoretically positioned the point of beginning at a place he indicated where singularity should be. With the cosmos the size it is and space so large compared to our smallness we have no chance in finding the centre of the Universe. The Universe started where singularity is and singularity is the sure indicator of the Universe. With all spinning objects holding singularity we then have located singularity in as much as finding the centre of the Universe. The Universe started with a dot forming. That answer arrive from taking mathematics back to a point of being the smallest possible position, far smaller than we may be able to calculate form.

Newton was Master of Motion formulating and yet, Newton missed the biggest motion achievement there can be. Newton missed what the atom is and why the atom is the Universe. Newtonians missed why a top is standing upright while spinning and why that will be space-time. If Newton tested Kepler better and not tried to denounce the work of Kepler as being mathematically poor and treated the work of Kepler undeservingly as mathematical incompetence, then

Newton had the best opportunity any one ever had to see that motion is space and space in moving is what forms gravity. From motion arrives the atom and the atom brings along the Universe. By spinning the top becomes an atom as much as the top shows what an atom is. If not for motion of space-time the atom would plunge the Universe into an everlasting Black Hole as it was before and as it will again be some day. The atom duplicate as much as the atom contract and that forms the gravity that keeps the Universe in the three dimensions we know.

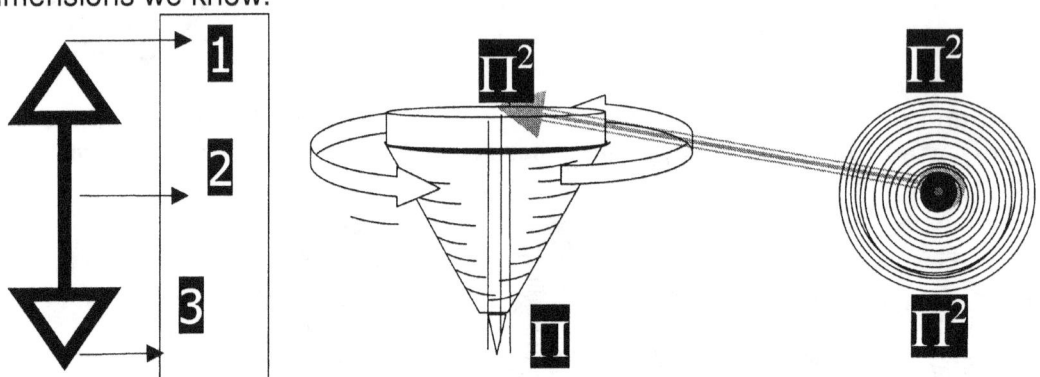

By motion the humble spinning top becomes defined as another Universe or another atom or another dignified star performing in the Universe as the Universe.

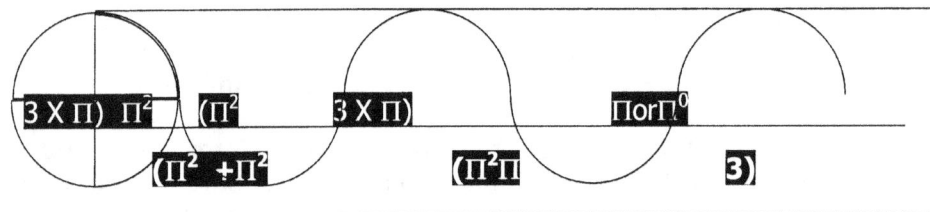

$\Pi \times 3 \times \Pi^2 (\Pi^2 + \Pi^2) = 1836$.

Our instincts, our logic and our calculating process all indicate that the sphere holds a centre point from where six evenly positioned point's position matter to be. Using The formula $F = G (M_1.m_2) / r^2$ it indicates to a force pulling objects closer, where each force is coming from each centre point the body in question has. The contraction must commit the two bodies towards a point in each case being spot on in the middle, not withstanding what direction the force is applying, the body will draw to the centre.

If the Universe spins around a centre point holding singularity, and singularity confirms the centre of the Universe, then every particle holds the centre of the Universe making the number of universal centres immeasurable many, and every atom and sub atom particle presented outside the atom in smaller bits, are all not pieces of the Universe but they are a Universe surrounded by many Universes. If every atomic particle no matter how small is holding the centre of the Universe, then the gravity is coming about from that point because that is where the gravity applying in the Universe are applying contraction.

It then is the atom in the most centre part where space and time meets singularity, that Einstein found a Universe collapsing to a single dimension, and every atom at a point post of the proton where gravity initiates in according with the proton dimensional colas of $(\Pi^2 + \Pi^2)(\Pi^2 \times \Pi \times 3) = 1836$

To circumvent the atom circle it takes 6188965056 times **k** to produce that number of **T²**, which duplicates that many a³. This means by the time the atom was completed the atom in ratio was confirming the duplication of **k** by a margin of 1:6188965056. That means when observing the motion there will be 6188965056 time units for every flicker one would notice. This differentiation is time delay forming space at the proton seen from singularity or there will be 1836 flickers seen from the electron. Being at the speed sees from the proton angle and there will be 300 000 flickers seen from our stance if the photon was one kilometre in length which we know it is not. That action brought on by singularity duplicating to form space-time is what makes the Universe seem solid. Once the measure of relevancy is broken time and space goes wasted. The solidness having flickering intervals is only applying in the eyes of the onlooker. Our Universe seems solid as long as we are on "the other side" matching the seemingly "shorter duration in time" and "extended space". Once the margin becomes fallible the space – time scenario becomes distorted and eventually destroyed. In the Universe we share, it seems the speed ratio or time duplication can withstand the combined effort of sixty protons where then the duplication of space falls prey to the destructing of space.

($\Pi=\Pi^3/\Pi^2$) or in Kepler's terms **k = a³ / T²** presents the duplication of space while

($\Pi^2=\Pi^3/\Pi$) or in Kepler's terms **T² = a³ / k** presents the destruction of space and

($\Pi^0 / \Pi = \Pi^2 / \Pi^3$) or in Kepler's terms **k⁰ / k = T² / a³** presenting the final act or the demise of space. **k⁰ / k = T² / a³** becomes totally dominating when there are a displacing ratio of **6** (materials Number) in the square of space **10** which then presumes the value of **60**. Space will remain duplicating as long as there is space available to convert to heat and as long as there is heat that can be converted to material and material available to transform to singularity. Nowhere is they're having a free ride coming to any part or dimension of Creation. There is built in only a lot of hard work and a dear price to pay for every inch gained or lost in growth.

In this mentioned ratio between the dismissing and the duplicating of space through motion stars form by accumulating material in a giant sphere and keeping the atoms secluded from outer space. All the atoms within the star that are forming the space within the star that is forming the star are as much the star as the star is all the atoms combined. Since all the atoms in the star works towards a mutual goal as to provide the star with the required security to provide the maintenance that brings about survival to components in the star is as much one atom in cosmic space as all the atoms are individually cosmic structures. When the flow of space is exceeded by a certain specific number of proton abilities to dismiss the space the dimensional walls keeping the space in form no longer can sustain the flow of space by substituting the demise with a flow of space. We have to remember that the initial motion was equal to the initial expanding that was in turn equal to the initial space **a³** that developed. The distance **k** that came about was the same value as space **a³** and the motion of the space T^2. The Universe divided innumerably as it remained one structure. Relevancies came about that excluded no possibilities and whatever one may think of being in the Universe came into place through relevancies between innumerable factors all acting in groups that remained one. There was no space but the space created in that cycle motion. There was the motion that provided the expanding in the straight line that was precisely the same value in the half circle and that brought about the triangle also to the identical value as the other two and was securing the space that was precisely the same as the other two factors. Reflecting on matters we still find this very same trend applying to light at the present time we live in. In the development space grew because the diminishing was only half the growth and size rather than direction became a major influence. The direction is the foundation and came in as a bases in the very first instant. The second repeating instant changed the scenario. With time progressing the distance k^1 will tend to lag behind as the rotation T^2 has to compromise for more space a^3 involved. With more space coming about the circle that had to produce the rotation was at the start equal to the expansion distance and there fore $T^2 = k = a^3$. Since then the ratio changed to $a^3 = T^2 k$. That is how space/time relates according to the original calculations Kepler introduced.

$1/k = T^2/a^3$ We know that there is a demise of space relating to the growth in space proving that when distance k reduces space a^3 will do the same and

$k = a^3/T^2$ when k expands it will produce space in relation to motion.

$T^2 = a^3/k$ When k demise the growth in distance k expand time T^2 will increase by the square but distance k will diminish by the single therefore time T^2 will grow faster than space a^3, which is the result of k will diminish.

At the beginning a trend was set that apply throughout Creation ever since. As the space increased the time ratio decreased since the distance in relevancy reduced in relation to the available space and that is the relevancy I simulate in the atoms ratio of space-time being $(\Pi^2+\Pi^2)(\Pi\Pi^2)(3) = 1836$. But as the star takes time in space back towards the earlier scenario conditions applying to the atom will reduce the space it holds and as such the time will bring the compromising aspect to the changing ratio. It is most important note that when the comet liquefied it was all the atoms within the comet that was liquefied. It was not a case of the comet destruct but the atoms within remained preserved. The comet is as all cosmic objects are, made up by all about the atoms within the structure and also the way the elements arrange their various positions to assimilate the sphere to prefect detail where the entire group of atoms as a group manifest in the form of a sphere spinning about an elected axis. As the group of atoms work in the way an ant ness will where the group form a unit, so does the atoms form a unit representing all the atoms as a group in the group where the group retained the position as one atom with one elected singularity in a charged centre. As the Universe formed by dividing innumerable times and yet remained a unit throughout every aspect, this leads on to groups forming galactica and stars where all the atoms within the star becomes the star and the star becomes anther single atom. The atomic relevancy apply as that quantity in the outer space and with Earth not being that much better developed the atomic relevancy in outer space and in the

Earth centre is $(\Pi^2+\Pi^2)(\Pi\Pi^2)(3) = 1836$ meaning the proton displaces space-time at a relevant value of 1836 times that of the electron, or the other way around is that the proton is 1836 times more intense going on route on to singularity than heat is going liquid through the strainer we call the electron. We can see that the atom formed in accordance with Kepler's prediction and followed the route Kepler introduced.

$\Pi^0 \Rightarrow \Pi$ **By motion Π became Π^3** The first motion implicated distance and space but space is one part of the requirement, which applies because the structure found independent space by separating from surrounding space through motion.

$\Pi^3 \Rightarrow \Pi^0\ \Pi^2$ Since singularity split the Universe, which in the case is the independent structure forms motion that also separate as it is separate on both sides of the divide because singularity is unable to move. The motion may seem as a combining factor but is doing the dividing on both sides of the divide because singularity is unable to move. The forming of space had to be completed by the forming of motion that provided the time factor and so space/time was presented for the first time.

$\Pi^0 \Rightarrow \Pi^2 + \Pi^0 \Rightarrow \Pi^2$ The immovability of singularity spawns space-time by producing space in motion on both sides of the divide as the proton duplicates ($\Pi^2+\Pi^2$). From singularity the double proton arrived in all particles through out the Universe in one motion by turning from one side of the universe to the other side. In this instant all four of the cosmic pillars came into affect that then became the building basis of all the cosmos.

$\Pi^0 \Rightarrow \Pi^2 + \Pi^0 \Rightarrow \Pi^2$ At the point where the initial motion came about from expanding or destroying singularity the plating of heat (in the same manner we see electromagnetic plating is done) where the process confirms the liquid heat into solid heat and in turn into singularity became contracting and by plating which confirmed the liquid onto the solid and into singularity it became it was feeding singularity and thus secured survival. The proton motion was no longer removing from singularity but was bringing into singularity. But this come at a cost where most of singularity in the cosmos failed securing the future by providing motion which is gravity and formed liquid that eventually became space. The double proton could not sustain or maintain the overheating due to friction that came about from the lack of space and space as a factor came about.

$\Pi^2\ \Pi^2\Downarrow$ The liquefying process mentioned above established at the very first a liquid particle that will hold the motion part in

$\Pi^2\ \Pi\Pi\Downarrow$ space to the value of ($\Pi^2\Pi$) where one factor (Π^2) confirmed space-time and the other factor could introduce motion (Π) and in that a third part became a factor forming the space within the atom. The differentiation between time space and distance became more and more apparent and influences as much as altered the composition of the developing Universe. The progress in space required factors to change and absorb the pace developing.

$\Pi^2\ \Pi 3$ Then only did the Big bang arrive as liquid heat proceeded in turning into space and space came into place as $\Pi^0+\Pi^0+\Pi^0 = 3$ which represent the 3dimensions of the six sided Universe we are within. Only with the arrival of heat forming part of the atom in providing space to the atom could the atom survive and bring about the Universe we know. Then only came the Big Bang. This became the atom $(\Pi^2+\Pi^2)(\Pi^2\Pi)(\Pi^0+\Pi^0+\Pi^0) = 1836$ and the atom formed stars that still act in accordance with and to the atomic relevancy

This became the atom $(\Pi^2+\Pi^2)(\Pi^2\Pi)(\Pi^0+\Pi^0+\Pi^0) = 1836$ and the atom formed stars that still act in accordance with and to the atomic relevancy

Outer space substantiate the atom as $(\Pi^2+\Pi^2)(\Pi^2\Pi)$ 3

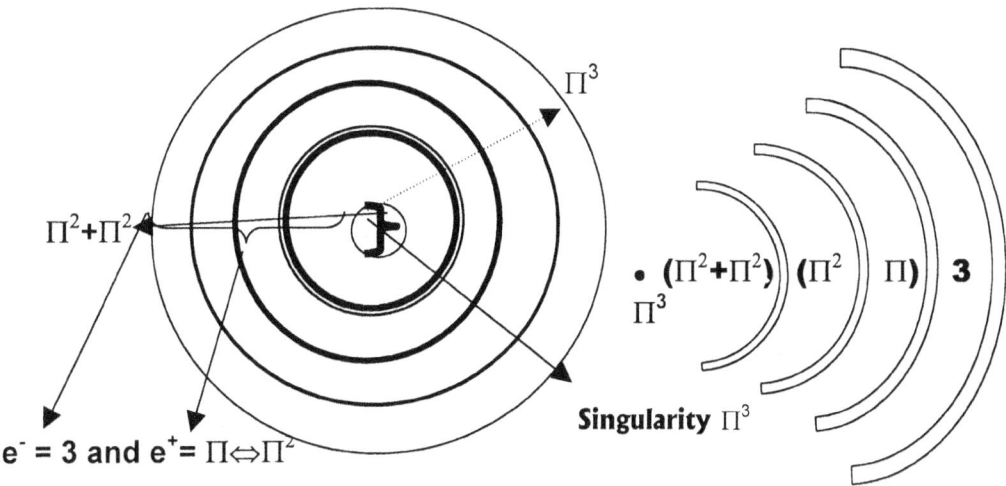

Every layer in the star represents one factor in the atom since the star is just another cosmic atom securing strings of atoms that as a unit aims for one goal and that is to secure one singularity within the star.

Mass has no part in any process but space-time demise and the relevancy brought about as such plays the only part. $(\Pi^2+\Pi^2)(\Pi\Pi^2)(3) = 1836$. At first with the first motion producing the first space, it took every proton in the entire Universe to suppress the space created in that motion. One movement found the ability to dispose all the space created but also created all the space there was. It was an innumerable number of protons forming the first atom. The first atom was the Universe and the atom became the Universe. The atom still is the Universe and will remain the Universe till the end. During this time relevancies came about that we named sub atomic particles. At first through motion the double proton came when seven points formed a relation with three and involving the Roche factor to form Π^2 on both sides of the divide. But the relation was in conjunction and not against so the total became an addition and not a result. $\Pi^2+\Pi^2$. Then the Universe was blessed with spinning space destroying double components allowing heat no escape. Some of the protons had to capitulate to survive. From this came the neutron $\Pi^2\Pi^2$ at fist presumably duplicating and later dividing $\Pi^2\Pi 3$. But every time it remained in relation to singularity extending, which is Π. In the Universe we enjoy the relevancy of the atom, which stands in direct relation to the atom holding space in the motion that duplicates the space. If the electron has this electron's mass of $9,109 \times 10^{-31}$ and the proton has that incredible bigger mass of $1,673 \times 10^{-27}$ making the proton a warping 1836 times more massive than the electron. Even in these differences in size and in gravity how can science connect size and mass. With this information and knowing where the subatomic particles are one can clearly see that gravity is about space dismissing ands what dismiss space more the action of the proton? The proton is the place in the Universe where I now am referring to which is where mass is created because that is where gravity is generated. Size just cannot be the cause of mass that produces gravity but rather more the other way around.

At this point one should delve deeper into the heat/cold and space density argument. Although it extends beyond the limits in information, which I wish to put on this book as far as my release of controversial and contradicting ideas I have colliding with the accepted and established norm straight on that mainstream science uphold I wish to elaborate somewhat on the big and small issue as well as the hot and cold issue. The Universe is without boundaries therefore there are no limits in what might be the hottest or the coldest. Neither can any one establish the limit of what might be big and what might be small. There is no biggest as much as there is no smallest. There is only some that is more developed or those that are less developed. But in saying even that I am extending what is and is not to what is not part of the Universe. It is far better stated in **Matter's Space In Time: The Thesis volume 5 or Matter's Space in Time: The Hypothesis I. S. B. N. ISBN 0-9584410-6-5** which incidentally is the same book since I have much more space devoted to the matter. Very briefly I shall commit myself by admitting to this: we are in the fourth developing stage in a cycle (as far as I can calculate) being the fourth in seven cycles of star development and in that some refer to one stage where it will hold prominence and the next will relate to another cyclic developing stage where the other will have prominence during the developing era. In the end all will conclude reaching the same finality but some will achieve it sooner that would others do. In that comes the "more or the less developed" but such a statement is not differentiating between two but merely identifying dissimilarities in singularity management under specific condition as the singularity maintain space-time in relevancies. The heat is everywhere and all over the same but confining the space to the concentration may help us to set a scale where such a scale only indicate a ratio of the balance we find in the density leaning to one or the other side. The more heat there is separated from space the colder the space must be holding the separation active. In the fine detail there is density in heat forming a liquid or there is density in space forming a gas but it always come down to the one or the other of both in balancing. When the Universe was 10^{34} K the Universe had everything that is currently a part of the Universe, which makes the Universe as big then as it now is small. Everything that was in it is still in it and will be in it and even a Black hole that has left the Universe is actively part of the Universe by consuming space-time through a proton action, which the Black hole placed outside the Universe. If we say the Universe was as big as an atom back when the Universe was 10^{34} K...then in proportions the Universe still spread space evenly and by equal measure as it does today. Space cannot add unless there is something willing to give away of the prominence it holds. Where space is abundant we find we find the concentration of heat to be low. The heat is not low or high but the concentration either favours the liquid factor or the space factor where if the heat being liquid is hot then the space being gas must be cold. It is a question of what is concentrated and dense and what is of little prominence. But when saying that while looking at our Universe I have to admit a lot of heat back then went the way of space now and I have to immediately correct myself the space out there is as hot as space can get while the space within the very centre spot of the Sun where there is no space and no motion is as cold as the Universe can get. If it was not the case fusion could no take place since fusion comes from freezing space between the two factors committed to fusion into the oblivious unknown. Space can only remove by cooling because pressure brings about heat and heat brings about space as the nuclear bomb so vividly proves. By removing plasma, which is a fancy, fill name for good old heat but sounds very intellectual, from space the space itself has to grow cold in order to allow so much distinction coming about. The cold can house the hot as the hot will accompany the cold and raising the one has to lower the other since it is clear that as space part it parts from heat and as heat lessons in density it joins space while it leaves heat. There is an unmistakable quantifying connection separating the two, which in itself is beyond separation. It is relevancies applying in favour of one aspect or another aspect. Gravity reduces space by employing motion and by reducing the space it makes the space cold as it removes heat from space by intensifying the motion of the space in the space. Gravity makes space loose from heat but also heat loose from space (if I will be excused for such low sophistication used to express myself). We measure the heat in the spot but we should agree that in that case we are measuring the space in the heat. The hotter the space will seem to us realistically the colder the containing space is. The relevancies have to push away from the centre or join at the centre. By removing the space and all the heat in the space to a point of no distinction. That means outer space is where heat joins space in the favour of space and singularity is where space joins heat in the favour of heat excluding space. In the past at some point cold and hot and heat and space was parted to the extent that the Sun's hot $18 \times 10^{6\ 0}$ was deep frozen because outer space was $10^{34\ 0}$ K and at that stage what is hot now was bitterly cold because the Suns corona must have been a solid freezer at only $6500^{\ 0}$ K which is 10^{30} below what was freezing outer space back then and what was bitterly cold then we regard the Sun now to be as extremely hot. The cold accommodated the hot keeping the Sun and all the stars in that range in a deep frozen state while the bitterly cold that outer space has to represent being the carrier of the heat limit was at that point in relation to our measure excruciating hot while the hot stored the cold but in the end it was the same as it is because it is the same Universe as it was.

One must not envisage this space duplication and space dismissing on the level where Einstein placed his vanishing and flat going Universe because it is on another plane. One must not see half a person following his other half into the future of a line of the same man continuously leading on in time. At the level we live and breath and the level we can see light and even the space we have and live in has become solid. A particle entering the atmosphere meets with a solid space that is so unbroken it puts the material back in time by billions of years by turning it into photons. The suspending of space by material in motion takes place at the most intricate of places locating where singularity brakes into space- time.

In outer space and near to outer space regions as the Earth is the atom confirms the security of space duplicating through time. The atom sustains a space-time duplication of the atom as it is combining the total value of the factors forming singularity. The factors related to the atom as a unit that is securing space–time relating to singularity to the form of $(\Pi^2 + \Pi^2 + \Pi^2 + \Pi + 3)$

× Π = 112.31. These factors are holding the atom apart from singularity by confirming space-time which is the atom in another relation other than the normal $(Π^2+Π^2)(Π^2Π3) = 1836$ and that is the point the sphere secures space in the six dimensions or space walls formed by the Universe when the Universe came to $7/10Π^6/6 = 112.162$. It is only when the requirements of maintaining singularity surpass the confirming the maintaining singularity in the atoms inner core stability that the double proton confirms in dismissing space-time without the required duplication also thereof will singularity dismiss the structural security that space-time in the double proton can provide. With me suspecting that there may be those that has super ambitious as space whirl creators living out there that plan to trick the Universe by building some double space whirl here this. In the event where there is a person with a super elastic imagination that foresee a situation where such a person can create or establish two of himself in the same space in the same time with some mathematical formula he has dreamed about, then such a person must first demolish his body down to fit a position where any triangle in his body will measure the size any half a circle that will take forming a straight line. Therefore all those out there planning on their next combining of space whirls they anticipate creating to split time or bridge time or just annoy time in space, remember that once you reach the space you occupy no more than a straight line you may proceed with your ambitions and plans.

The atom like the star is a sphere and on the inside of the sphere, there is the point where space dismisses into singularity because of the way that nature designed the sphere. The sphere secures a point where no space is possible therefore; no motion can duplicate such a space. Then as space-time develops stronger locations flowing to the other border of the sphere, space-time becomes more defined and secured. On the outside, the number of spots per volume of space-time developed is far reduced to the space claimed because the vacancy in space can only be represented by singularity that is the securing aspect of space-time. The duplication and the dismissing is such a small part of the space-time singularity secures that in the dynamics where a triangle must hold a different value from that of the half circle, which in turn has to hold a different value to the line, the difference there is in the forms secures the definite solidness of space in time. One must not see an object breaking composition because at a displacement factor of $(Π^2+Π^2) × Π = 62$ the walls of space comes tumbling down on the flow of time, but only at a point where the governing singularity developed to a point that the displacement can secure the survival of the appointed located singularity the star developed throughout its life span.

The breaking down of space by the inability to create motion is present in every atom and particle just as the duplicating of space is in every atom and molecule. Since all stars are group representatives of innumerable atoms within the space-time confinement the governing singularity lays claim to the dismissing of or duplicating of space-time is present. It is the ratio in the atom or stars favouring the presence of one or the other that allows the atom or star the characteristic it shows. If the duplication of space –time is prevalent as it is with Xenon, then notwithstanding how massive the protons grouping is, it still seems that space will duplicate much more frivolous than would the diminishing of space-time be a feature of the element. It is the ratio that gravity establish placing relevancies against relevancies that produce space-time or even the lack there of as in the case of a Black Hole. Where the Black Hole lost all ability to motion with in because of its securing of singularity maintaining does the proton then ejects into outer space way outside the star or star that now became an overdeveloped atom to a position in outer space because the motion of the double proton is no longer valid within the well developed singularity governing the Black Hole.

The proton position

The electron has the outer edge of the atom to fill, which is far greater in volumetric size than the inside of the atom where space is at a premium compared to the outside. Even just looking at this picture should tell everyone all there is to tell about gravity. It cannot be size that implicates mass. Referring to the massive potential size discrepancy that **became synonymous with the quantum factor, the name used as a quantum state is indicating unbelievable space that is beyond any explaining the size the atom holds and space going the way of the atom from the electron down to the proton. The name alone given to this state of space became synonymous with incredibly big, unexplainably huge and beyond measure of understanding**

The electron position

The position the proton has within the circular atom cannot promote any idea that the proton can be that more massive than what the electron is. There is just not enough room down there in the centre of the atom. Therefore one has to search and see what there is down there that would allow the proton to be that more massive. The only thing in total abundance is the lack of space and the only way that the proton can be that much more massive is if it destroyed 1836 times the space than what the electron manages or on the other hand if the electron can duplicate 1836 times more space than what the proton can achieve. What this does tell is that the electron duplicates space 1836 times more than the proton or the proton reduces space 1836 time more efficient than does the electron. This atomic relevancy we at present hold were not always applying and as space grew in progressive development the atom had to adapt in relation to the requirements of singularity.

This reality is frozen in time in all stars when the star seeks independence from the galactica centre as it drifts outwards. There are two definitions we can use when looking at such a growth. We can look at the space not holding material that grew in size in which the stars froze their development by remaining behind all because of a lesser developing singularity or we can focus on the stars growing and with that push the push the space much more into expanding. The star froze cosmic development as it came about in its search for independence and at that it holds the atom to the value the cosmos had before and during the time it became independent. The cosmos grew in space progress just as much as the star was left behind. In that there are young and not that young stars but it has no implication on the particles in numbers volumetric holding mass and as such the object produces gravity. If an object had the intensity of I kg in outer space, the object will be 3 times more intense on the outer edge of the Sun or it will be Π^2 more intense on the very inside That object will be 1836 times as intense in a neutron star or 6188965056 million tons in a Black Hole. This comes from the manner that the star manage to destroy space and redirect the space to fluid heat or the solidity of frozen space as matter really is. In the Black hole it reduces much further as it claims the singularity, which the object had, and destroy all space and all time there ever was.

The eventual combined value forming the total combination of spinning protons produces a velocity of flow of space in relation to the time thereof towards the centre and to compromise for the flow being insufficient the space will compensate by becoming denser heat once again. The heat that forms is in the centre of the star and the compromising of space by forming heat comes as the result of space flowing into the star. The star has to have an iron inner core to generate the flow of space towards the concentrated centre. If the sun did not have an iron core, the sun would not have gravity. That again contradicts the science teachings. This is the way space flows towards the centre of the star and that would influence the space/time displacing within the star that would affect a star: If it was merely centrifugal force keeping objects in gravity, one would surmise that the most massive should be swinging on the very inside of the star. That is not the case.

Every star (even a midget such as the sun) is a gas giant going down to a Black Hole The inner limit in the star centre space is Π^3

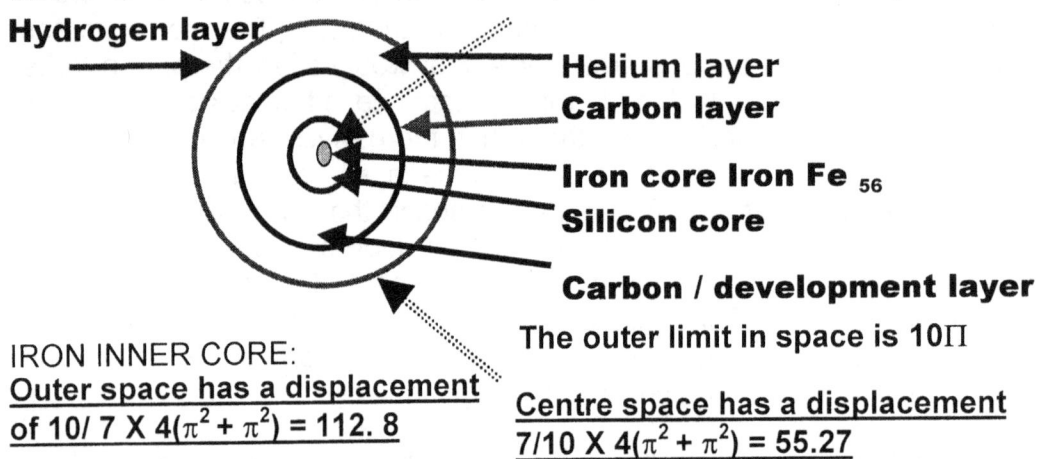

Hydrogen layer

Helium layer
Carbon layer

Iron core Iron Fe $_{56}$
Silicon core

Carbon / development layer

The outer limit in space is 10Π

IRON INNER CORE:
Outer space has a displacement of $10/7 \times 4(\pi^2 + \pi^2) = 112.8$

Centre space has a displacement $7/10 \times 4(\pi^2 + \pi^2) = 55.27$

Every layer is a star in development that might be part of the original star, but once the layer has resolved in the following phase the star characteristics changes drastically. Every layer is a star almost independent from the rest of the star and yet it is in support of the rest of the star.

By going dark the star did not die because only that which holds life has the ability to be in any position to can die... and Newtonian boilers with mediaeval coal stoves of course. The star did not diminish the space and the space did not outgrow the star. It is a relevancy where the one factor represents a compromise to substitute for the other factors changing the relevancies applying. It is the space that grew as much as it is the star development that fell away and started in initiating independence from further development by securing sufficient heat within the inner core to provide the motion that will produce such independence from its surroundings. The diminishing of space takes place in every atom in such an atom inner core per ratio of the number of protons acting as one unit. The result of the product is the accumulated to form the total value of the star. In every atom there is the dismissing of space that is fed from the top or outside of the star that spirals the diminishing value downwards towards the circle and into a centre.

IRON INNER CORE:

The inner core has to be Fe_{56} to produce gravity. This is what reduces space in conjunction wit singularity where the atoms produce a dismissing value that the space-time can sustain with enabling the flow of heat through space. In the one limit of the six sided Universe no element can sustain duplicating above the value of $10/7 \times 4(\pi^2 + \pi^2) = 112.8$ and above $7/10 \times 4(\pi^2 + \pi^2) = 55.27$ within the star inner core dismissing space beyond that capability will no longer contribute to duplicating space-time of the atoms involved. Only the iron atom producing and maintaining a displacement value of 55 – 56 can produce gravity by being on the edge of demising space time while maintaining duplicating which is gravity and in our Universe only stars with an iron inner core has the ability to bring about gravity. Gravity can only achieve a displacing relevancy at $7/10 \times 4(\pi^2 + \pi^2) = 55.27$, and that produces a potential difference that brings about gravity within the inner star core where gravity accumulates. This then relates directly to the second value of the Titius Bode value of $10/7 \times 4(\pi^2 + \pi^2) = 112.8$ that limits outer space in the three-dimensional and six sided boundaries of what forms our Universe as outer space forming the value of $7/10 \, \Pi^6 / 6 = 112,162$. That is the outer relation to the inner relation set by the core in ratio to the outer space securing a position for the star identity in the space limits and is an indicator of the balance in space-time displacing potential of the star. In every star there is this flow towards the centre firstly of every individual atom but also as a combined unit flowing towards the centre of the star and the dismissing of space in every atom centre brings about the forming of a relation as a group within one unit structure we call a star. This flow is there because we gave it the name of gravity and gravity is the result of all the atom protons dismissing space and as such then has a linking that is invisible to the naked eye. In young stars the core ability is yet to develop and in such stars the gradual reducing come about as layers support the effort little developed inner core. The space reduced becomes a unifying effort from all the atoms in all layers from the outer ($hydrogen_1$ and $helium_2$) through the carbon / oxygen centre and the silicon layer down to the iron core and even going down further into space-time obscurity where the atoms as a group combining their effort acting as one atom. An atom securing one proton will provide much more space a much better field to flow. That leaves space the opportunity to support the demising of space of all the individual protons by substituting the loss with an new supply of space that becomes converted to heat than would an atom supporting 56 protons within one containing centre. The stepping down assists the young stars with a weak heat envelope sustaining the spin effort of the yet underdeveloped inner core.

The dismissing has little effect in the immediate vicinity of the atom as the removing of space is compensated by the supply of space from positions where more space is available. The flow of space from outer space will substitute the dismissing of space. But in the centre of the star where all the heat accumulates and gravity is at the very prime, where there is no space left such diminishing will have an accumulating effect gathered from all the atoms accumulated efforts that cannot be substituted by the flow since the flow comes from every proton in the atom within the star housing all atoms and protons as one unity. The eventual combined value forming the total combination of spinning protons produces a velocity of flow of space in relation to the time thereof towards the centre and to compromise for the flow being insufficient the space will compensate by becoming denser heat once again. The heat that forms is in the centre of the star and the compromising of space by forming heat comes as the result of space flowing into the star. The star has to have an iron inner core to generate the flow of space towards the concentrated centre. If the Sun did not have an iron core the Sun would not have gravity. That again contradicts the science teachings This is the way space flow towards the centre of the star and that would influence the space/time displacing within the star that would affect a star:

The demand on space flowing will be much more beneficial to the flow where all the atoms comprises of hydrogen and helium such as we found large super giants have. In the event where fifty percent of the star holds iron$_{56}$ and the rest is composed of silicon $_{26}$, the demand on space flow will be at a prime and the heat envelope that will support such gravity flow coming from such demand will not likely allow any fluids, as the photon is to escape from the star. The concentration of protons overshadows the flow opportunity by far and the star will become darker. Such a star going darker does not die because it is not life or a coal stove that can go out. There is no time period where it consumed all the available fuel as is the case with coal stoves that was used during and just past the Elizabethan age. The fuel stars use is available in unlimited quantities forming volumes that only time extending to eternity can consume.

This dismissing of space puts a relevancy onto the atom and that allows the atom some space/time conditions that diminish the occupying space the atom claims. This puts totally different equations on the atom. Restricting the occupying space the atom claims indicate the relevancy of space demise within the star that will culminate in the centre of the star and will control all aspects of the star. $(\Pi^2+\Pi^2)(\Pi^2\Pi)3$ **=1836.12 =1**. The super "gas" giant is nothing but a bowl of fluid the compare more or less to conditions on the Sun. A periodic cycle will be about the same as the planets give or take a few years per century. The atoms would display the same gravity relevancy with the electron establishing a dividing value between the atom and the atmosphere of that particular star. The relevant function of the star will then be 3. By aligning with three it can only have enormous space that the giants hold because stars in that league is all about duplicating space-time.

In the view space has on material all material is one gaping black hole ready to consume as much space as the star could manage and the space restraining would allow. But from the offset space has a black hole locked up and growing in every star striving for independence. This is the result of the first moment in space forming time. The first expanding distance **k** was equal to the space as well as the time factor and the reducing of the time factor brought on reducing of the space by half and the distance by half. This is the result of space-time being double the distance since space is just as much as the distance and the motion of the time is.

By eliminating either or, either or remained a factor and the Universe came about. Then the second moment in space arrived and the distance created was half the space and half the time that doubled. That refers to being small and that still finds much prominence in stars developing. In the Universe there is a balance between $\mathbf{k=a^3/T^2}$ and $\mathbf{k^{-1}=T^2/a^3}$. On the one side particles are adding time to material by removing time from time space by making outer space less dense. Then in stars that are very young the trend of $k = a^3 / T^2$ carries on because it is taking time from a gas to a much denser and colder liquid. This is concentrating but it is concentrating heat in space and not concentrating space. Our atmosphere is concentrated heat and not concentrated space. However at one point the duplicating $k=a^3/T^2$ is overcome by concentration of time $k^{-1}=T^2/a^3$ and the star goes dark. A red giant and yellow stars are all about duplicating $k=a^3/T^2$. That is why they are blubbering liquid heat sloshing as they move along.

From what I gather the red Giant is the equivalent of a gas structure such as Jupiter only containing a lot more space filled fluids. Strains placed on the atom in the Red star, which is just a liquid bowl of contained heat, separating atoms will be precious little. There will be many protons but the protons will be spaced individually where each proton claims individual space securing individual survival since the space dismissed is relatively little and the space duplicated are relatively very much favouring the duplication and not the destruction of the space.

100kg ⇒ 100 kg

Red giant Betelgeuse
Dia. 1400000000 km
Diameter relevancy is
35.2 km.

$(\Pi^2+\Pi^2)(\Pi^2\Pi)3$

A 100 kg in outer space will register as a **100 kg** in the star because the atom will have to release very little space

Consider a sloshing mass of cosmic blubber stretching in diameter from the Sun centre all the way to Jupiter. It is a relative cool supergiant, about 1200 times as bright as the Sun, 24 million times as volumes, 15 times as massive and has a surface temperature of about one and a half times the solar value. The density of Betelgeuse is extremely low, and has been aptly described as "red-hot vacuum" which is a sharp contrast when compared to white dwarfs, which are extremely small and dense. If mass was the factor considered to produce gravity, and by mass I refer to the volumetric amount of material in one container, this was a Black Hole's grandpa. Yet it is a sloshing lump of liquid blubber with less gravity per proportionate cubic meter than any other star. From this it is very apparent size and amount of material does not constitute to gravity. Gravity is seated in density, which is seated in size reducing.

In this star the core has developed considerably concentrating heat to the inner regions. Within the inner regions the heat concentrating will sustain the proton in a manner it will show much more growth within the denser heat surroundings and as the proton grows in stature the proton finds a stronger ability to dissolve space and with that ability it can therefore dismiss more space with the increase in mass as a result. The more the mass the more it produces more massive protons that can displace more space.

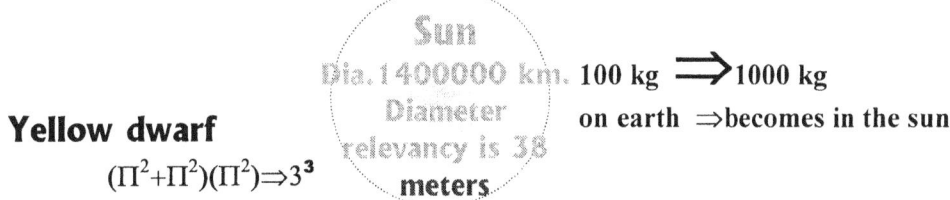

Yellow dwarf

$(\Pi^2+\Pi^2)(\Pi^2) \Rightarrow 3^3$

Sun
Dia. 1400000 km.
Diameter relevancy is 38 meters

100 kg $\Rightarrow$ 1000 kg
on earth $\Rightarrow$ becomes in the sun

In this class star by this time has extended the heat concentration to a level that the star on the outer regions are as brilliant and strong as the electron concentration and the inside centre is starting to diminish any space that can hold heat. The space in the centre is so critical reduced it can no longer allow the flow of light on the inside while on the outside the star is one unbroken solid electron and the protons are diminishing that space also at a rate. From this point the star will become darker as the protons grow more massive finding the ability to demolish even space holding heat so concentrated the heat is pure electrons and photons filling space as material would. It is discarding the electron from the atom. The more massive protons must create much better conditions for fusion.

White dwarf

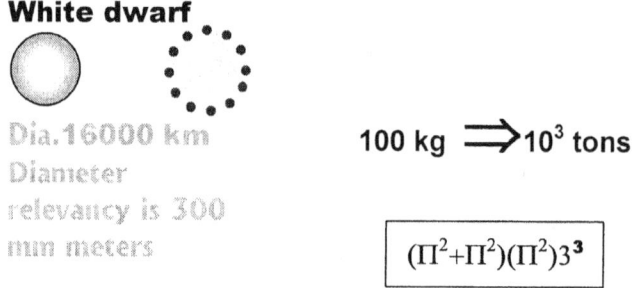

Dia.16000 km
Diameter
relevancy is 300
mm meters

100 kg $\Rightarrow$ 10^3 tons

$(\Pi^2+\Pi^2)(\Pi^2)3^3$

Within the Black hole only singularity remains. It is the single dimension of space less ness where the singularity retained the single unit of all singularity it once contained. That is the epitome where the Universe id heading, the final goal that all stars securing singularity will finally reach and there is no unsuccessful candidates that will float in darkness in space forever and another day. The star has completed the journey and is now in the effort of demising outer space as an entirety.

Black hole
Dia.9.8 km.
Diameter relevancy 1.5mm

100kg $\Rightarrow$ 10^{19} tons

The mass increase is only an educated guess but it is most likely that a mass of **100kg** will become **10^{19} tons** in the Galactica but it is much more accurate than the official number presented as a calculated value.

The Black hole reached the point of singularity. Singularity is a point where **k** = Ω placing **a³** = Ω and also **T²** = Ω. I feel sorry for the mathematicians but this time they cannot take it out the cupboard and play with it for a while. It is too small and only fits into thought. On Earth we may give it a meaningless value of being one molecular mass. At first the space surrounding the atom holding the heat relevancies diminish as the space within the star compromises in favour of heat accumulating. This is quite in coming into place within the gas structures and is starting to apply in the solid stars.

In the star development it is about containing while at the same time gathering. It is about removing space to accommodate cold. The fact of the star is to remove the volumetric size if the star and therefore remove heat as the star reverse the Big Bang growth. The star is counter acting what the galactica does by growing. A star is about shrinking space while space is about growing in size. The only way that is possible in nature is if outer space is still overheating rapidly while at the same time by the same measure is removing size by condensing and that is only achievable if cooling is adopted. The only way a star can reduce its volumetric size is by reducing the size of its atoms because the atoms form the star.

In condition in outer space where the heat is fully intergraded with space the atom has the relevance we experience on Earth. It is as far divided from the Big Bang as possible. The purpose of the star is to bridge that gap and bring space-time back to singularity by securing singularity position within the star. In that the atomic relevance changes drastically from what it is in outer space being $(\Pi^2+\Pi^2)(\Pi^2\Pi)3$ =1836.12 =1 to what the stars space-time functions will allow.

$(\Pi^2+\Pi^2)(\Pi)3^3 = 1674.3$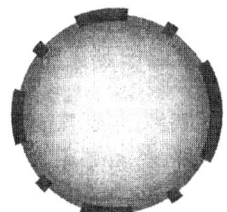

The cyclic period relating to our value that the sun displays is Π because from "hot" to "cold it measures in time about twenty one years according to the butterfly diagram of space and heat contracting and expanding. With that in mind, a cyclic period in the sun is about Π to the outside and therefore we can deduct Π from the atomic relevancy as we experience it to be on Earth and restrict the space occupation to $(\Pi^2+\Pi^2)(\Pi^2)3^2$. With having, a displacement of Π only puts space in the value of the electron acting as a neutron substitute in the electron as part of the neutron function. There is very minor fusion applying in the centre and growth is mainly coming from reducing space by concentrating heat that enlarges the productivity of the proton.

The sun has a relevancy of $(\Pi^2+\Pi^2)(\Pi^2)3^2 = 1753.36 = 5.70 \times 10^{-4}$

Blue giant $3\Pi^2 + (3)^3 = 56.6$

$(\Pi^2+\Pi^2)(\Pi^2\Pi)3 / 3\Pi^2 + (3)^3 = 32.438$ which then is dominating 10Π as well as Π^3 where 10Π form the outer space limit and Π^3 form the inner star limit. At that point, the outer space has no more influence on the star and the star will not emit any light.

Blue giant going dark $(\Pi^2+\Pi^2)(\Pi^2) =$

Neutron star $(\Pi^2+\Pi^2)(\Pi^2\Pi)3 / 10\Pi = 58.4455$

The pulsating star $(\Pi^2+\Pi^2)(\Pi^2\Pi)3 / (\Pi^3) = 59.21$ and space inside the inner sector of the star collapses the dimensional walls of the atom from the proton side. After this, it becomes a new ball game all over again and in this class, there are two separate classes.

After which only the proton action of $(\Pi^2+\Pi^2) = 19.73$

Black hole producing the ultimate relevancy $(\Pi^3) = 1$ or $(1836.12)^3 = 6190178657.$

This spot indicates the existing of a theoretical star that is just a spot of no energy sitting in space too small to observe and too powerful to ignore.

Removing volumetric size can therefore only be about removing the measure the atom has. The atom has certain specific standards, which it holds while the atom is contained in outer space or some feeble place as far as gravity goes such as the Earth. Every part of the atom has a specific function and the function serves a purpose. The first factor of any star is to remove space and space must be equal to heat. The only way any object can possibly grow is by heating and in that expanding. The only way any object can possibly shrink is by cooling into a freezing frozen spot. It is not what we sense that matters in heat and cold but what we come to realise about hot and cold that matters.

In the neutron star the space dismissing reached a level where no heat as such is available any more because conditions inside that star has gone pre Big Bang conditions. The space applying is not sufficient to separate atoms any longer and the reducing of space has limited the motion of space to almost a standstill. There are two forms of motion being very apart. The heating allows the neutrons the ability that Photons had before by dismissing space through motion and in motion the neutron can expand the space by duplicating the space into which it expands. In that manner it finds a way to escape much like the spacecraft does when launched. The protons however reduce all space, which may be inside the atom by discarding the neutron from the atom leaving only room for the protons housed in space less ness

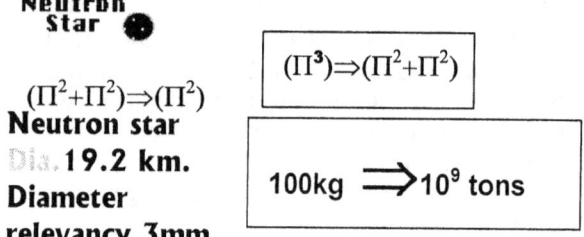

Neutron Star

$(\Pi^2+\Pi^2) \Rightarrow (\Pi^2)$

Neutron star
Dis. **19.2 km.**
Diameter
relevancy 3mm

$(\Pi^3) \Rightarrow (\Pi^2+\Pi^2)$

$100 kg \Rightarrow 10^9$ tons

The initial doubling was possible by consuming and concentrating heat of less fortunate and less movable singularity. That became the second motion where some atomic particles completed gravity by reducing and confirming heat and the other part went liquid and establishes room in which to move with the ability to move. Then came the third phase we call the Big Bang when heat turned into space everywhere. Bu the first was coming as a result of atoms grouping in particles as well as groups that brought maintaining to singularity and other singularity destructed as it melted into liquid. Stars in galactica and atoms in stars then formed as in each case the atoms elected a governing centre and the stars with their elected governing centre elected a centre governing the galactica in quite the opposite way that the stars govern them selves. The relevancy applies because we are attached to the singularity within the Earth. Being connected to the star for instance would have allowed a much different relevancy to apply but then again we would have been much different and very lifeless. I have no idea how any relevancy would be when I find a connection to some other relevancy than the Earth. Seen from a Black hole the Universe will be much different than seen from the Earth and in that we are unable to presume what would apply. It is crucial that humans only be concerned with the relevancies from where humans are allowed to work. Our relevancy would also be much different if we were not under the stringent control of the Sun, which again help us to define our relevant position in space-time. It is a culmination of circumstances that apply to place us in the position that we can value the atom at $(\Pi^2+\Pi^2)(\Pi^2\Pi)3 = 1836.12 = 1$

This spot indicates the existing of a theoretical star that is just a spot of no energy sitting in space to small to observe and to powerful to ignore. The whole position came about from speeds not matching and the larger velocity overtaking the comets lower velocity at a time when the lower velocity was challenging the higher velocity for supremacy. The overheating was the result of the comet firstly not finding enough space to duplicate but secondly by overheating the comet has the chance to enhance its velocity and find a manner to beat the velocity Jupiter excels on space by depleting space. With the magic force applying the comet would not pass Jupiter and would not later find a way to brake up.

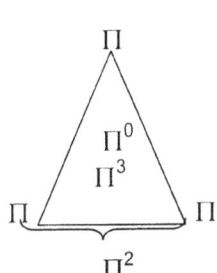

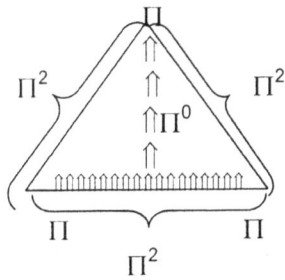

In every atom hides a Universe that remains part of the original Universe that we apparently came from. That which we see is

The reason why I changed Kepler's formula was that I established singularity change existing formulas to match, because I did detect the evidence of singularity within the formula. Although Kepler used different symbols the value of the symbols were dimensional alike. All came from Π^0. If everything came from Π^0 it should be going to Π. That places everything according to Π in one or the other side of the universe in accordance with Π. From that there has to be Π^3 which is the triangle being equal to Π^2 the half circle and Π the single dimensional straight line. Whichever way Π takes form Π^2 would be the norm.

Where ever matter is it will hold $\Pi\Pi\Pi$ relative to $\Pi^2\Pi$ and that makes nonsense of Einstein's Flat Universe, (however I must correct myself) where Einstein wanted to place the universe that he saw to be *fl*at. Where the universe truly is (one or three or seven points away from singularity) within the atom and more correctly where the proton's Π^2 in the double links to time and meets at the edge of the Universe at $\Pi\Pi\Pi = \Pi^3$ that is within the core sector of densified space-time within every atom. At that point the universe goes flat, but that is only because the universe is in every atom and the protons move so fast.

only extending, more a product of what came afterwards and definitely not representing that which was first. From singularity the space-time of which we are only a part is the contact there is between various points holding the original singularity governing the space-time in matching frequency in time. All that are is in singularity.

There would be no compromise with Jupiter having so much stringer gravity than that of the comet the comet with its own application of force would have participated in the collision. In summarizing thus far the following. Mass does not establish gravity. There is no magical graviton. There is no grabbing and there is no pulling of material on each other in any form. Mass does not inflict upon forming gravity but come as a result of motion differences in space secured by material a^3 in relation to space being secured by motion T^2 in relation to a specific centre **k** from where singularity holds the Universe true to form. Gravity produces mass but mass is only the result of gravity applying a single factor of the three that combines to form gravity. Mass does not produce gravity and the means to measure that mainstream science use is flawed not only by argument or concept but in the way science indicate by the formulas they then use in calculating the measure gravity's effect. The reason why the formulas work and they do work well is that on Earth only one of the three factors apply and then the application is in reverse of the cosmos space-time flow in general seeing the manner in the way we experience gravity. In outer space however there is a Universe of difference in the gravity applying there. In outer space there are three factors. One is the duplication of space by motion and the second is the dismissing of space by the lack of motion within the centre and the third is the balance between the two motions.

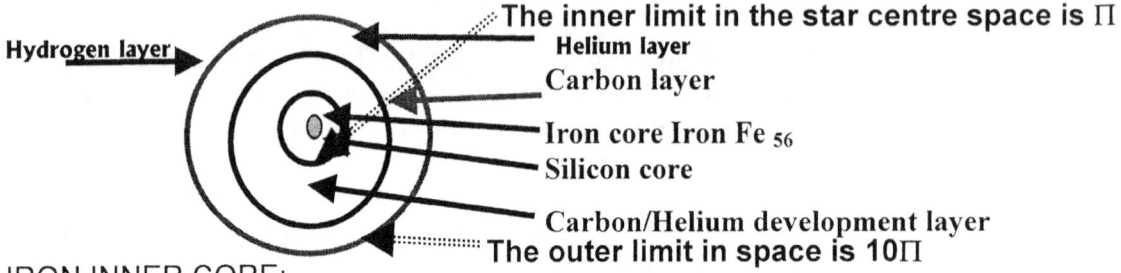

IRON INNER CORE:

Gravity is about space concentrated to form heat, which is stored in motion that produces gravity. Any one not in agreement then convince every one by comparing the neutron star with the massive red giant and be convinced. Mainstream science came up with some cockeyed way of introducing the square of the speed of light to hide the failing of their explaining as to why and how stars evolve. To calculate a Black hole they go and throw C^2 next to the dividing radius and throw the square onto the C that presents the speed of light instead of keeping it at the original position of squaring the radius/diameter as it normally apply. We the mindless shall be baffled when they the Members of Mainstream science sit back and feel smart in the way they manage to once again baffle all of us and the "us" include everyone include everyone with such an exceptional weak mind hat will believe any and every story they can think up with no ability to screen fact from trash with their incorrectness as they are outsmarting us the brainless bunch into some form of accepting as those honourable members cheat once more to prove their incorrect views correct. After all who will ever fly down a Black hole and return to support or deny their calculations. When I suggest the Gravity of the Black hole is a speed because all gravity is speed and speed is space of one particular kind overcoming in relation to the motion of another specific kind versus the time it takes to bring in motion deference's. Gravity is a ratio between space that is displaced or dismissed and space duplicated. It is space over time. It is $k^0 = a^3 / T^2 k$ bringing about space divided by the time effecting the space. Then the speed that light has is gravity. The gravity of the light can be gravity as much as it at that very same time can be antigravity.

What the hell has C^2 got to do with a Black hole because you can pop what ever nuclear device far away from a Black hole and the effect of such a nuclear blast would be at the most and at the worst very much insignificant. The motion the Black hole brings about destroys any other motion coming from the Black hole. The light produced by the nuclear explosion will not even escape form the gravity of the Black hole. When it became apparent to Mainstream science that the radius of stars reduces as the stars develop through progress, someone was supposed to say: hey there is a dead rat I smell. I for one have been on some mission about this matter for how many years now but for my saying so I am regarded as a mindless mutt being the clown in the courtyard having no friends and only foe. The reason why I changed Kepler's formula was that I established singularity change existing formulas to match, because I did detect the evidence of singularity within the formula. Although Kepler used different symbols the value of the symbols were dimensional alike. All came from Π^0. If everything came from Π^0 it should be going to Π. That places everything according to Π in one or the other side of the Universe in accordance with Π. From that there has to be Π^3 which is the triangle being equal to Π^2 the half circle and Π the single dimensional straight line. Whichever way Π takes form Π^2 would be the norm.

BACK THEN when the Universe was new

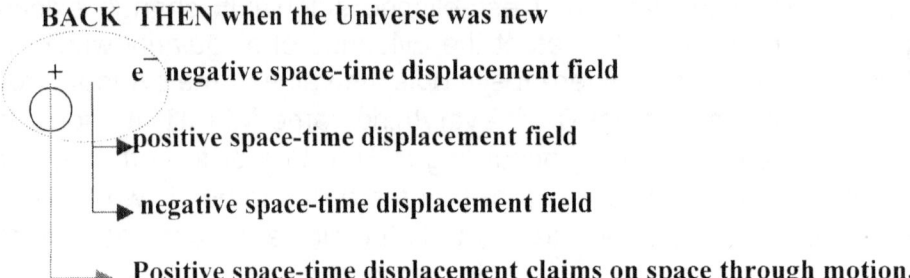

If space outside the atom grew to where we now see the Universe the space within the atom also grew substantially and if a star demolishes the space it has, such a star must then also reduce the atomic space because after all that is what the star is all about, dismissing of all space including the space surrounding as well as the space inside the atom.

PRESENTLY we refer to the sizes we find space has in the sun as quantum meaning they are inexplicably big

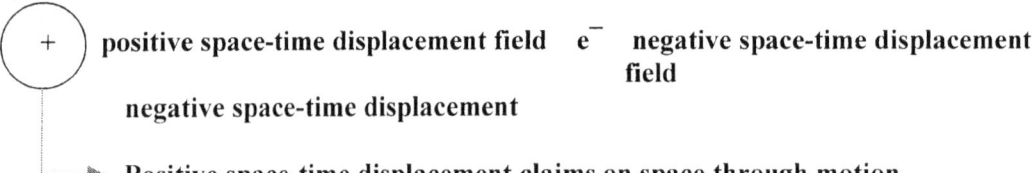

IN FUTURE TO COME they are going to get a lot bigger than the quantum size now present.

positive space-time displacement field e^-

negative space-time displacement   negative verplasing field

⟶ Positive space-time displacement claims on space through motion.

The manner, in which the schematic layout presents it self as follows.

ATOM NUCLEUS   ELECTRON

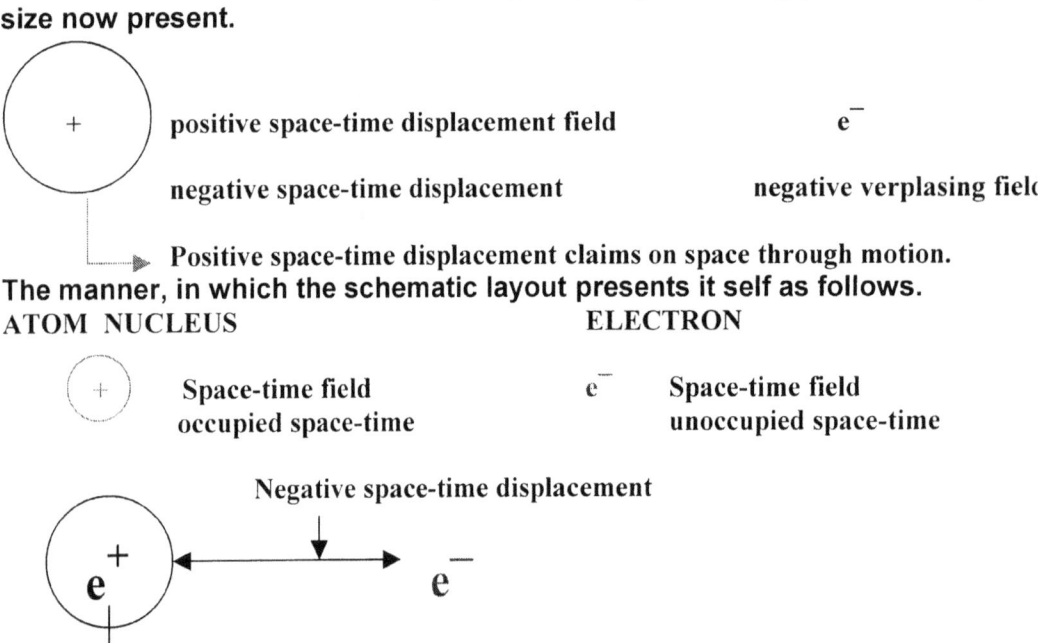

The Universe is in a cycle whereby time in space as well as space-time grows $k=a^3/T^2$. With the aid of strong stars the cycle is swapping in favour of concentration $k^{-1}=T^2/a^3$ as the atomic control goes onto the governing control, however while the growth of material and space is prevailing the atom will gather material within the atomic limits $k=a^3/T^2$ by employing the role of time conserver $k^{-1} = T^2/a^3$. That places the dynamics of concentrating in the atom for preservation while the atom is performing the act of duplicating.

The Hawking Black Hole

It is not only outer space that grows because $k = a^3 / T^2$ is as much the cosmic value as the value within the atom. That means that $k = a^3 / T^2$ is also in place within the atom and that shows the space within the atom grows as the Universe grows because the atom represents the Universe that is in growth. As gravity brings space-time reduction from the centre of the proton so must the growth come from the centre to the atomic proton cluster. As the atom expands in space-time, the proton can also grow dimensionally bigger through the neutron growing in stature. It expands in captured space–time by pushing the electron walls to allow the atom more space to occupy. It is pushing the electron to achieve a distance every time in the same manner that the body lets nails and hair grow. There are three factors of space-time where space-time is released. Cosmic unity and space and heat parted as singularity released the space heat holds by forming motion which produces time to set boundaries and relevancies applying.

The Hawking Black Hole found to be in the centre of some large Galactica, which is generated by the total motion of all the spinning stars, material and mass surrounding the centre with the governing singularity. We also must consider that due to out position we have in the Iron period, we can only see half of the material since the larger stars using more advanced inner core would seem dark to us. What should enable us to form a concept or to form and idea of what constitute Black Hole we can imagine a star where all the mass within such a galactica was in one star. This was long before our individual Big Bang because at this period we find ourselves to be in there were seven Big Bang events going around that forms the Universe we now see. The star we see as a Black hole comes from a time before the iron period we now experience was due. The star then developed as all stars presently do and eventually fused all the atoms that formed the star into one gravity generating monster that absorbs space –time far beyond the speed of light. Having the ability to generate gravity beyond the speed of light puts the singularity core beyond what we think of as our personal Big Bang where our light became visible. In our little Universe we call Earth the atomic relevancy at present is $(\Pi^2+\Pi^2)(\Pi^2\Pi)3$ but in the Black Hole the relevancy already returned back to Π^0. In all aspects of the Universe there are the motionless which forms the role of the solid. We may think of the stars in the galactica to be solid but since those stars form motion around the centre of the galactica all space-time in motion including such stars are liquid as far as that centre is concerned. In comparison to this state of affairs there is liquid and only liquid can move. All stars spinning around the centre holding the Hawking Black Hole is therefore liquid and the motion by principle using the Coanda affect generates a gravitational centre that has the velocity to generate the gravity we will find in the Black Hole.

In every atom hides a Universe that remains part of the original Universe that we apparently came from. That which we see is only extending, more a product of what came afterwards and definitely not representing that which was first. From singularity the space-time of which we are only a part is the contact there is between various points holding the original singularity governing the space-time in matching frequency in time. All that are is in singularity.

From singularity comes the motion and the space we call space-time. Singularity is dimensionless, time less and space less and because of all this features it carries the value of Π^0. By expanding does singularity apply a relation coming about that reform singularity from Π^0 to Π. Only when extending Π^0 to Π the extending creates motion and the motion creates space that then doubles through motion applying which cuts the space in motion in half by matching the space as a duplicate. Motion creates another dimension or another level reforming singularity from Π^0 to Π or from Π to Π^2 or from Π^2 to Π^3.

The motion comes as a result of different motion claiming space within space in relation to individual positions they hold relating to singularity. If everything started off small it must include everything and not part of everything. How big was the atom as a unit when a star was the size of an atom. The relevancies apply all the way through and not just when science needs them to explain certain aspects gone array. Gravity to us in the way we experience gravity being on Earth and part of Earth yet also apart from Earth we find the connection we have called gravity to be $\Pi^3 \Rightarrow \Pi^2 \Pi$. Explained it would read that it would be the space we are in Π^3 that we claim within the Earth is confined to the space the Earth claims Π^2 extending from outer space Π the centre Π^0. But the rules applying to the roles changed a little since Π^2 apply to the motion singularity use to retrieve space and Π indicates lateral shift.

It is not only outer space that grows because **k = a³ / T²** is as much the cosmic value as the value within the atom. That means that **k = a³ / T²** is also in place within the atom and that shows the space within the atom grows as the Universe grows because the atom represents the Universe that is in growth. As gravity brings space-time reduction from the centre of the proton so must the growth come from the centre to the atomic proton cluster. As the atom expands in space-time the proton can also grow dimensionally bigger through the neutron growing in stature. It expands in captured space –time by pushing the electron walls to allow the atom more space to occupy. It is pushing the electron to achieve a distance every time in the same manner that the body let nails and hair grow.

As the atom expand it pushes all space into expanding because it takes the heat from space where the heat in the space retains the growth of the space by allowing the distribution of space go in favour of space and against the density of heat losing relevance. By relieving space of heat the density of s[ace grow by comparable measure to the density heat loses in space therefore the Universe is growing in space by the measure the Universe is losing space

There are three factors of space-time where space-time is released cosmic unity and space and heat parted as singularity released the space heat holds by forming motion which produce time to set boundaries and relevancies applying.'

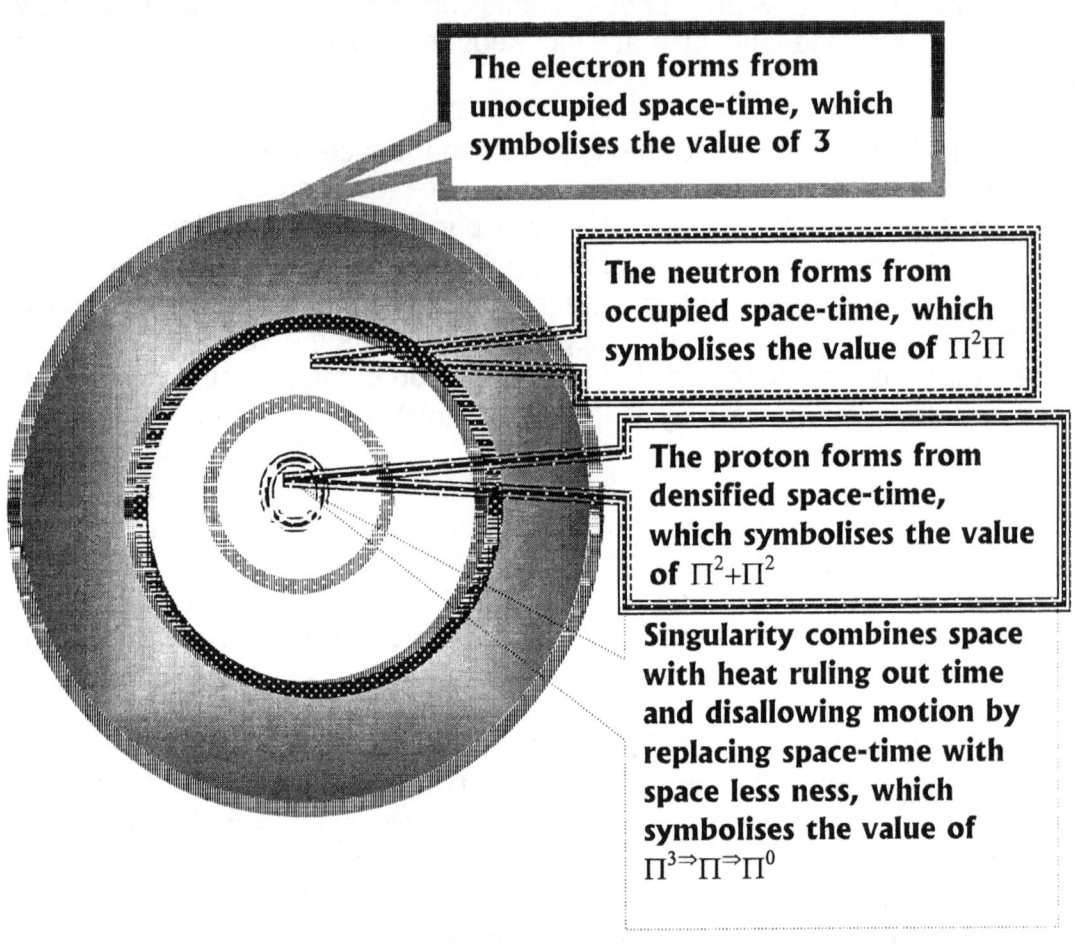

The electron forms from unoccupied space-time, which symbolises the value of 3

The neutron forms from occupied space-time, which symbolises the value of $\Pi^2\Pi$

The proton forms from densified space-time, which symbolises the value of $\Pi^2+\Pi^2$

Singularity combines space with heat ruling out time and disallowing motion by replacing space-time with space less ness, which symbolises the value of $\Pi^3 \Rightarrow \Pi \Rightarrow \Pi^0$

Unoccupied space: This forms the atomic relevance of 3, which is where the ratio of space moves to the ratio of liquid in space. This is bringing motion in contact with **unoccupied space**.

Occupied space: Then forms the atomic relevance of $\Pi^2\Pi$, which is where the ratio of liquid space moves to the ratio of solid space. This is bringing motion in contact with **occupied space**.

Densified space: That forms the atomic relevance of $\Pi^2+\Pi^2$, which is where the ratio of solid space moves to the ratio of **densified space** or motionless space. This is removing motion by disallowing contact with space and then forming space less ness.

Space less ness: That forms the atomic relevance of $\Pi^2+\Pi^2$, which is where the ratio of solid space moves to the ratio of densified or motionless space. This is confining motion in a position being part of eternity.

This forms the atomic relevance of $(\Pi^2+\Pi^2)(\Pi^2\Pi)3$

Every galactica grow individually in accordance to the singularity at what every position in relation to singularity influence in space-time. Stars accumulate heat from space and become dark space less giants. By duplicating the space of any particle sharing space within a larger cosmos structure such as an atom inside a star or a human inside the Earth there are two relations applying. At this point I must indicate that I entirely disagree with Stephen Hawking on the matter of a Black Hole being in the centre of a Galactica.

That cannot be possible since the galactica presents more mass than the entire galactica present with hidden matter included. This I say because the Black hole has gone the full circle and has developed back to singularity while all galactica is somewhere in the process. In the centre of the galactica there is the very opposite of the Black Hole. In that centre singularity is so much in charge it has not yet released space-time but is feeding on space-time form outside its realm of influence. To maintain form the singularity governing is removing space-time from the Universe and charging the yet to be developed space-time within the structure that is still in a state of pre Big Bang conditions. Since that is not the same as a developed Black Hole gone through the cycle material will escape in a very small amount since there is an out flow of space-time too as the displacement will require space duplication that causes ventilation. In **MATTER'S TIME IN SPACE:** The Hypotheses ISBN 0-9584410 –6-5 there are 550 pages of explanations about how the galactica formation is distributed.

Not only does space grow but stars holding space that holds atoms also grows in relation to the space that grows with the atoms growing and so does the inside of the atoms forming the growing stars. How else would space grow if there is no material pushing and converting unoccupied space-time into occupied space-time.?

Stars expand from space and singularity has singularity. The levels but along with the expanding through singularity development claiming heat developing matter in relation to the progress securing value in relation to the Alfa expanding of space is represented on all expanding is the diminishing of space as well and such diminishing shows relevance to growth by indicating reducing of space density and increasing of mass in truly massive stars.

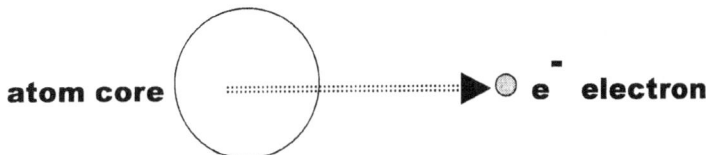

With matter growing the rate the growth would advance the growth in space must be at least $(\Pi^2/4)$ times the growth of the diameter matter holds to singularity.

With outer space carrying the blackness in progressive multiplying, the very essence of space being space within the atom too must be in growth claiming more space. Of all the above factors Mainstream science only acknowledge the growth of space in as Much as calling it the Hubble Constant. However that is not where the growth affects ends because it originates as much from any individual atom as it comes from Alfa singularity. Space does not expand because the space is only reducing the heat in density while producing density in space. Stars expand through singularity development claiming heat from space and developing matter in relation to the progress singularity has securing value in relation to the Alfa singularity. The expanding of space is represented on all levels but along with the expanding is the diminishing of space as well and such diminishing shows relevance to growth by indicating reducing of space density and increasing of mass in truly massive stars.

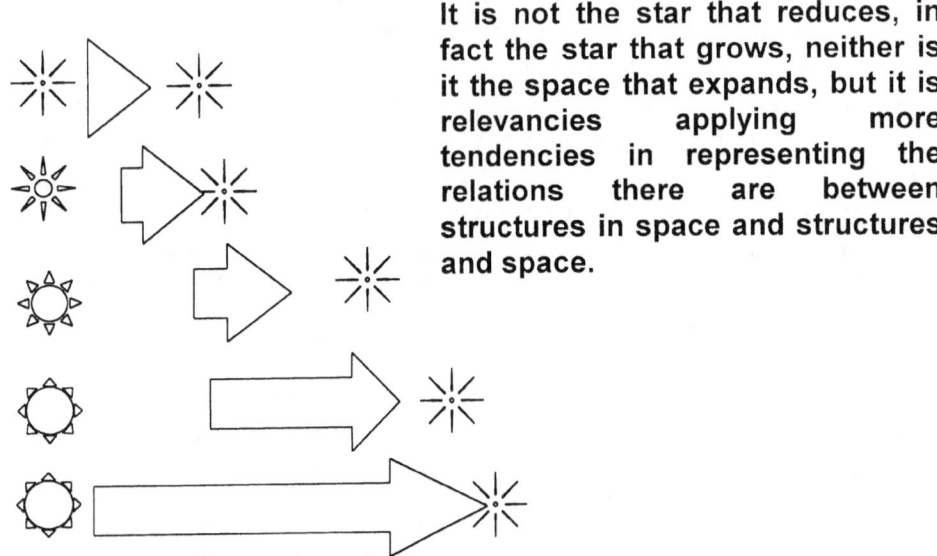

It is not the star that reduces, in fact the star that grows, neither is it the space that expands, but it is relevancies applying more tendencies in representing the relations there are between structures in space and structures and space.

With outer space carrying the blackness in progressive multiplying, the very essence of space being space within, the atom too must be in growth claiming more space. Of all the above factors Mainstream science only acknowledge the growth of space in as much as calling it the Hubble Constant. However, that is not where the growth affects ends because it originates as much from any individual atom as it comes from Alfa singularity. Space does not expand because the space is only reducing the heat in density while producing density in space.

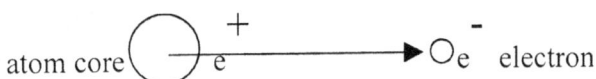

The singularity within the atom grows, which supports the growth of the singularity within the star. This goes on up to a point where the star starts to decline in volumetric size by contraction overcoming duplication. By the reducing of time, heat is gathered in the atom as material and since the time zone called outer space is losing density and therefore becoming bigger in relevance, the atom is doing the same. After all it is the atom that is expanding and not outer space, which by singularity improving within the atom it is shifting outer space in relevance. The heat, which is put in the atom as material by gravity contracting the heat as a cooling substance is in affect time delayed forming heat and is put there to support the ever-growing singularity. Before anyone shouts about my contradicting myself, the singularity within the atom is growing in prominence and not size because size it has not. Once the atoms are gathered in a star and the star singularity in governing is growing, the singularity in governing in the star forces the demise of the atomic individuality by promoting the governing singularity prominence. In size the star is declining but in stature the star is growing. Actually it is very simple because the more outer space seems to have solace the less the density outer space would seem to have. The more space outer space has the more space there would be for matter to be within. The more space outer space has the more space there would be to move about. Most of all the Universe forms a balance and the density decline in outer space has to go somewhere, so it is going where all heat goes…into material to be confined in space by material. The balance is $k = a^3 / T^2$ and $k^{-1} = T^2 / a^3$

With matter growing the rate the growth would advance the growth in space must be at least ($\Pi^4 / 4$) times the growth of the diameter matter holds to singularity.

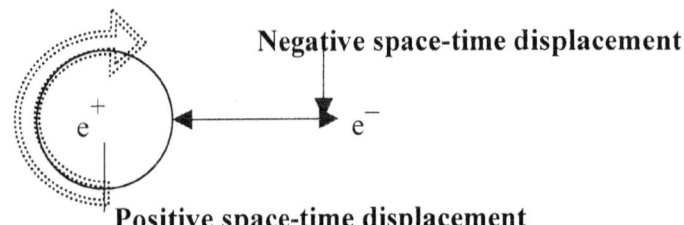

ATOM NUCLEUS   SPACE GROWTH   ELECTRON

(+) Space-time field e^- Space-time field
     occupied space-time   unoccupied space-time

This side of the electron...and...   **That side of the electron**

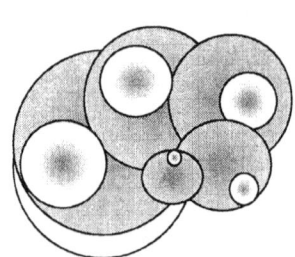

Negative space-time displacement

Positive space-time displacement

Space-time within the atom is the motion that the neutron establishes removing one dimension from the three times two dimension of space leaving singularity extending motion $\Pi^2\Pi$. That too is what the earth atmosphere re-enacts.

The more space the star converts to heat in relation to the most little space the star has to duplicate places untold compensating demands on the atom to release the space the atom claims and hold. The closer the motion of the star duplicates singularity in relation to the position of singularity the more the motion will find an opportunity to remove the occupied space within the atom as it converts such motion responding as mass increasing.

In the first factor of singularity a line indicates direction and hydrogen becomes a volatile product. But singularity also provides space a^3 and duplicates by motion **k** to destroy by rotating T^2. In dismissing the space the proton grows by accumulating space that the other side lost. The growth depicting of the dot I use to symbolize the proton's growth are highly exaggerated I have to admit, but that is only to bring across the idea I wish to convey. Looking at carbon$_6$ one would think that proton numbers would bring about mass, which we then associate with density between particles. But then comes Nitrogen with seven proton pairs and oxygen with eight proton pairs, Fluorine nine and Neon having ten. These mentioned are significantly highly volatile which means they truly extend duplicating of space.

As much as the atom expands in outer space by gathering heat and accumulating the heat as material, the star is taxed to reverse the process. The star removes heat from outer space by contracting, which in fact is condensing space. As it removes space it accumulates density in the atom and that helps the absorbing of heat by singularity. Every atom on Earth holds a completely different role in this because every atom stands different to heat and cold in maintaining singularity. The star may present a singularity in governing that sets rules on how the atom functions within the star but in that sets an environment that the atom finds suitability or not. In a white dwarf hydrogen has no role except to be accepted as just more heat flowing in. The carbon would have a relevancy of 7/10 Π, leaving the hydrogen atom without a feasible Universe. That is the reason why after a Super Nova event the carbon that pores out is so numerable. It is definitely not like Chandrasekhar tried to convince the world that life came from there. Such a notion is childlike stupid. People believing that has the child like mentality that Chandrasekhar find exploitable. To consider life as a natural part of the cosmos indicates towards a clearly defined incapability to understand the cosmos.

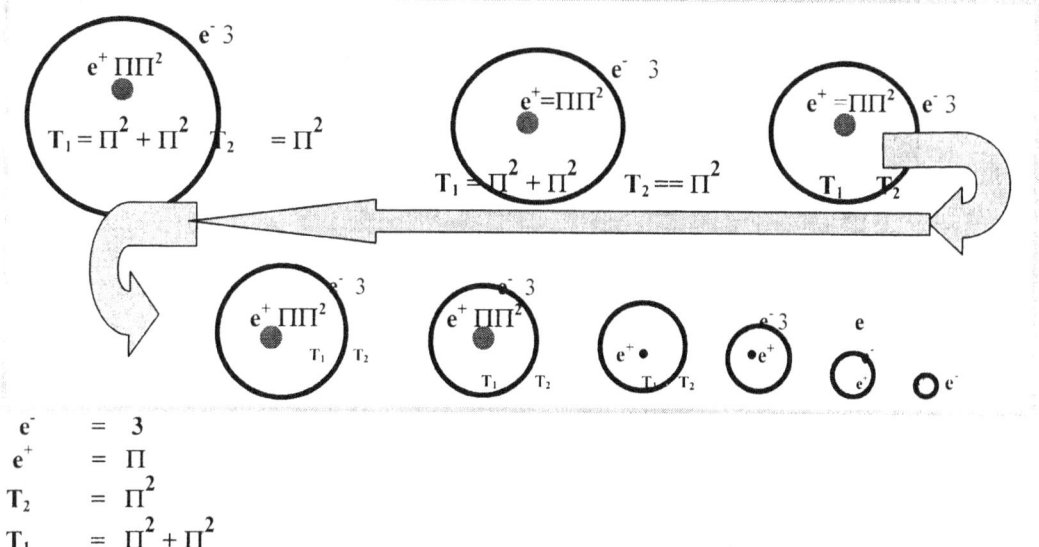

$e^- = 3$
$e^+ = \Pi$
$T_2 = \Pi^2$
$T_1 = \Pi^2 + \Pi^2$

In the first factor of singularity a line indicate direction and hydrogen becomes a volatile product. But singularity also provides space a^3 and duplicate by motion **k** to destroy by rotating T^2. In dismissing the space the proton grow by accumulating space that the other side lost. The growth depicting of the dot I use to symbolize the protons growth are highly exaggerated I have to admit, but that is only to bring across the idea I wish to convey. Looking at carbon$_6$ one would think that proton numbers would bring about mass, which we then associate with density between particles. But then comes Nitrogen with seven proton pairs and oxygen with eight proton pairs, Fluorine nine and Neon having ten. These mentioned are significantly highly volatile which means they truly extend duplicating of space. This as a group forms a relation with heat unlike any other. If mass did the trick these must have been the group having the second least density, but they form the group with the least density as a five point group. The next group holding the Pythagoras five or the Lagrangian five plus two or three, four or five enabling their relation with heat to be quite remarkable. This must be some indication of events during the period just preceding the Big Bang at say 10^{-7}, 10^{-6}, 10^{-5}, the time when the fuel that would ignite the Big Bang turning heat into space turned material into heat. The relation of five plus on, plus on and five plus one plus one plus one are just too uncanny to ignore. Since the star is the total configuration of atoms characteristics the atoms will tell us what we should know about every layer from what is applying in such a layer too what characteristics such a layer would show when it provides the function of what it has to for fill within the star.

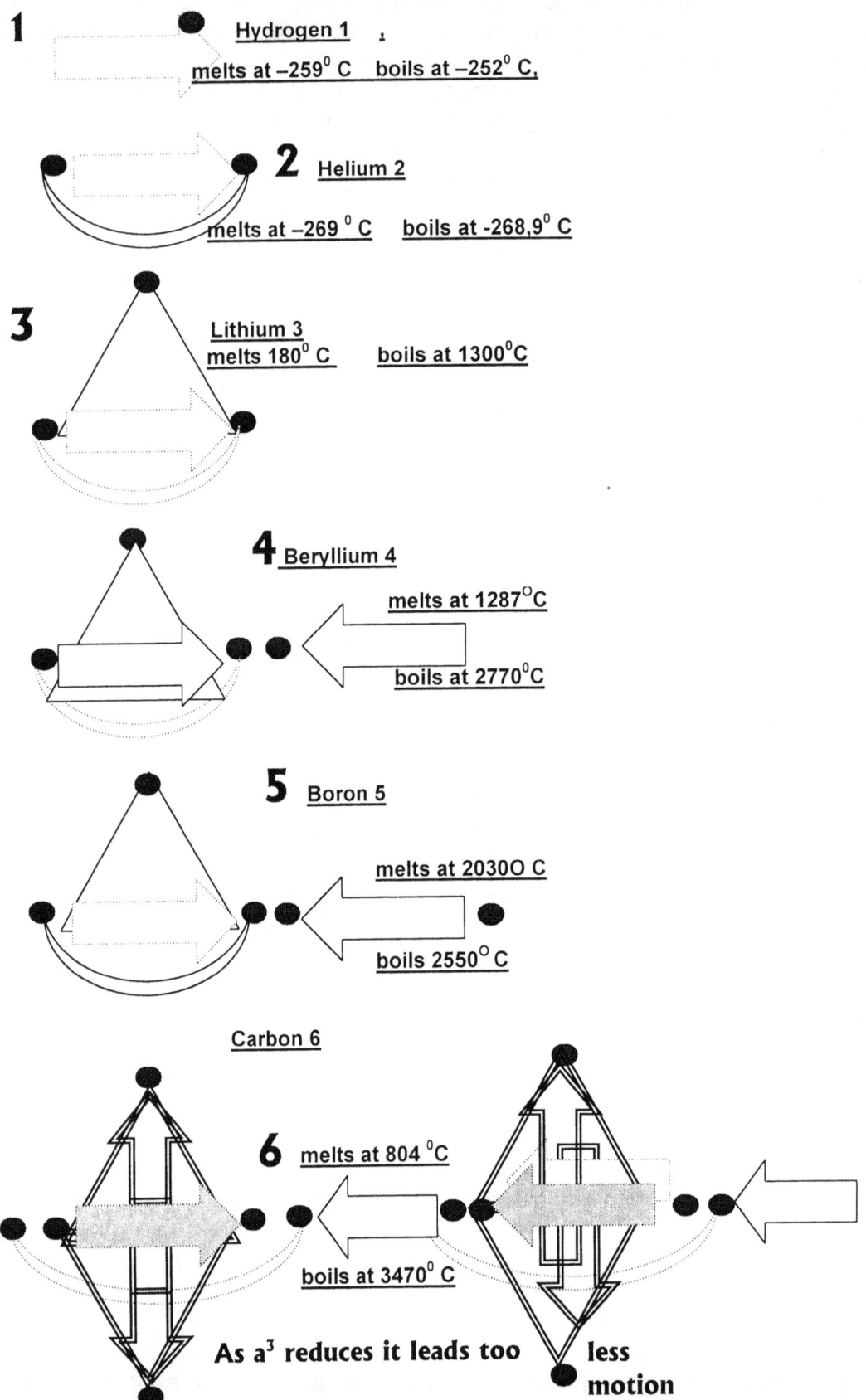

1 Hydrogen 1
melts at –259° C   boils at –252° C,

2 Helium 2
melts at –269 ° C   boils at -268,9° C

3 Lithium 3
melts 180° C   boils at 1300°C

4 Beryllium 4
melts at 1287°C
boils at 2770°C

5 Boron 5
melts at 2030O C
boils 2550° C

Carbon 6
6 melts at 804 °C
boils at 3470° C

As a³ reduces it leads too less motion

In all it is not mass contributing to gravity but gravity establishing mass. Mass has no influence on gravity but mass is the creation of gravity.

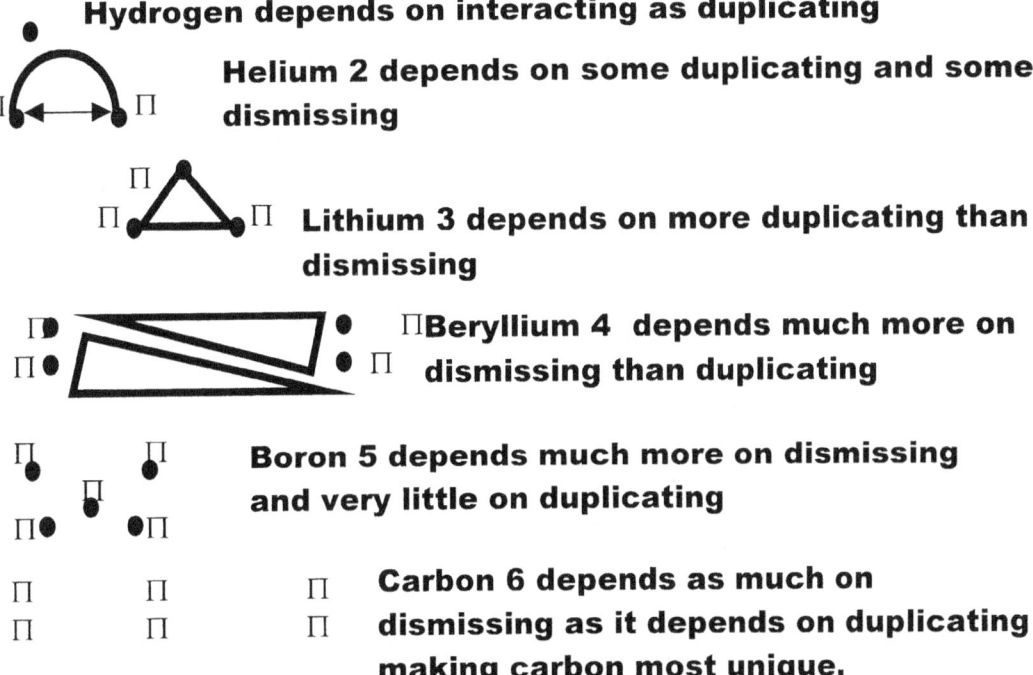

Hydrogen depends on interacting as duplicating

Helium 2 depends on some duplicating and some dismissing

Lithium 3 depends on more duplicating than dismissing

Beryllium 4 depends much more on dismissing than duplicating

Boron 5 depends much more on dismissing and very little on duplicating

Carbon 6 depends as much on dismissing as it depends on duplicating making carbon most unique.

It is the way the atom formed before the atom took on space-time. It is in the formation, that space-time relates to motion. We have some elements being quite massive but also lighter than air and others are quit light but as dense as they come. This can only be a contribution from the way the atom relates to heat, which make the atom volatile (movable) or dense (motionless). Those elements being volatile are also very movable and in that we find the role that such elements play in the star. Stars that are predominantly made up of hydrogen and helium with very slight support from the metallic inner core are those stars that duplicate by producing motion. However the point I wish to press is that mass and being massive and being heavy do not support the fact that some elements have more gravity they produce because their protons are more numerous than others. The fact that mass generates gravity is a myth.

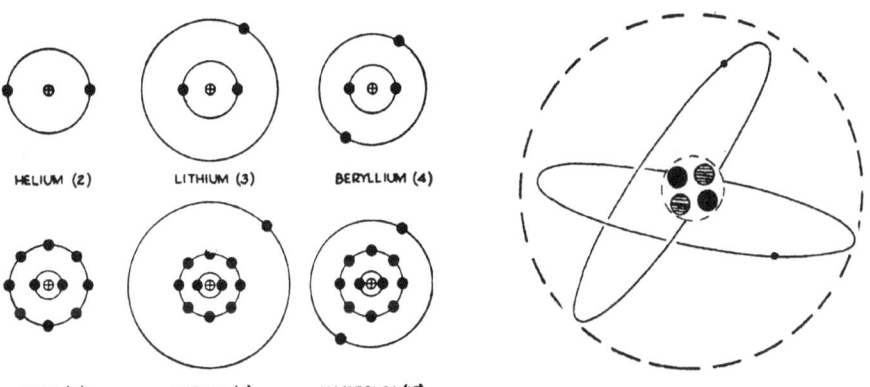

The relation that an atom has with heat stems from the number of protons in the nucleus of the proton cocoon.

One will find that whatever group one chooses there are gasses and there are solids. If mass was attracting mass then the strongest mass must be attracted to the strongest mass and the least mass must float in the air. $F = G(M.m)\, r^2$ hardly can even begin to explain the fact that there is a gas that is more massive than iron but floats in the breeze just as hydrogen which is the least massive element.

Nitrogen 7	melts at -210°C	boils at -195.8° C
Oxygen 8	melts at -218.8°C	boils at -183° C
Fluorine 9	melts at -219.6° C	boils at -188.2° C
Neon 10	melts at -248.59° C	boils at -246° C
Sodium 11	melts at 97.85° C	boils at 892° C
Magnesium 12	melts at 650° C	boils at 1107°
Aluminum 13	melts at 660° C	boils at 2450°
Silicon 14	melts at 1412° C	boils at 2680° C
Phosphorus 15	melts at 44.25° C	boils at 280° C
Sulphur 16	melts 119° C	boils at 444.6C
Chlorine 17	melts at -101	boils at -34.7 C
Argon 18	melts at -189.4° C	boils at -185.8° C
Potassium 19	melts at 63.2° C	boils at 760° C
Calcium 20	melts at 838° C	boils at 1440° C

Ignoring these facts, Mainstream science will hardly answer the problem we do not understand and such ignoring brings strong doubts about the quality and sincerity of science.

Excluding Argon, which is six (carbon's number) times two and suddenly that is a less dense material. The four times five plus... group are the following:

Scandium 21	melts at - 157° C	boils at -152° C
Titanium 22	melts at 1670° C	boils at 3260° C
Vanadium 23	melts at 1902° C	boils at 3400° C
Chromium 24	melts at 1857° C	boils at 2665° C
Manganese 25	melts at 1244° C	boils at 2150° C

Iron being the five times five plus one is the only generator of electricity and therefore the producer of gravity making five times five plus one the ultimate relevancy to heat in reducing space. Still Krypton is much more massive and turns out to be a gas.

Krypton 36	melts at 1539° C	boils at 2730° C
Iron 26	melts at 1536.5° C	boils at 3000° C
Cobalt 27	melts at 1495° C	boils at 2900° C
Nickel 28	melts at 1453° C	boils at 2730° C
Palladium 46	melts at 1552° C	boils at 3980° C
Silver 47	melts at 1412° C	boils at 2680° C
Cadmium 48	melts at 321.03° C	boils at 765° C
Xenon 54	melts at -111.79° C	boils a -108° C

How can science promote their image of establishing honesty when they are confronted by such truths but choose to ignore the truth so long as a lie will bring them some respectability.

Following the process and seeing the influence of singularity should bring about a pattern that may lead one to a pattern of how the required heat formed and how the intended heat transformed to space. Density depends more on proton number arrangement producing specific form in relevancy as to merely and only having mass as factor that contributes to the forming and development of stars in the cosmos. The evidence is so clear that mass has nothing to do with gravity but density has everything to do with gravity. Density is the volume of space in numbers used to fill material in ratio with numbers of space per volume not filled with space. It is matter versus space in every sense there are.

In the first factor of singularity a line indicates direction and hydrogen becomes a volatile product. But singularity also provides space a^3 and duplicates by motion **k** to destroy by rotating T^2. In dismissing the space the proton grows by accumulating space that the other side lost. The growth depicting of the dot I use to symbolize the proton's growth are highly exaggerated I have to admit, but that is only to bring across the idea I wish to convey. Looking at carbon$_6$ one would think that proton numbers would bring about mass, which we then associate with density between particles. But then comes Nitrogen with seven proton pairs and oxygen with eight proton pairs, Fluorine nine and Neon having ten. These mentioned are significantly highly volatile which means they truly extend duplicating of space.

This as a group forms a relation with heat unlike any other. If mass did the trick these must have been the group having the second least density, but they form the group with the least density as a five point group. The next group holding the Pythagoras five or the Lagrangian five plus two or three, four or five enabling their relation with heat to be quite remarkable. This must be some indication of events during the period just preceding the Big Bang at say 10^{-7}, 10^{-6}, 10^{-5}, the time when the fuel that would ignite the Big Bang turning heat into space turned material into heat. The relation of five plus one, plus one and five, plus one plus one plus one is just too uncanny to ignore.

Following the process and seeing the influence of singularity should bring about a pattern that may lead one to a pattern of how the required heat formed and how the intended heat transformed to space. Density depends more on proton number arrangement producing specific form in relevancy as to merely and only having mass as factor that contributes to the forming and development of stars in the cosmos. The evidence is so clear that mass has nothing to do with gravity but density has everything to do with gravity. Density is the volume of space in numbers used to fill material in ratio with numbers of space per volume not filled with space. It is matter versus space in every sense there are. This came about before the Big Bang took place and before space was formerly space and time was formally motion. It was a time when singularity set relevancies moving from Π^O to Π

In that manner we know that that was the way particles formed combinations just after the arriving of moment-Alfa. Singularity brought the Universe but also singularity brought the divisions between the many Universes that followed the immeasurable many Universes that came after the flooding of Universes to follow the leaders. The term "moment-Alfa" is the way I refer to the moment when singularity changed, not when space formed or time began or space exploded but even before anything including mathematics became definitive. At this point mathematics renders it useless. There was no space or time to calculate because relevancies came in place. Form took shape but space there still was not because Π^O moved to Π. Every slightest point in space became an opportunity of establishing a Universe with most different functions and ingredients there might form. This is apparent from the fact that it still takes place at the present moment by motion attaching new singularity through duplication and through duplication releases previously attached singularity from serving the purpose of duplicating by motion.

When the cosmos came to motion, motion was not yet defined. When the cosmos brought about motion, the first motion was relevancies. Cold parted from hot. Eternity parted from infinity. Motion parted from motion absence. Infinity broke the laboriousness of eternity for the duration of infinity. The spot became the **Dot** becoming

The Spot

This was the era of distinction, when separation brought an all-possible new Universe

From what the spot was to what the dot now is might be just a mathematical implication of going from 1^0 to 1^1 but in reality that first motion was the creating of and establishing of an entire Universe with all possibilities now in it. Never again can that much growth become a reality, although to us the growth is beyond what we ever can notice. But it is because the growth is so massive and we are so small that we are unable to notice such almighty growth.

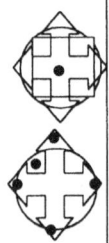

When the spot Π^0 became functional and established all relevancies possible, heat parted from cold as eternity parted from infinity. The expansion was not clear motion but more a parting of relevancies where a centre formed a relevancy because the centre could not provide motion. Without being capable of motion, the centre established four points, which also served singularity. From the inverse square law we know that the centre doubled by producing the four points holding singularity.

By exciting the centre spot, the centre spot came to be because of the heat that formed in relevancy as heat parted from the cold bringing about the division that followed and that was the motion that formed. Therefore the heat had to move but being singularity it could not get singularity to move. In an attempt to establish growth, singularity activated six spots of which four was having motion drawn into relevance four spots that was providing what was to be motion and three that was to be securing the position the centre holds. There were four forming a ring around singularity with two forming in locations we will refer to as above and as below or north and south.

The three in line was in singularity not being able to move but the four was also in singularity and just as incapable of moving. All the points came as relevancies applying the forming of more of what was to come but only the four committed to time were expected to move. The four points that came as a result of discrepancies that became time that produced form and that established the relation with the one but had to perform the motion by expanding was as much incapable of motion as the centre was that charged the four with motion in the first place. As they were incapable of motion, it still required a tendency to apply motion that did separate Π^O from Π. This not only involved form but it involved all relevancies that did come or may in the future come about as a result of the attempt to commit motion. If mass was a factor contributing to gravity the cosmos would have frozen back to singularity without ever releasing singularity to relevancy.

Mass does not establish gravity. There is no magical graviton. In the beginning there was no mass but boy was there gravity! The only means that the cosmos could find a way to break from the grip of eternal eternity was to expand into relevancies. Such a feat can only go to task by forming opposing hot and cold. Becoming hot produces more of what is heating. That implies motion or a moving away from where it was by generating more of what is available. Only where hot released from cold could whatever was repeated once again and duplicate what was before into what then is more. Secured by motion T^2 in relation to a specific centre **k** from where singularity holds the Universe true to form. The **k** was an intention to place apart and by today's standards will not even qualify any noticing.

All that are is in singularity. From singularity comes the motion and the space we call space-time. Singularity is dimensionless, time less and space less and because of all this features, it carries the value of Π^0. By expanding, singularity applies a relation coming about that reforms singularity from Π^0 to Π. Only when extending Π^0 to Π, the extending creates motion and the motion creates space that then doubles through motion applying which cuts the space in motion in half by matching the space as a duplicate. Motion creates another dimension or another level reforming singularity from Π^0 to Π or from Π to Π^2 or from Π^2 to Π^3

As said before we now know Π came about since Π is achieving form and not space. Only **r** can establish space as size will accumulate and as it had with everything else singularity had **r** covered by one as in being $r^0 = 1$. By reducing the circle radius **r** by half continuously will lead to an infinite small circle and an infinite number holding r would place **r** to the power of one as a factor. Then as a factor **r** would not contest any change when change is introduced into any future equation but Π will remain because the circle as a form remains even being infinitely small. By reducing r indefinitely to the tune of half each time, r would become infinitely small, beyond human calculating means, however as mentioned in the case of the smallest dot holding one spot, r would become insignificant beyond human comprehension even, but never reaching zero and still Π would remain intact and dictating form. To amplify by dimension a value has to be set to r but if r remained covered by singularity all alterations that could possibly come about was in the form, which was Π.

This expanding can be a problem one can wrestle with for one lifetime and never reach any conclusion. How can something grow without getting more that what was before? Then it hit me like a ton of bricks. The answer is in heat but not heat, as we know heat. It is heat in getting relevancies between outer limits. Only heat could break the monotony of singularity. Heat in the form we now know heat as heat is now. Since the Big Bang heat is material transforming from one state to another state.

The change that took place involved singularity but singularity was 1^0 and being ^{01} could not grow. The growth came about. Heat rose from singularity, but if heat rose from singularity. Singularity as a factor changed from 1^0 to 1^1, which means a relevancy came in place that no one could detect. It is true that 1^1 are still one, but one could then escape from singularity by producing factors other than 1. Heat came about but only as a relevancy to utter cold. If there is heat, there is cold or if there is no heat there can be no cold. Space came into forming a relevancy that brought form. Since it is a relevancy and not a generation by accumulation, the form produced was Π. The spot formed a dot by heat and cold establishing relevancies and from that singularity was broken to allow all other forms of relevancies to come about. The cosmos did not start because of gravity. The cosmos started with heat and cold coming into a relevancy and in the cosmos there is no hot as much as there is no cold. The cosmos broke, put from the confinement of singularity by establishing a singularity in a relation of heat and cold. The heat that came about was beyond measure because the cold that held the heat was also beyond measure. The immeasurable heat was on the outside of the dot that formed and the cold was on the inside of the dot that formed. The cold contracted because in nature cold contracts. The heat expanded into a dimension of form and heat by expansion is in nature about motion. Motion is duplicating that which is and heat is what is duplicating by motion. But only heat by expansion was possible because in affect singularity cannot move. The motion became contraction, as the motion was the result of heat expanding which was forming four points in the rim of the dot. The expanding of the points created motion in relevance of a centre that formed because of the motion, which established an immovable centre as the Coanda effect, placed more dots in relation to more dots that formed

Every dot was Π and every dot formed Π^3 because of the expanding heat, which produced Π^2. With that a new relevancy came about forming a centre in between the four points of expansion that was resulting in time. But since the points were in themselves singularity, which is immovable and space-less, they still heated forming a cold centre with the heat bringing about motion. It became a repetition where infinity broke eternity by producing a centre because of space (or rather form) forming the motion to enable the space to form in relation to the heat applying motion. This brought about a Cosmos being conceived.

The spot forms a full circle, but the line running through the circle is forever present because that is the future radius of the circle that will one day develop the circle, which is equal to the present diameter. The fact of the presence of such a possible line in such a possible circle dividing the possible circle into two parts makes the centre line equal to the half circle. The line forms the half circle but not only that the line presents the half circle as much as the line is the half circle. The line then is 180^0 and the half circle is 180^0 because in singularity the two factors are the same.

The same value is of course $\Pi^0 = 1$. The issue of concern is o understand that singularity cannot move. Singularity has no space. Singularity is no only part of the Universe but singularity is the Universe. By establishing motion singularity has to be charged with the time delay we find space to be. The space is time taking a period or a duration while moving from one singularity point to another singularity point while conducting the heat and the accumulation of heat that built up due to the retarding of the time to conduct the heat forms the space that is conductor to bring about the motion of the space.

It takes heat time to entice singularity and singularity can only entice. Singularity cannot move and neither can singularity form space. By enticing from one relevancy to another there is a bridging of heat that has to be crossed in order to send the gravity or the enticing or the relevancy to depart the space and reconnect the space to the next singularity. Bridging all the accumulated various time delays that formed an accumulation of heat through time distorting brings us the space we see and have. However there is no true space or motion but it is eternal motionless space is singularity charging time to provoke heat into forming space.

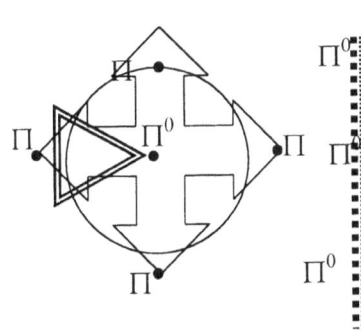

Three points formed a line covering singularity where the centre singularity recovered heat to grow and two points served as an axis to allow the rotation and to assist the duplication. There is one centre connecting the duplication of three as well as the recovery of one (the fourth one) that is applying the tie aspect. Therefore, motion consists of three positions in relation to a centre, which forms as space in relevancy to the motion and the space receive a controlling centre.

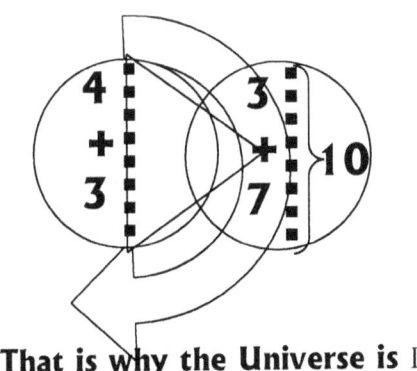

That is why the Universe is Π

The duplication comes about as singularity is exciting another singularity in precise relevancy of 3 to 3 to 1, but the points charged is as space less and as motionless as only singularity is. The heat it requires to carry the exciting between points forming space and the space excites heat and the time delay it takes to excite singularity between points forms space-time.

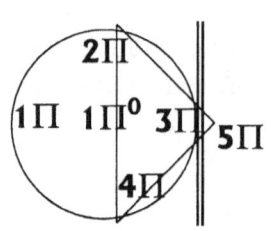

Where motion conducts electrical charging which is equal to gravity the charging of motion is to entice duplication of singularity. This is the basis, the heart and the sole ingredient of the Coanda principle that includes the Roche limit ($\Pi^2/4$). The charging of gravity $((7/10) + (7/10)) / (10/7) = \Pi^2$ and the charging of space-time $\Pi^3 = \Pi^2\Pi$ is all due to the relevancy brought on by the Coanda principle. The value of motion came from singularity exciting singularity and that is the duplication while the duplication or motion presents the space.

The development came into eras as the relevancies brought about new relevancies that spawned even newer relevancies that all remained in touch with the original singularity centres. Every one focused a new time delay that eventually brought about space and every distortion of time brought more. That concentrated between singularity points that charged the points to form space. When the charging became overdue in some sectors it erupted in forming the Big Bang. By the time the Big Bang erupted there was such a huge backlog in heat and time corrupted and delayed the next result was the employing of space as a commodity in the Universe. The relevancy was C the gravity was C^2 and the space was C^3. That left what was inside atom still spinning faster than the speed of light applying the relevancy of **k** = C where the electron applied the relevancy of $\mathbf{T}^2 = C^2$ and that formed the atom which then became the cube of the speed of light $\mathbf{a}^3 = C^3$. That left the atom at the relevant size of what the speed of light permitted at the time but since the Universe from that the relevancy expanded as the Atom grew in space to the extent it has now. The purpose of the star is to recapture the space the atom grew into and from there dismiss the space by spinning faster than whet the speed of light will be on the outside of the star.

Before the Big bang the lot was form without dimensions playing any part. It was the point lining up and forming positions that was spinning faster than the speed of light can ever achieve. However every point today still serve the role it took on at that stage and serves in the position that it had during the time it had no space with eternal time.

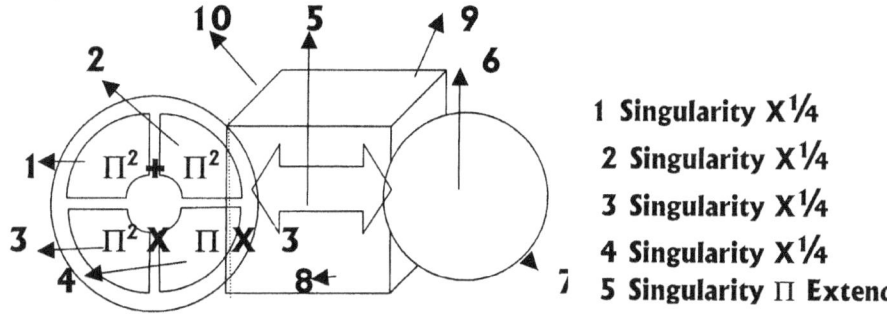

1 Singularity X¼
2 Singularity X¼
3 Singularity X¼
4 Singularity X¼
5 Singularity Π Extend

1, 2, 3) Singularity and supporting
(4) Time
(5) Space -time
(6) Matter
(7) Formation
(8,9,10) Dimension

During this cosmic period of developing singularity by overheating the relevancies that still dictate today came about. The relevancies form a chain that cannot be broken and apply through out the Universe. Material came in the place when the heat delay compacted to a point the compacted product formed an excluding zone of individual proportions. But as the time-delayed points formed each had a role in regard to the others and that today we call atoms. Every atom secludes its Universe by motion that is feeding, preserving and excluding the Universe maintained from all other Universes maintained.

This form came about when only form was present in the cosmos. It was in a time era where form featured in relevancies that would lead to one day becoming the atom. The atom forms a dual purpose of duplicating as well as dismissing and some prefers the one better to the other. This relevancy came in place when time was not time and space was form. Time is forever eternity being interrupted by form in infinity to bring about eternity ticking as infinity ticks. Before that singularity took on stages in forming relevancies between duplicating and dismissing space-time, which incidentally was not yet truly space-time in the sense we think of as space-time. At first a dot moved from the spot leaving the spot but taking with the spot as part of the dot to remain in the dot. The two never separated but the one allowed the other to be.

As the dot confirmed a discrepancy between infinity and eternity by defining infinity as an interruption of eternity cold and hot parted a union.

The dot that formed was not space but a relaying of time to form a new point of singularity where eternity was interrupted by infinity. Time took form from 1^0 to 1^1 or from Π^0 to Π. It brought form into differentiating between interrupted eternities with infinity doing the interrupting

Then a true distinct relevance came about that positioned a time differentiation outside the realm of time by four. In this realisation we can assume that space had some meaning at this point and the formula used to investigate suggests just that.

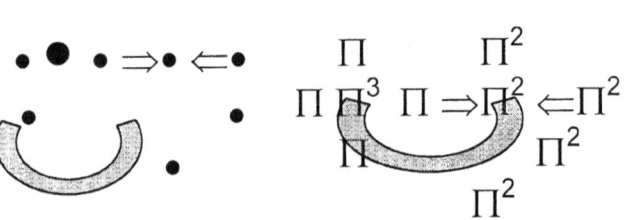

Even in grouping, there are characteristics, which make a certain group of atoms more perceptible to duplicating and others more perceptible to dismissing

The lagging of exciting one point in relation to another point takes time. It takes time to send the message across to get singularity at that point excited. It takes effort to bridge from the dominating singularity to the independent singularity and that effort slows time down. The crossing of the divide is space formed by pushing time into duplicating. When time brought in a five points to the four points it took time to be, that fifth point became more than only form, it became space because it was one point outside the Universe of four or of form. One must see the three points established as motion duplicating singularity in relation to one dismissing singularity. This always has to strike a balance in order to establish space-time. It began as a relevancy and developed into space-time flowing or space-time displacement.

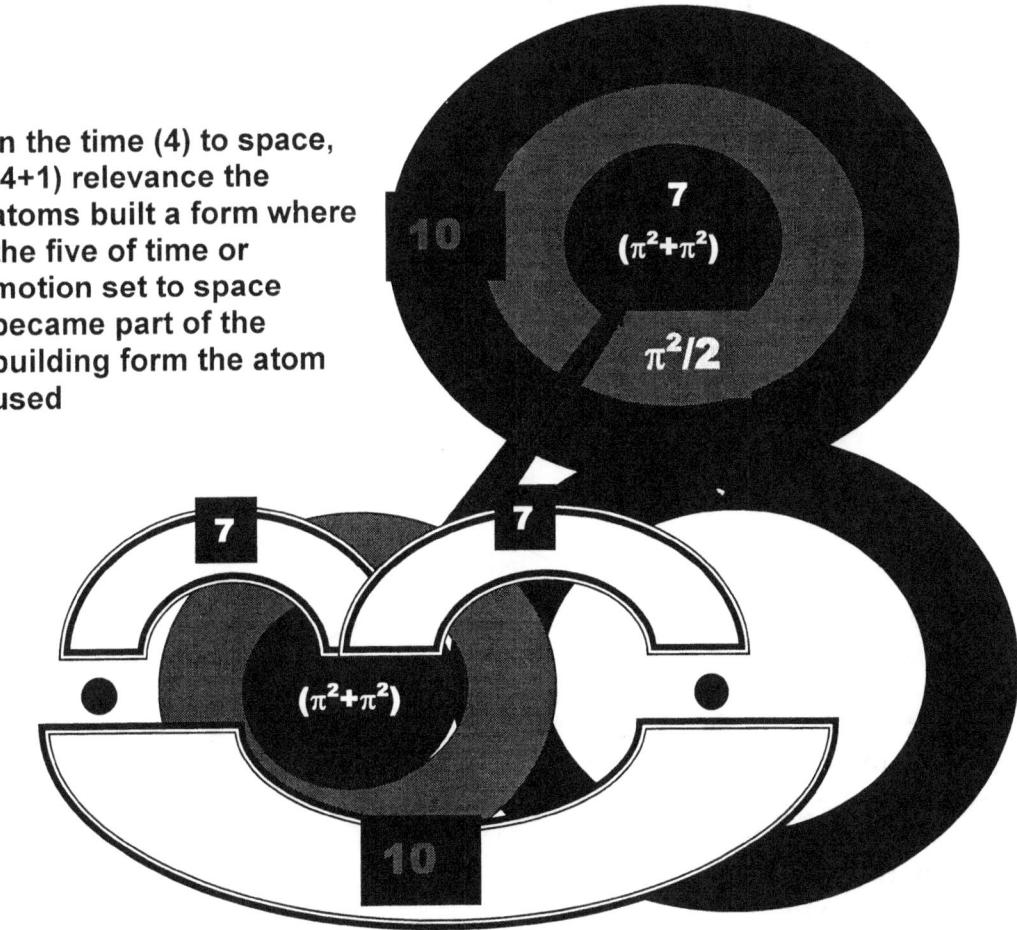

In the time (4) to space, (4+1) relevance the atoms built a form where the five of time or motion set to space became part of the building form the atom used

What the Coanda effect proves is that the rotating motion is acclimating a centre that exemplifies all phenomena in nature as we use nature to our advantage. All of nature including gravity uses the same method of motion forming around a circle in rotation and in the centre of the circle a point of no motion holding no space comes about. This is what Kepler taught us when he taught us $a^3 = k\, T^2$. With the Coanda effect forming the basic principle of all natural phenomena we can see from that, that the motion of liquid in the presence of a solid forms a centre that excites as it establishes singularity. From that rotation, space flows to a controlling centre but because of the lack of motion in that centre, there is a lack of space in that centre. Therefore, there is proof of a flow towards such an established centre and there is control from that point of singularity. In every case, the singularity controlling space-time sets standards for space dismissing in relation to space duplicating.

The duplicating stands in regard to the flow that the liquidity of the atom in relation to the solidity of the atom can reproduce. This forms density and mass but mass has little influence on the scenario.

There is a balance between the duplication in relation to the dismissing of space and the relation extends to the number of atomic elements present which then creates the balance applying within the star. As the liquid heat subsides in the centre of the star and the heat density is dissolved by the dismissing-prone elements the motion or moving ability of the star as a unit fades away as the star becomes static and solid with less space providing the star with less motion.

Since the star is the total configuration of the atom's characteristics, the atoms will tell us what we should know about every layer from what is applying in such a layer to what characteristics such a layer would show when it provides the function of what it has to for fill within the star.

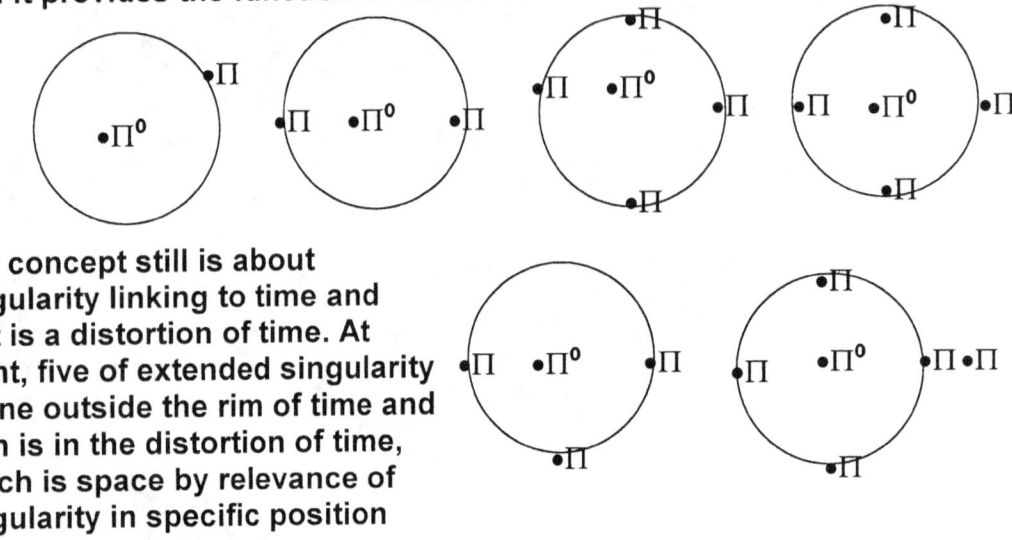

The concept still is about singularity linking to time and that is a distortion of time. At point, five of extended singularity is one outside the rim of time and then is in the distortion of time, which is space by relevance of singularity in specific position according to time.

Let us investigate and try to find a way by using logic how a star applies gravity. Therefore it is not the number of dots that is important. It is not the size of the number of dots occupying the position or the size of the space the dots occupy that is prominent. It is the relation in the dismissing of space and the duplicating of space that becomes important. The less space there is the more the favour will be to reduce the space because of the advantage the dots have in securing space-time that will prevent overheating. On the other hand the more space secured will also prevent overheating and therefore those will opt to duplicate space in order to find space to secure and prevent overheating.

Since the Earth has no singularity demand that is much better developed than the Universe sustains, we find on Earth a relevancy of Π to $(\Pi^2+\Pi^2)(\Pi^2\Pi)3$ is adequate. But in bigger units the space-time displacing relating to space duplication presents much more demands on atomic structures occupying space within the star containing through set boundaries. In the presumed to be bigger stars there is much space filled with atoms occupying much space. In the stars more massive but holding lesser space the atoms must also hold lesser space but they also hold more protons by number in the lesser space.

The space the particles hold is directly in relation to the particles the containing structure duplicate. The more space that is relevant to the structure that the star duplicate by motion is then in turn once again relevant to the space the structure destroys by proton action in space less units. The more space the particle claims in relation to the space the container holds that relates to the space the container duplicate is relative to the space the containing structure destroy. From that mass derives value. As individual occupying space the atom is an individual container by own merits and as such duplicate space in this regard within the specific confinements of atoms.

This we will classify as normal applying structure values the atom has in outer space or in structures with very little atmosphere. Please note there is no pressure involved because the motion involved creates conditions naturally instead of unnatural pumping that causes pressure. Pressure is an artificial creation as part of life but has no role in the natural cosmos. Pressure is a condition where the retaining of particles has to be confined in a patrician made of material where the outer wall does the retaining of the substance within. This obviously cannot be in a star because the "pressure" is regulated from a condition applying and space-time controlling inner centre that needs no solid walls to contain whatever is inside. With that one can see there is a Universal difference between the concept of pressure forming due to human action inside a container and what comes about as secluded space-time within a star.

As the demand of singularity in such units grow stronger some relevancies within the atom come into play and I developed a system whereby I can arrange the space-time merits of space-time curtailing within the confinement of the star borders applying in the star to place such a demand in relation to singularity where the ultimate demand sets the standards.

In the Sun, for instance, which is a minuscule small star a relevancy in the outer region might be 3^3 relating to singularity and with the atom having, a sustaining displacement of $(\Pi^2+\Pi^2)(\Pi^2\Pi)3$ there is no danger of the atom demising. The electron in the Sun will have a diminishing factor of 27 whereas the atom can sustain $(\Pi^2+\Pi^2)(\Pi^2\Pi)3 = 1836$. The relation in the atom degenerated by 27 leaving the atom a sustaining value of the electron plus the neutron applying space-time without involving any of the neutron aspects at all. That is the mass of the space-time that the electron will consume in the space reducing flow of space-time.

The flow is the result of heat distributed where the heat is delivers to the dismissing sector and producing of the duplicating of space by mass within the star that then forms a favouring of duplication in comparison to dismissing. The star is a bright little boy shining by dismissing pebbles of light-photons into space. When a demand on space-time displacement reaches an accumulated general displacing or movement to the value of what 56.6 protons can achieve in a general flow of conducting space-time that would be the requirement for such accumulated displacement within that space forming the motion of the space or the time aspect of space.

The star accumulated more heat by consumption applying direct dismissing without accumulating space-time in liquid form beforehand therefore there is no heat remaining to dispose of by producing light. When the general displacing flow of space-time within that sector of the star or the star in total reaches 56.6 displacement the natural state of absolute solidifying becomes the norm within the star. From then on, the star will exclude all electron functions and stop shining as the demand on space-time duplication and diminishing reduced the atom to space without a heat envelope that will be electrons or a liquid/gas jacket. Only the nucleus will be able to sustain the diminishing and the reducing of space by increasing of time. The entire star becomes a solid structure by reducing space-time directly freezing the space-time from a gaseous state to a solid state. By motion, speeding up the tempo of the flow of space-time the liquid state of space-time is by passed going from gas to solidity in one motion. The atom would shrink to such little space it will have space within the star that only the centre nucleus will fit. More reducing by applying motion in creating space differentiation will leave a star with so little space the space will be insufficient to secure a position for the neutrons and the star will then have the name of being a neutron star. Going even further will find the proton rejected from the star.

Every atom holds (I am guessing), as many dots as the Sun has subatomic particles per atoms and that would still be a very conservative guess. Every dot is a controlling centre selecting a regional centre where every regional centre selects a centre. This goes on as long as there are spots forming groups as individuals unable to survive independent. The others that was unable to group formed heat that became space, which became the broken dots. The dots form groups to survive and as a group, the survival depends on doing what the group has to do to remain cool. In another book, I reserve one chapter to explain the phenomenon what I called the Lagrangian atom. These dots arrange in a manner that they could favour either the space duplicating aspect or the space dismissing aspect.

This can only be the result of the fact that even in the case of the Sun, the inner space is almost entirely liquid heat and the liquid heat produces sufficient space to dismiss as the centre that holds the heavy metal particles, where all the dismissing is done. The liquidity provides motion while the solidity removes motion in the centre of the star. The dismissing going on is in the space factor where the space leads to a denser heat within that space because there are insufficient material to accommodate all the heat by the dismissing factor T^2. In that case motion far outweighs dismissing $k>T^2$ but a time comes in every star that the dismissing takes absolute charge. $k<T^2$ That is when the star goes dark. The Earth is mainly about duplication of space much more than dismissing of space and so is every structure in the solar system.

I would suggest we think of stars in the following terms. A star that generates and transmits a lot of light is weak on gravity because their progress started recently. They command a lot of space-time but the demand they have to keep their cooling acceptable is very low. In that they can generate a lot of light but with the demand on cooling low and the gravity in the centre not very developed, those stars cast a lot of light back into outer space. It is just because of the size the stars hold that tell the that the stars are still young and have a weak developed governing singularity. The stars will have very prominent hydrogen and helium layers, with the inner core not very prominent. The control of the star is still very much in the individual atoms and in that the motion the atoms have to produce in order to maintain their individual singularity will only come about through motion. The atom has to make contact with as much space-time through motion as possible since it has a very poor ability in contracting space –time in support of the cooling system.

I would suggest we think of stars in the following terms. A star that generates and transmits a lot of light is weak on gravity because their progress started recently. They command a lot of space-time but the demand they have to keep their cooling acceptable is very low. In that they can generate a lot of light but with the demand on cooling low and the gravity in the centre not very developed, those stars cast a lot of light back into outer space. It is just because of the size the stars hold that tell the that the stars are still young and have a weak developed governing singularity. The stars will have very prominent hydrogen and helium layers, with the inner core not very prominent. The control of the star is still very much in the individual atoms and in that the motion the atoms have to produce in order to maintain their individual singularity will only come about through motion. The atom has to make contact with as much space-time through motion as possible since it has a very poor ability in contracting space –time in support of the cooling system.

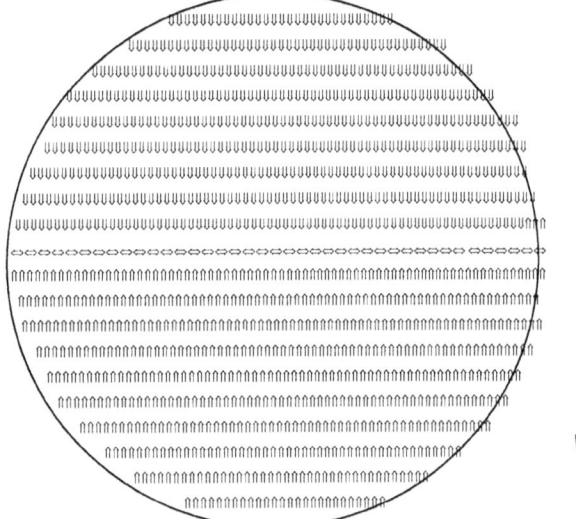

The entire motion and the entire contraction of every atom culminates as one effort and this produces a single combining effort which is then displaced to the center of the sphere of the star where singularity is normally nurtured as a result of the shape of the sphere

The contracting action is at present the only part of gravity that Newtonian science credit as gravity. There is a lot more to gravity than such simplicity. Every atom in a star is pushing the atom in front by filling the space the atom in front vacated. Every atom in front of every atom behind is pulling the atom behind as the atom behind is urged to fill the space that the atom in front vacated. That is motion, which is the most complex issue one can find in the Universe. Since every atom is driven by singularity and no singularity are able to move it bring about that every singularity must remove and rebuild the space every atom fills or vacate as the atom moves along. There is a building of an entire Universe going on in every split second and this split second is so fast we cannot name it. By naming it there will be so many time units gone by, by the time we said the name, the Universe might not even be recognisable. We might call it energy but I hate to call it energy because energy is a lot like Holy water. It can come from anywhere and you can use it for everything and in the end it does not even become something durable because its use eventually comes to nothing.

Each atom is demolished and then rebuilt again in the space it occupies in the time it occupied the space and because singularity cannot move singularity is charged with the constructing of the space-time every time the space-time requires a new position. This we accept as motion. This is gravity. This is so intense the human mind will never realise the true magnificence of the process. As the star develops the star sacrifice expendable individual atoms in creating one centralised governing singularity that becomes charged with all gravity and cooling. The atoms bond individual proton clusters up to a point where the gravity within the atom find the space-time unable to support the three dimensional structure. That is where Einstein's Universe goes flat! With the governing singularity becoming absolute then somewhere in the centre a time arrive when the centre find the support it requires to dispose of space-time totally within the centre. In this the governing singularity slowly takes charge of the star. However since singularity disposes of space-time that inner core dissolves light directly in order to maintain competence in securing the cold the star requires in soldiering on. The more dominant the centre singularity becomes in governing the entire star, the less the role would be of atoms maintaining individuality. At a point the need for motion becomes quite little and the star dispense the helium and hydrogen layers in favour of a building of carbon and other heat preserving elements.

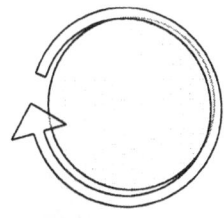

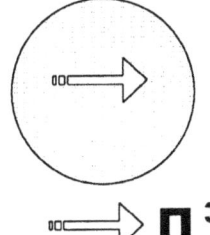

Π^2 this factor by the square proves to be the motion in charge of producing the gravity or time aspect.

$\Rightarrow \Pi^3$

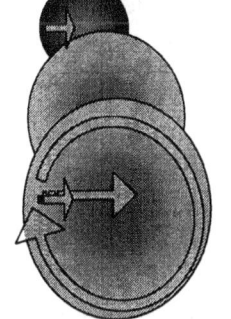

The time aspect Π^2 in relation to the relevance Π of the centre establish a confined and identifiable space Π^3

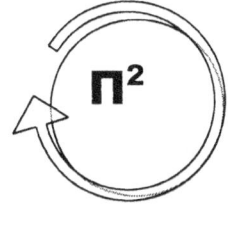

It was when the centre retaining ability became unable to match the motion and the relation between containing and motion was beyond control that the Big Bang produced space as we think of as space. The relevancy of Π increased to C, (the speed of light) which placed gravity at C^2 and the entire Universe at C^3, which is the value of the electron. That act put the Universe in 3D as it is today.

It was when the centre retaining ability became unable to match the motion and the relation between containing and motion was beyond control that the Big Bang produced space as we think of as space. The relevancy of Π increased to C, (the speed of light) which placed gravity at C^2 and the entire Universe at C^3, which is the value of the electron. That act put the Universe in 3D as it is today

```
                    ⋔⋔⋔⋔⋔⋔⋔⋔⋔⋔⋔⋔⋔⋔⋔
                 ⋔⋔⋔⋔⋔⋔⋔⋔⋔⋔⋔⋔⋔⋔⋔⋔⋔⋔⋔⋔⋔
              ⋔⋔⋔⋔⋔⋔⋔⋔⋔⋔⋔⋔⋔⋔⋔⋔⋔⋔⋔⋔⋔⋔⋔⋔⋔⋔⋔
           ⋔⋔⋔⋔⋔⋔⋔⋔⋔⋔⋔⋔⋔⋔⋔⋔⋔⋔⋔⋔⋔⋔⋔⋔⋔⋔⋔⋔⋔⋔⋔
         ⋔⋔⋔⋔⋔⋔⋔⋔⋔⋔⋔⋔⋔⋔⋔⋔⋔⋔⋔⋔⋔⋔⋔⋔⋔⋔⋔⋔⋔⋔⋔⋔⋔⋔
       ⋔⋔⋔⋔⋔⋔⋔⋔⋔⋔⋔⋔⋔⋔⋔⋔⋔⋔⋔⋔⋔⋔⋔⋔⋔⋔⋔⋔⋔⋔⋔⋔⋔⋔⋔⋔⋔⋔
     ⋔⋔⋔⋔⋔⋔⋔⋔⋔⋔⋔⋔⋔⋔⋔⋔⋔⋔⋔⋔⋔⋔⋔⋔⋔⋔⋔⋔⋔⋔⋔⋔⋔⋔⋔⋔⋔⋔⋔⋔⋔
    ⋔⋔⋔⋔⋔⋔⋔⋔⋔⋔⋔⋔⋔⋔⋔⋔⋔⋔⋔⋔⋔⋔⋔⋔⋔⋔⋔⋔⋔⋔⋔⋔⋔⋔⋔⋔⋔⋔⋔⋔⋔⋔⋔
   ⋔⋔⋔⋔⋔⋔⋔⋔⋔⋔⋔⋔⋔⋔⋔⋔⋔⋔⋔⋔⋔⋔⋔⋔⋔⋔⋔⋔⋔⋔⋔⋔⋔⋔⋔⋔⋔⋔⋔⋔⋔⋔⋔⋔
   ⋔⋔⋔⋔⋔⋔⋔⋔⋔⋔⋔⋔⋔⋔⋔⋔⋔⋔⋔⋔⋔⋔⋔⋔⋔⋔⋔⋔⋔⋔⋔⋔⋔⋔⋔⋔⋔⋔⋔⋔⋔⋔⋔⋔⋔
  ⋔⋔⋔⋔⋔⋔⋔⋔⋔⋔⋔⋔⋔⋔⋔⋔⋔⋔⋔⋔⋔⋔⋔⋔⋔⋔⋔⋔⋔⋔⋔⋔⋔⋔⋔⋔⋔⋔⋔⋔⋔⋔⋔⋔⋔⋔
 ⋔⋔⋔⋔⋔⋔⋔⋔⋔⋔⋔⋔⋔⋔⋔⋔⋔⋔⋔⋔⋔⋔⋔⋔⋔⋔⋔⋔⋔⋔⋔⋔⋔⋔⋔⋔⋔⋔⋔⋔⋔⋔⋔⋔⋔⋔⋔⋔
 ⋔⋔⋔⋔⋔⋔⋔⋔⋔⋔⋔⋔⋔⋔⋔⋔⋔⋔⋔⋔⋔⋔⋔⋔⋔⋔⋔⋔⋔⋔⋔⋔⋔⋔⋔⋔⋔⋔⋔⋔⋔⋔⋔⋔⋔⋔⋔⋔
⇔⇔⇔⇔⇔⇔⇔⇔⇔⇔⇔⇔⇔⇔⇔⇔⇔⇔⇔⇔⇔⇔⇔⇔⇔⇔⇔⇔⇔⇔⇔⇔⇔⇔⇔⇔⇔⇔⇔⇔⇔⇔⇔⇔⇔⇔⇔⇔⇔⇔
 ⋃⋃⋃⋃⋃⋃⋃⋃⋃⋃⋃⋃⋃⋃⋃⋃⋃⋃⋃⋃⋃⋃⋃⋃⋃⋃⋃⋃⋃⋃⋃⋃⋃⋃⋃⋃⋃⋃⋃⋃⋃⋃⋃⋃⋃⋃⋃⋃
  ⋃⋃⋃⋃⋃⋃⋃⋃⋃⋃⋃⋃⋃⋃⋃⋃⋃⋃⋃⋃⋃⋃⋃⋃⋃⋃⋃⋃⋃⋃⋃⋃⋃⋃⋃⋃⋃⋃⋃⋃⋃⋃⋃⋃⋃⋃⋃
   ⋃⋃⋃⋃⋃⋃⋃⋃⋃⋃⋃⋃⋃⋃⋃⋃⋃⋃⋃⋃⋃⋃⋃⋃⋃⋃⋃⋃⋃⋃⋃⋃⋃⋃⋃⋃⋃⋃⋃⋃⋃⋃⋃⋃
   ⋃⋃⋃⋃⋃⋃⋃⋃⋃⋃⋃⋃⋃⋃⋃⋃⋃⋃⋃⋃⋃⋃⋃⋃⋃⋃⋃⋃⋃⋃⋃⋃⋃⋃⋃⋃⋃⋃⋃⋃⋃⋃⋃⋃
    ⋃⋃⋃⋃⋃⋃⋃⋃⋃⋃⋃⋃⋃⋃⋃⋃⋃⋃⋃⋃⋃⋃⋃⋃⋃⋃⋃⋃⋃⋃⋃⋃⋃⋃⋃⋃⋃⋃⋃⋃⋃⋃
     ⋃⋃⋃⋃⋃⋃⋃⋃⋃⋃⋃⋃⋃⋃⋃⋃⋃⋃⋃⋃⋃⋃⋃⋃⋃⋃⋃⋃⋃⋃⋃⋃⋃⋃⋃⋃⋃⋃⋃
       ⋃⋃⋃⋃⋃⋃⋃⋃⋃⋃⋃⋃⋃⋃⋃⋃⋃⋃⋃⋃⋃⋃⋃⋃⋃⋃⋃⋃⋃⋃⋃⋃⋃⋃⋃
        ⋃⋃⋃⋃⋃⋃⋃⋃⋃⋃⋃⋃⋃⋃⋃⋃⋃⋃⋃⋃⋃⋃⋃⋃⋃⋃⋃⋃⋃⋃⋃⋃
          ⋃⋃⋃⋃⋃⋃⋃⋃⋃⋃⋃⋃⋃⋃⋃⋃⋃⋃⋃⋃⋃⋃⋃⋃⋃⋃⋃⋃
            ⋃⋃⋃⋃⋃⋃⋃⋃⋃⋃⋃⋃⋃⋃⋃⋃⋃⋃⋃⋃⋃⋃⋃
               ⋃⋃⋃⋃⋃⋃⋃⋃⋃⋃⋃⋃
```

> The atom restricts dismissing of space by the containing structure to the atoms relevancy being Π^0 in singularity bringing on Π relating to $(\Pi^2+\Pi^2)(\Pi^2\Pi)3$. As the layers swap there aligns between duplicating and dismissing the atomic relevancy adapt to comply

Since the star performs as an accumulated atom where innumerable atoms inside the confinement of the star combine to select one centre spot forming singularity that represents the star, I have chosen the to use the same symbols that I found in atoms to describe the relations in space –time to singularity within the space-time of the star. I refer to a star as a cosmic atom in other books.

Early stars still in the envelope of heat within the centre of the Galactica have only space duplication and growth through the cover of such enormous heat. These class stars are not visible but are shrouded in a blanket of heat covered by light. The atoms forming the stars are small and under developed. They remain cool because they contrast with the heat surrounding the star where the star material supports the cool space and does not form part of the liquid heat forming the outer limit. I would like to draw your attention once again to the fact that the sun at one stage was a cool $18 \times 10^{6\ 0}$ on the inside and a freezing cold at $6500^{\ 0}$ on the outside while all the time outer space was a blistering $10^{34\ 0}$. This was considered the coldest place in the Universe because the sun was still part of the deep frozen space inside the blanket of heat. Look at any galactica and see in the centre there are stars surrounded by a blanket of heat with stars conversed by heat sitting like a duck frozen in this pond of liquid heat.

As every atom is a Universe apart from another Universe by virtue of protecting and maintaining singularity the atom provides the star its reason for being there. In that regard every atom forms part of a group that combines in one governing singularity formed in the centre of a star and all the atoms has one purpose, that purpose is to maintain the governing singularity. This feature makes the star form a part of a Universe one Universe apart from all other Universes. Every atom is the Universe standing secluded by the electron apart from space-time.

Being part of the 3D we have the inclination to think of something and then we also includse in that something the space we experience. Thinking of the spinning top we will think of the edge (A) forming the end of the line, the position of seven. This cconcept we have is part of the 3D Universe. To understand cosmoogy we have to rteturn to the biginning of cosmology. The relations then was when Π^0 formed Π.

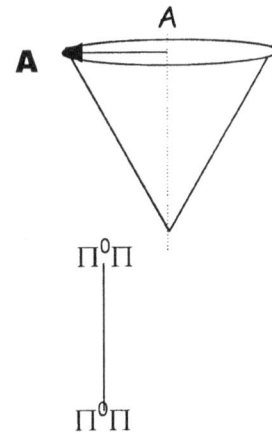

The moving of Π^0 to Π involved relegation and not motion as we consider motion. It was Π^0 getting a side and that is all. There was no true side but only a form that came into place. Singularity (A) received singularity (**A**) and no more of anything but the shift to comply with having a relevancy forming in relation to singularity. The dots had no sides, had no length or diameter. There was not measurable space or measurable time involved. The time could have been a micro, micro second as much a trillion millennium because time had no relevance. It was eternity interrupted by infinity, as it still is the case, however the line that eternity followed was no line because there was no space to hold the line. The line was momentarily interrupted by infinity, however with no one there, there was no one to notice. The lines were not lines but relations to sides being formed.

The relevancy that had the power to set Π apart from Π^0 is the only relevancy that still has the power, to set particles apart or join particles. It is heat in variation from cold. In order to excite singularity, singularity must establish a basis of heat that sets such a heat basis apart from cold. From there the form the atom will take on, however, the atom was still enumerable eternities to the development side.

The relevancy that had the power to set Π apart from Π^0 is the only relevancy that still has the power, to set particles apart or join particles. It is heat in variation from cold. In order to excite singularity, singularity must establish a basis of heat that sets such a heat basis apart from cold. From there the form the atom will take on, however, the atom was still enumerable eternities to the development side.

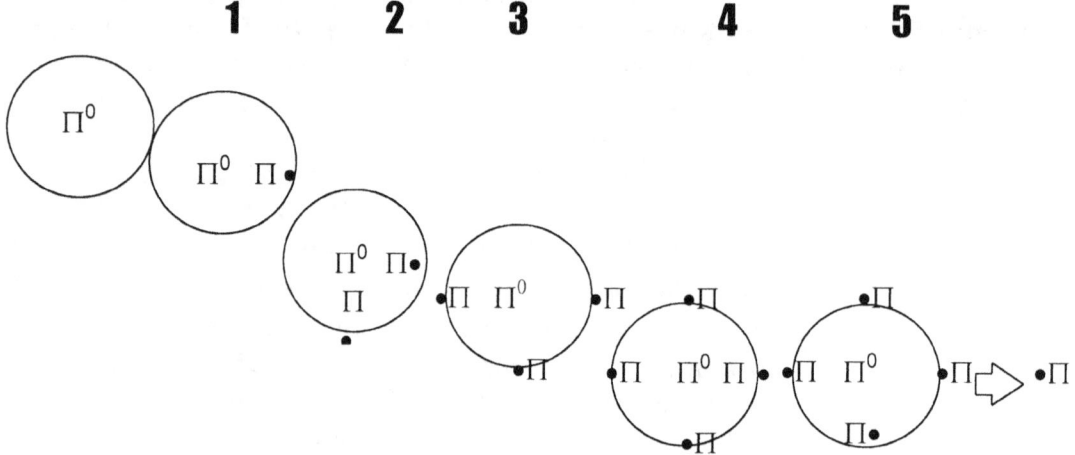

Long after the establishing of the ground principles did the Universe introduce these measure in space. It was only when the distortion of time became tio much to control by singularity in direct control that space burst onto the scene as part of sin gularity in influence of space

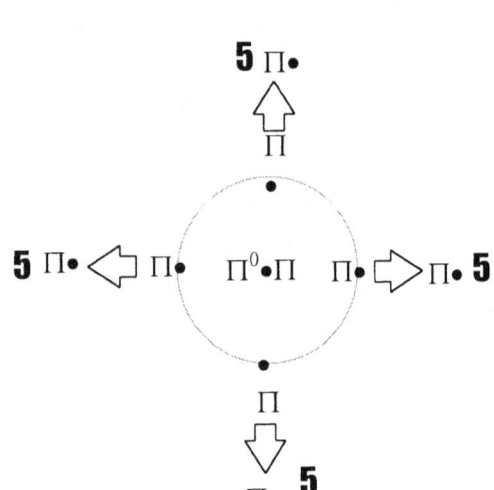

The first four dots form time and the three the foundation of singularity bringing the points in support of the sphere to seven. Forming the distortion one of time that, which is one to the outside of time, is the fifth dot. There is a fifth dot for every dot distorting time, which brings the total number of fives to four. Controlling singularity is singularity Π^0, which is 1 / 10 less than the expanded singularity, which is one. The singularity expanded is the two points of five away from the original going down (or up depending on which way one looks at it) from the original. Every point there came about it was four dots relating to each other in singularity as the fifth, which totalled twenty and with the full account of singularity it was 21.9991. Place that is relation to the next seven points that will originate form that and another Π is about.

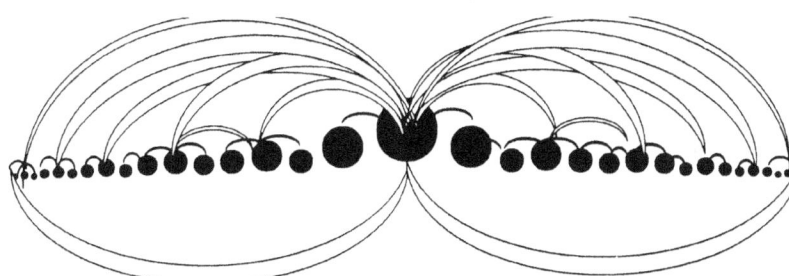

We have the time delay that occurred back when space was not yet developed since tiem was eternal (ro at thet iem pretty close to it whan compared to what is todyay applying. From the density at first and the very apparent less numerous to the lesser density required in the forming and the far greater numourous in quanteties it is pretty apparent that the expandingthat went on with the Big bang started long before the Big bang. The Densiy that was required to compress the multi proton cocoons becameless as heat turned to space. With more space available the proton cocoon numbers dwindled and and the individual atoms became more numorous in numbers. Still it came about as a result of time delay and the Big Bang was in space development long before the Big Bang went Bang.

Element	Relative number of ato
Hydrogen	1,000,000,000,000
Helium	90,000,000,000
Carbon	350,000,000
Nitrogen	85,000,000
Oxygen	590,000,000
Sodium	1,500,000
Magnesium	30,000,000
Aluminium	2,500,000
Silicon	35,000,000
Phosphorus	270,000
Sulphur	16,000,000
Potassium	110,000
Calcium	2,100,000
Chromium	300,000
Iron	3,200,000
Nickel	120,000

The same method now used to produce more space in widening circles as the atom finds more heat was the process that became the Big bang. However every star has its set of rules and every element abide by another set of rules as interpreted in the range that the star dictates.

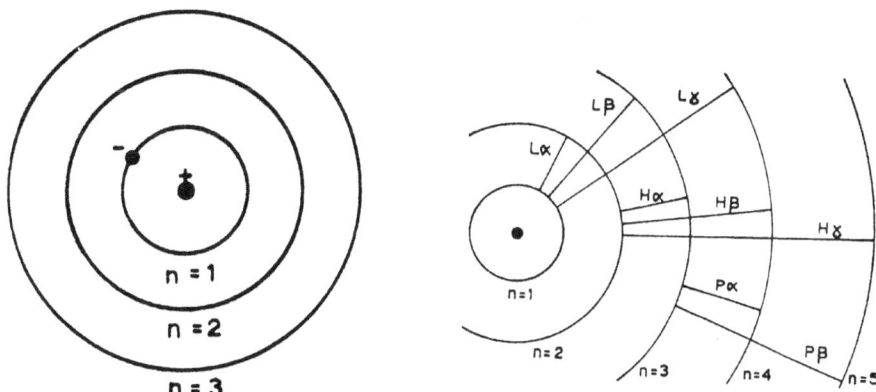

Whenever an electron jumps from a higher into the lower (innermost) orbit, the atom gives out radiation at a wavelength corresponding to a spectral line of the Lyman series. Jumps down into the second lowest level contribute to the Balmer series. The greater the jump, the closer the emitted radiation is to the limit of the series, which is reached when an electron enters from outside the atom. Outward jumps involve the absorption of energy and give rise to absorption lines. Even today the forming of heat increase becomes space and proves that material is time delay. This was the process that was started by the spot becoming the dot and the only difference is the increase of space to the demise of time.

In this manner and by sets of this ratio a variation of atoms and atomic groups came about that set the rules for the future Universe to comply with. The forming must not be viewed as a line or a ring that came about because that which formed had no sides yet and had no formation. There was only a forming of relevancies running from one too seven.

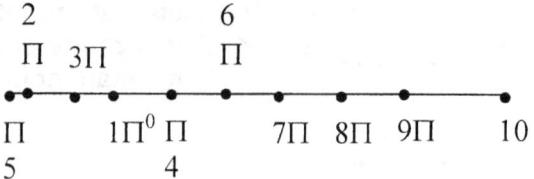

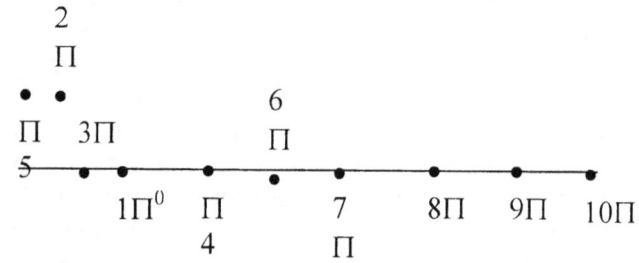

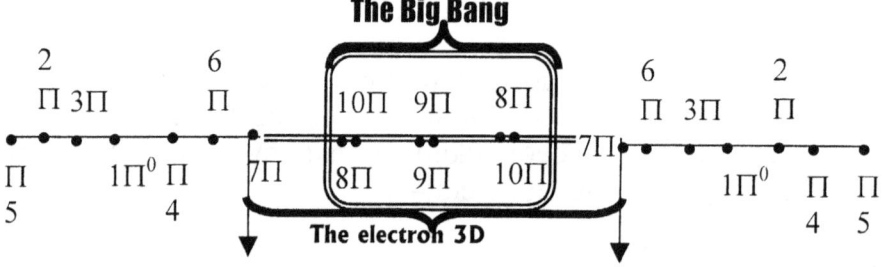

When the Universe set relevance, the motion was way beyond the speed of light. The speed of light only arrives at the third dimension coming about. This was a Universe in relevancies

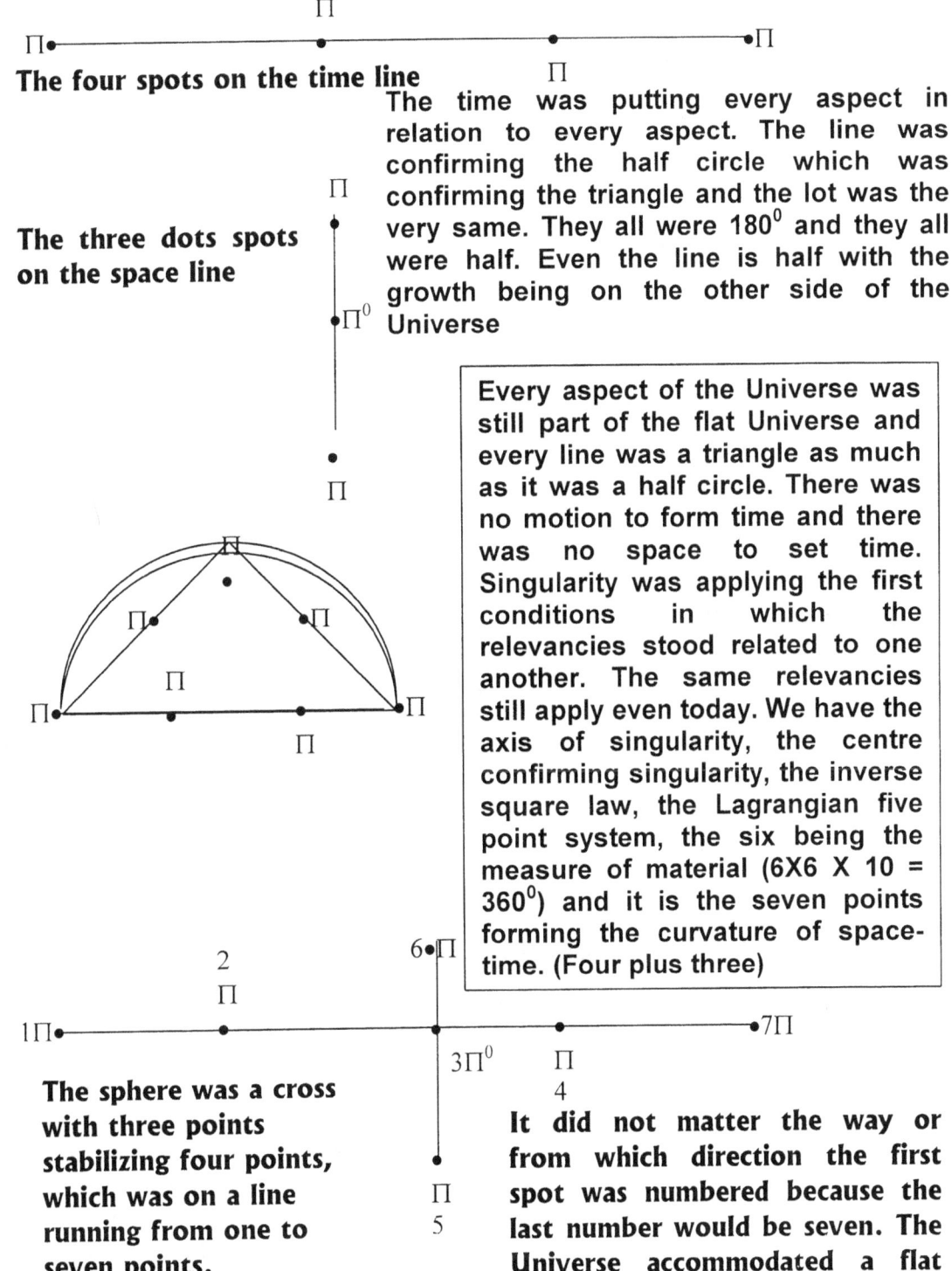

The four spots on the time line

The three dots spots on the space line

The time was putting every aspect in relation to every aspect. The line was confirming the half circle which was confirming the triangle and the lot was the very same. They all were 180^0 and they all were half. Even the line is half with the growth being on the other side of the Universe

Every aspect of the Universe was still part of the flat Universe and every line was a triangle as much as it was a half circle. There was no motion to form time and there was no space to set time. Singularity was applying the first conditions in which the relevancies stood related to one another. The same relevancies still apply even today. We have the axis of singularity, the centre confirming singularity, the inverse square law, the Lagrangian five point system, the six being the measure of material (6X6 X 10 = 360^0) and it is the seven points forming the curvature of space-time. (Four plus three)

The sphere was a cross with three points stabilizing four points, which was on a line running from one to seven points.

It did not matter the way or from which direction the first spot was numbered because the last number would be seven. The Universe accommodated a flat structure and not one with certain sides removed to serve the explaining of the Educated.

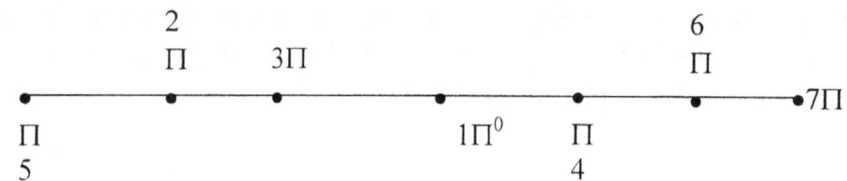

1•Π⁰ Singularity governing

2Π• Singularity in relevance

3Π• Singularity in relevance

•4Π Singularity forming motion to become time

•5Π   Singularity forming time distortion becoming space.

•6Π Singularity forming time distortion becoming material in space

•7Π Singularity forming time distortion becoming material ending space

Long after the establishing of the ground principles did the Universe introduce these measure in space. It was only when the distortion of time became tio much to control by singularity in direct control that space burst onto the scene as part of sin gularity in influence of space

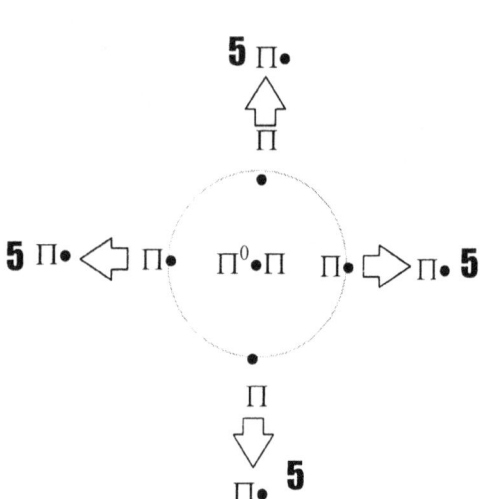

The first four dots form time and the three the foundation of singularity bringing the points in support of the sphere to seven. Forming the distortion one of time that, which is one to the outside of time, is the fifth dot. There is a fifth dot for every dot distorting time, which brings the total number of fives to four. Controlling singularity is singularity $Π^0$, which is 1 / 10 less than the expanded singularity, which is one. The singularity expanded is the two points of five away from the original going down (or up depending on which way one looks at it) from the original. Every point there came about it was four dots relating to each other in singularity as the fifth, which totalled twenty and with the full account of singularity it was 21.9991. Place that is relation to the next seven points that will originate form that and another Π is about.

Second Over All Overview 651

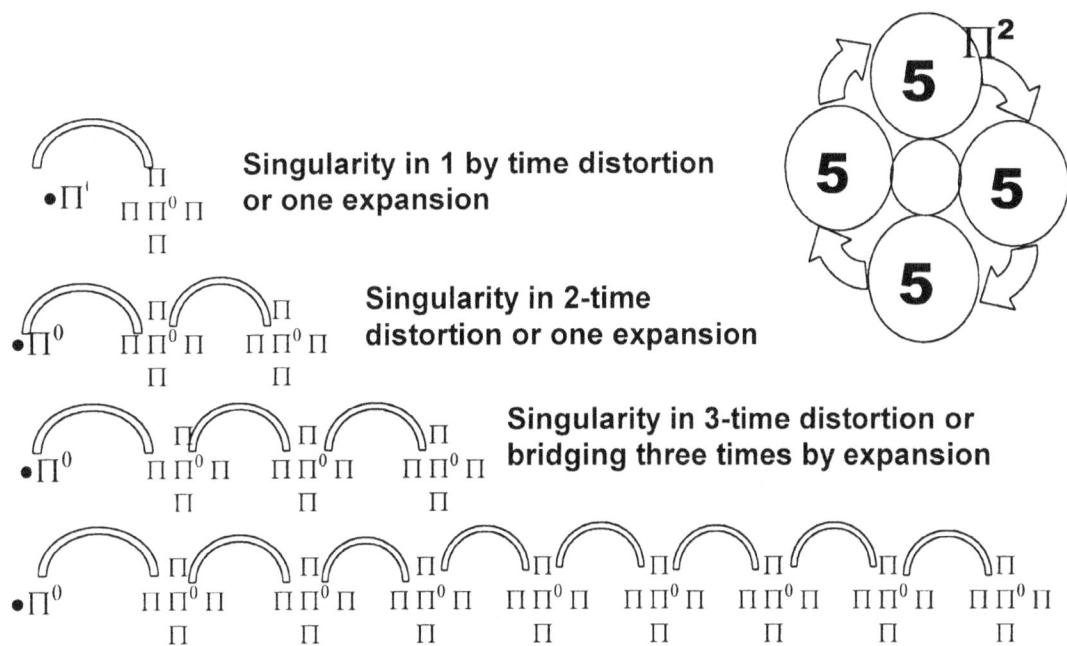

Singularity in multiple time distortion or multiple expansions forming space by time delay

It is by means of the Coanda effect that proton numbers formed in individual cocoons. Then we can gauge how the process of atomic cluster development started and how it progressed.

1 H	IIA											IIIA	IVA	VA	VIA	VIIA	2 He
3 Li	4 Be				Transition metals							5 B	6 C	7 N	8 O	9 F	10 Ne
11 Na	12 Mg	IIIB	IVB	VB	VIB	VIIB	VIIIB			IB	IIB	13 Al	14 Si	15 P	16 S	17 Cl	18 Ar
19 K	20 Ca	21 Sc	22 Ti	23 V	24 Cr	25 Mn	26 Fe	27 Co	28 Ni	29 Cu	30 Zn	31 Ga	32 Ge	33 As	34 Se	35 Br	36 Kr
37 Rb	38 Sr	39 Y	40 Zr	41 Nb	42 Mo	43 Tc	44 Ru	45 Rh	46 Pd	47 Ag	48 Cd	49 In	50 Sn	51 Sb	52 Te	53 I	54 Xe
55 Cs	56 Ba	57-71 *	72 Hf	73 Ta	74 W	75 Re	76 Os	77 Ir	78 Pt	79 Au	80 Hg	81 Tl	82 Pb	83 Bi	84 Po	85 At	86 Rn
87 Fr	88 Ra	89-103 †	104 Rf	105 Ha	106 Sg	107 Ns	108 Hs	109 Mt	110	111	112						

Following the line one can see how the development took place as the intensity of time being heat in liquid form lost density to space acquired and as the intensity reduced and the density removed in favour of space, the compactness of atomic proton numbers in the nucleus declined. Then came a point where the protons were not forming and the motion could no longer support further compacting. Because there were no feeding of heat the overheating set in as the gravity manufactured by the motion of liquid time that the cold no longer contain the growth. The atoms that was formed held the motion that formed them by an electron containing but the rest went on the explode into space. The Bang got Big and the forming of atoms came to a halt.

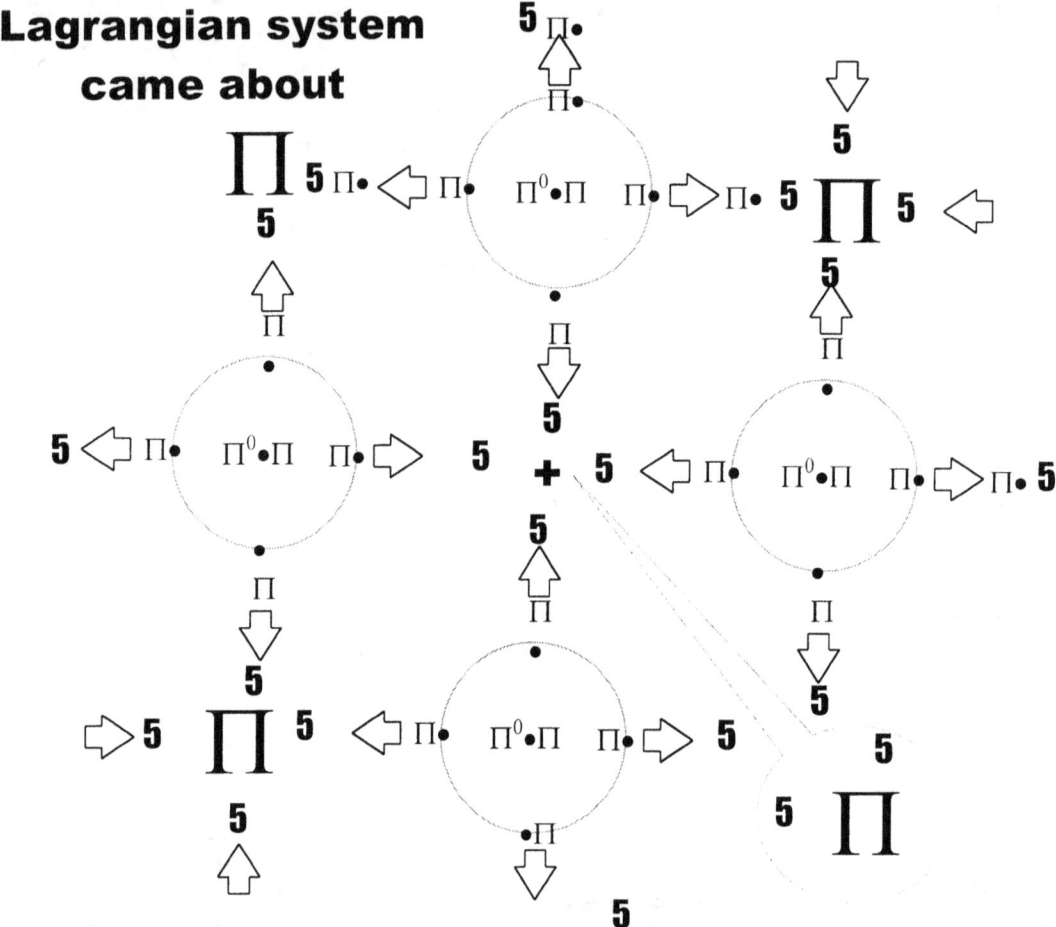

In that way dots formed more dots by becoming spots to the more dots. Every four dots formed a centre from where four dots became time aand three dots became singularity supporting time. The measure f the dot thaty formed was Π. The distance the dot was apart as well as of the dot was Π. The dots were Π apart.

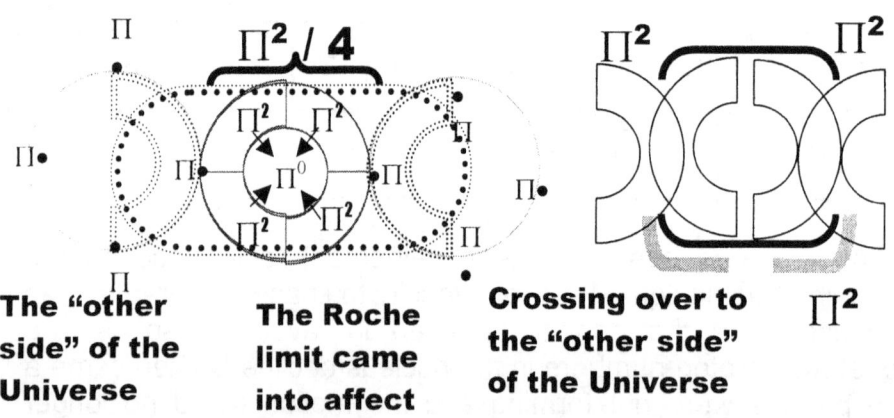

The "other side" of the Universe

The Roche limit came into affect

Crossing over to the "other side" of the Universe

Second Over All Overview — 653

TIME STARED WITH ETERNITY
TIME STANDING IN ETERNITY R = O / T = Ω
TIME STARTED WITH THE APPLICATION OF Π=1

$$\Pi = 1 = \text{AT } R^0 / T^0 = 1$$

Π=		TIME ZERO + TIME		$T = 10/7\pi^2/2(\pi^2+\pi^2) = 139$	10
Π=		IN MOTION STARTING		$T = 7(\pi^2+\pi^2) = 138$ TB.	1
Π=		AT $R^0 / T^0 = 1 + R^\Omega / T^\alpha = 1$		$T = 7/10\, \pi^2/2(\pi^2+\pi^2) = 136$	7

Π=	THE	124	$T = 2\pi(\pi^2+\pi^2) = 124.0251$ TB. 1
Π=	PLANK	119	$T\, 2(3)(\pi^2+\pi^2) = 118.435$ TB. 3
Π=	ERA	112	$T = 10 \div 7(4(\pi^2+\pi^2)) = 112.795$

Π=	THE BIG BANG	112	$T = 7/10\, \pi^6/60 = 112.162$ COSMIC TIME
Π=	FORMING		$T = \pi^2(\pi^2+\pi^2) = 107.278$ TB= 1
Π=	THE ATOMS	102	$T = 3\pi((\pi^2+\pi^2) = 102.88$ TB= 3

Π= 98	THE	$T = 3^2(\pi^2+\pi^2) = 98.69$ TB= 6
Π= 88	QUARKS ERA	$T = 10/7\pi(\pi^2+\pi^2) = 88.6$
Π= 84	THE LEAD	$T = 10/7(3(\Pi^2+\Pi^2) = 84.6$ TB=3
Π= 78	ERA	$T = 4(\Pi^2+\Pi^2) = 78.9562$ TB= 3
Π= 69	THE	$T = 10/7\,(\pi/2)^2(\pi^2+\pi^2) = 69.57$
Π= 55	CURRENT ERA	$T = 7/10\,(4((\Pi^2+\Pi^2) = 55.3$ TB =6

Π= 28	ERAS LAYING	$T = 10/7(2((\Pi^2) = 28.2$
Π= 14	TO THE FAR	$T = 7/10(2(((\Pi^2+\Pi) = 14.1$
Π= 5	AND DISTANT	$T = 7/10((\Pi+\Pi) = 5.34$
Π= 2	FUTURE	$T = 10/7(\Pi)$ TB. 96 = 2.2 TB= 96
Π= 1	APPROACH TO	$T = 7/10(\Pi/2) = 1$ TB= 192

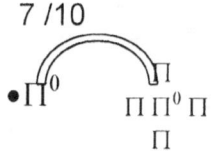
With every time configuration involving the Titius Bode law, the duplication produce a distortion doubling the delay due to the $1.4 / 1.42 = \Pi^2$

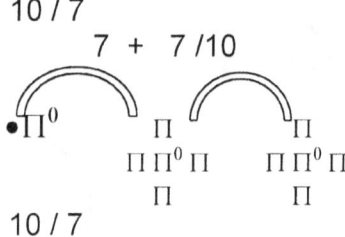
That puts time in delay by the square, as it would befit time with time being motion. However the biggest implication is that time slows down by the double every time a new relevancy appears. By the time of the arriving of the Big Bang time delay was beyond singularity control.

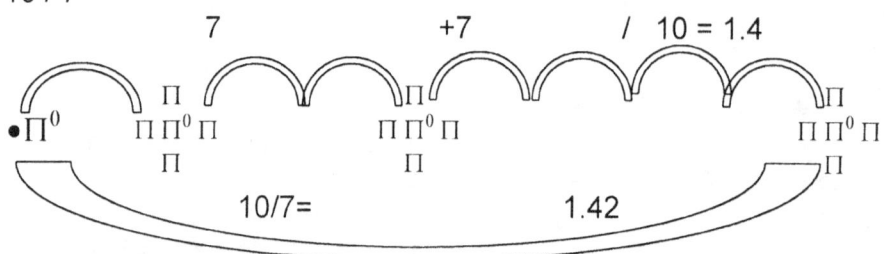

$(7/10)+ (7/10) = 1.4$ and $10 / 7 = 1.42$
$1.4 / 1.42 = 0.986 \times 10 =$ gravity

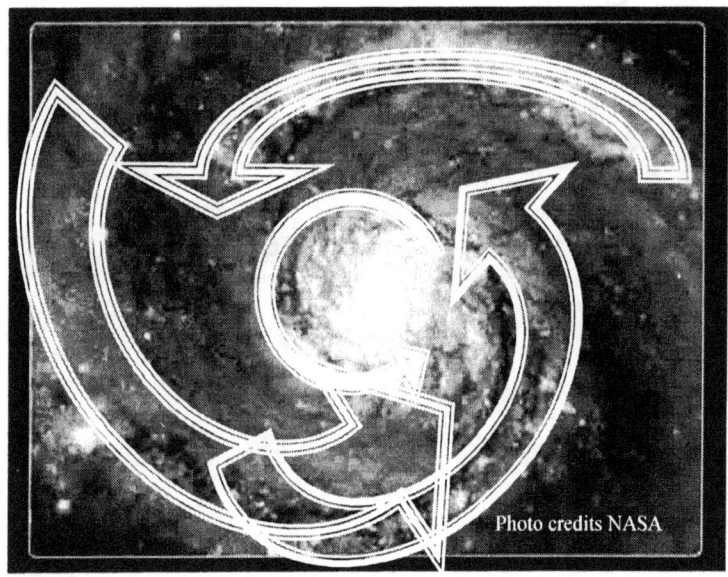

In all galactica there is a spiral effect turning the one way or another but the cycle runs towards the inside dome. A one can see there is a spiralling motion T^2k that leads towards a defined space a^3 on the inside.

In all we have two forms of gravity that developed in simulation of each other. The galactica indicate the controlling gravity encircling of motion around a specific centre.

Then we have the duplication or motion of singularity that is unable to move. Because of this inability to move time had to invent space to move time about.

This brings evidence of at least two things. One is we know how the Universe started and two we know how gravity works. What we now see being as spread out and so wide that we can never really see all of it once was pretty close. Yet, the formation remained the same throughout. Once anything is a part of the Universe and is in a part of the Universe that then can go nowhere again because the Universe do not allow shifting and swapping. We can see how it started because what the galactica contains was what the galactica at first contained. It was liquid heat covering lumps of solid material and the solid material is lumps of heat contained and preserve by motion using the Coanda effect. It is heat preserved by motion with heat preserving what is preserved by motion. How did it start? In the photo on the next page we can see the moon heating but we know the moon cannot heat. We know it is the Sun heating the moon and in all the talk as usual we miss the point there is to be made. When anything heats it expands. But more so, if anything expands it must heat up. Time is on an eternal line with infinite dimension, which we named singularity. The line is so small it has no sides and the line is so long it can be generated by anything with heat that unleashes motion. However we also know that to find drive we need heat. What the heat is and where the heat came from I leave to another book called "Starstuffin" as well "Seven days Of Creation". The fact is that heat expands or makes what there is available being more and more available. Looking at the photo we can see how much the stature of the moon grows with the heat coming from the moon. Fair enough you may say it is light but what is light other than heat very densely packed and being very small and compact while being contained as heat. Should you hold the opinion that what we see coming from the moon is merely light, well then how can light be anything else than heat and how can heat be anything else than light. Put what you see and the expanding that you see into a box that you cannot see because there is no space to see. Put that lot into a box where the expanding has no place to go because there is no place in the Universe but the small part that was there before expanding came about. Then you see that's how it started! It is only heat with the ability to expand and establish a new Universe.

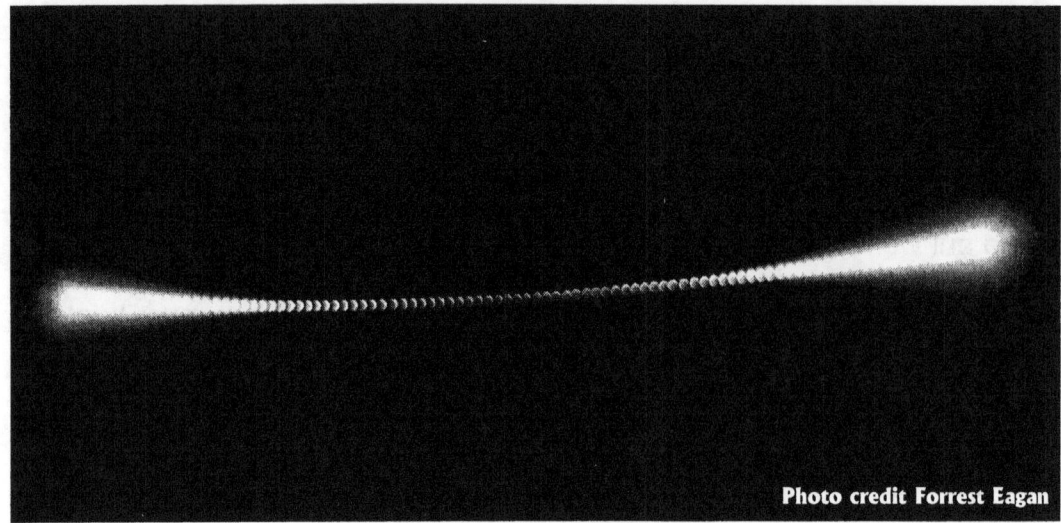

Photo credit Forrest Eagan

● First came the centre. From the centre advanced four in equal presenting motion

 If gravity were about contraction creation would have ended at this point. However it did not and therefore it is wise to surmise that Creation is about duplication more than it is about contraction.

If in doubt about my statement go and see the outside where expansion brought changes since creation started and then rethink your doubt about my statement. And even in contraction the contraction is about duplicating the centre in preventing overheating

The next act that came after separation came in place was the conception of time. Time brought about four points in singularity. As one can see from the picture of the moon dimming and brightening again the heat or light clearly indicate that it seems to fill more of what is filled when the heat is less prominent.

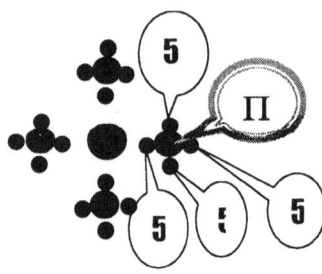The heat brought motion about. Singularity has an inability to move, therefore the heat had to perform a sequence to provoke the insinuation of possible motion. At the point where five points sprouted by four to develop a new Π in relation to the seven points the sphere established such relating brought on an indication of possible motion.

From this meeting of space-time developed space as time spawned space. The retarding of time took heat longer to reach the spawning points, which became the retarding or the distortion of time. That space still is and that is what is growing, as material while the Blackness of the night is time extending or growing.

Every time duplication came about ten points was in affect of which seven was retained by the sphere in as much as the proton $(\Pi^2+\Pi^2)(\Pi^2\Pi)=7$ positions in singularity.

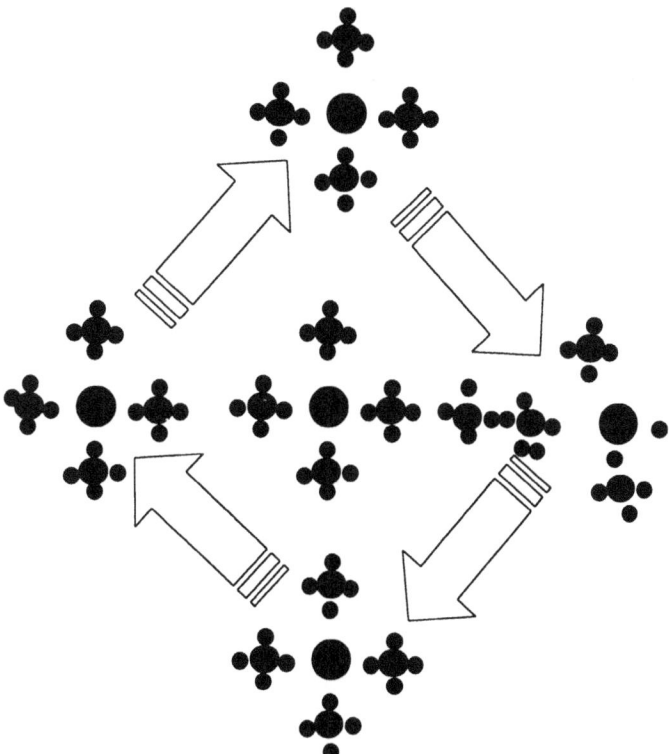

From such duplication of singularity heat is providing motion, which is providing duplicating of singularity. Singularity is not moving, as singularity has no space to move. In allowing motion that forms gravity the relevancy of affecting new singularity and by heat enticing the moving of the enticing of singularity provide a sustaining of what would be motion. As the exciting spreads the singularity manifest by continuing tradition and roles move about. The roving singularity becomes the holding singularity and as five forms for spots a new pi elevates to become the roving singularity being in control of the newly established containing singularity. In this way the Universe formed.

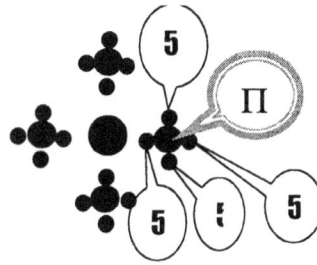

The heat brought motion about. Singularity has an inability to move, therefore the heat had to perform a sequence to provoke the insinuation of possible motion. At the point where five points sprouted by four to develop a new Π in relation to the seven points the sphere established such relating brought on an indication of possible motion.

From this meeting of space-time developed space as time spawned space. The retarding of time took heat longer to reach the spawning points, which became the retarding or the distortion of time. That space still is and that is what is growing, as material while the Blackness of the night is time extending or growing.

Second Over All Overview   658

Every time duplication came about ten points was in affect of which seven was retained by the sphere in as much as the proton $(\Pi^2+\Pi^2)(\Pi^2\Pi)=7$ positions in singularity. The seven was confined leaving three points not detained and unattached. The factors of motion being 3 Π^2 or was absorbed by the motion the proton instigated in conjunction with the proton $\Pi^2 \Rightarrow \Pi^2$ implicating $\Rightarrow \Pi^2$ the neutron. The relevant implicating factor became the second adding to the atom forming Π. That left the retaining space as 3 or Π^0 x

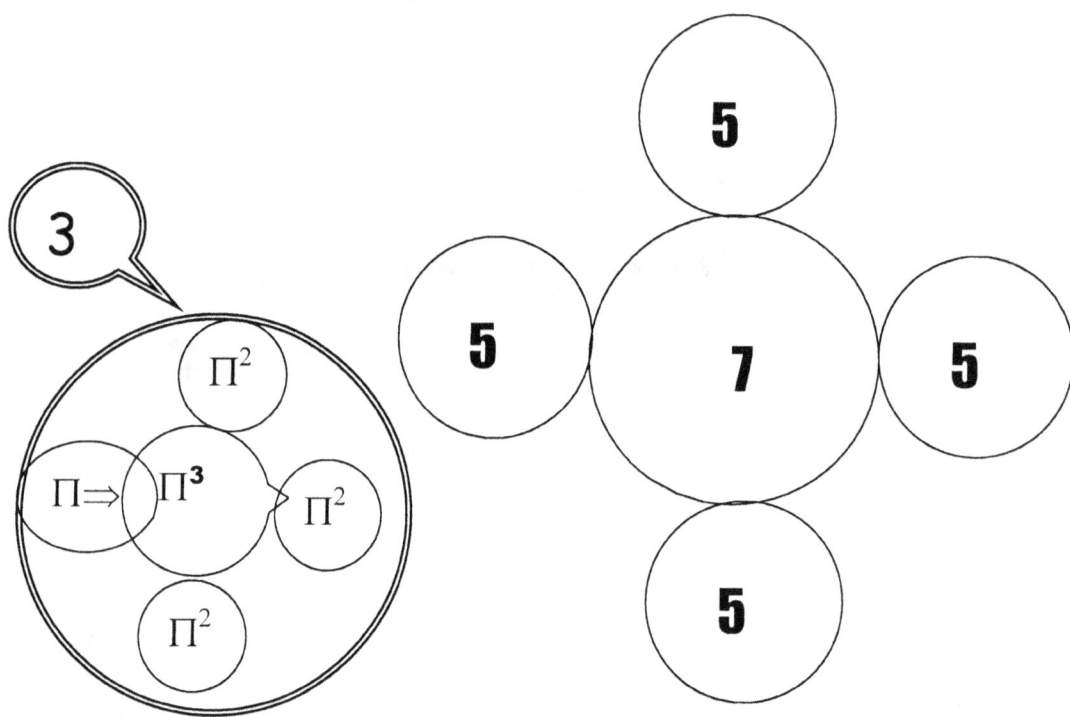

The Black Hole proves to be the ultimate atom and is also the ultimate profile of the Coanda principle. It is stretching as far as singularity can stretch space-time in the reducing of the cosmos by gravity. What I introduce is what Kepler introduced although Kepler may have used somewhat different factors in the manner he introduced what I am about to introduce. The black Hole shows the Coanda principle in its ultimate role. Any person doubting my theory about space –time moving towards a centre and thereby creating gravity must then explain the Black Hole working. But please for sanity's sake let's forget the theory of gravity going mad in a super nova event and then becoming a monstrous Black Hole. The super nova is as natural having the process where gravity is in the process of moving from the atom accumulation to the governing centre singularity. However in the case of the super nova things went wrong because the super nova occurrence is where the transfer from movement or gravity within the star went out of control. The motion was to slow and the build up of accumulated heat exceeded the motion and over came the contraction of the governing singularity. In the process heat went back to gas as time increase by millions of folds.

If it expands it would go straight but if it cannot expand because there is no space it has to go straight while going around in a circle. Knowing our Universe doesn't like changing we know it kept what it started with back then pretty much the same.

The dot does not spin therefore the dot is unmovable. The dots immovability renders them the opportunity to destroy the space by incorporating the space into the dot. The series of dots flow in a direction around a democratically elected centre that means the dots are moving as a group. The group motion becomes the space duplication by motion. But if the dots remain in one location, then the space they accumulated react and the space they accumulated shift on spot on because the space they accumulated and use to return the previous standard plus one in space-time relocate each relative position in relation to the centre one spot. The motion factor T^2 has singularity rotate the space in relation to the motion of time whereas k has singularity rotate in relation to a select centre and the selected centres selected a centre that provide the group of selected centres a rotating centre as well as a rotating direction in which the group of selected centres will move in a line. The space created by the motion spins as it rotates but then when the dot secures the space by rendering the space secured through the immovability of the dot, the dot changes the relevance it had in the group and as part of the group the dot spins. This then is k. The factor k confirms the duplication of space when the factor T^2 confirms the space as dismissed and when the interacting takes place it confirms space-time through motion duplicating or motion to dismiss space-time. There is forever a balance and this balance we call the Universe. The containing structure represents in a duplication relation applied within the boundaries of the star and in that it applies a norm set on pre-determent conditions that is as old as space-time self applied by the governing singularity that space is supporting and maintaining. That will demand the mass brought about by all that the space contains. This relevance means that without a specified container (star forming boundaries) the space producing specific duplication and destroying of space in the area known as outer space, the outer space container will apply a diminishing relevancy of space-time displacing that which can support a maximum number of 112 protons working as a unit and in conjunction dismissing space can withstand. We know that is a theory because the atoms in space can sustain much less than 112. Inside containers being stars the direct relevancy of singularity applying puts much more strain on the surviving abilities of atoms. In outer space the atom has an own relevancy of seven and the space demand on the atom is only three that it must maintain in order to duplicate. But in stars the containing star places a demand of the containing seven plus the space creating three in relation with the time applying inside a star, which are four.

The seven was confined leaving three points not detained and unattached. The factors of motion being 3 Π^2 or was absorbed by the motion the proton instigated in conjunction with the proton $\Pi^2 \Rightarrow \Pi^2$ implicating $\Rightarrow \Pi^2$ the neutron.

The relevant implicating factor became the second adding to the atom forming Π. That left the retaining space as 3 or $Π^0$ x the motion of the photon being $Π^2$. Since the Earth has no singularity demand that is much better developed than the Universe sustains we find on Earth a relevancy of Π to $(Π^2+Π^2)(Π^2Π)3$ is adequate. But in bigger units the space-time displacing relating to space duplication presents much more demand on atomic structures occupying space within the star containing through set boundaries. In the presumed to be bigger stars there is much space filled with atoms occupying much space. In the stars more massive but holding lesser space the atoms must also hold lesser space but they also hold more protons by number in the lesser space. The space the particles hold is directly in relation to the particles the containing structure duplicate. The more space that is relevant to the structure that the star duplicate by motion is then in turn once again relevant to the space the structure destroys by proton action in space less units. The more space the particle claims in relation to the space the container hold that relates to the space the container duplicate is relative to the space the containing structure destroy. From that mass derives value. As individual occupying space the atom is an individual container by own merits and as such duplicate space in this regard within the specific confinements of atoms.

The atom restricts dismissing of space by the containing structure to the atoms relevancy being $Π^0$ in singularity bringing on Π relating to $(Π^2+Π^2)(Π^2Π)3$. This we will classify as normal applying structure values the atom has in outer space or in structures with very little atmosphere. Please note there is no pressure involved because the motion involved creates conditions naturally instead of unnatural pumping that causes pressure. Pressure is an artificial creation as part of life but has no role in the natural cosmos. Pressure is a condition where the retaining of particles has to be confined in a patrician made of material where the outer wall does the retaining of the substance within. This obviously cannot be in a star because the "pressure" is regulated from a condition applying and space-time controlling inner centre that needs no solid walls to contain whatever is inside. With that one can see there is a Universal difference between the concept of pressure forming due to human action inside a container and what comes about as secluded space-time within a star. As the demand of singularity in such units grow stronger some relevancies within the atom come into play and I developed a system whereby I can arrange the space-time merits of space-time curtailing within the confinement of the star borders applying in the star to place such a demand in relation to singularity where the ultimate demand sets the standards. In the Sun for instance which is a minuscule small star a relevancy in the outer region might be 3^3 relating to singularity and with the atom having a sustaining displacement of $(Π^2+Π^2)(Π^2Π)3$ there is no danger of the atom demising. The electron in the Sun will have a diminishing factor of 27 whereas the atom can sustain $(Π^2+Π^2)(Π^2Π)3 = 1836$.

The relation in the atom degenerated by 27 leaving the atom a sustaining value of the electron plus the neutron applying space-time without involving any of the neutron aspect at all. That is the mass the electron will consume in the space reducing and producing mass within the star that then forms a favouring of duplication. The star is a bright little boy shining by dismissing pebbles of light-photons into space. When a demand on space-time displacement reach an accumulated general displacing to the value of 56.6 protons of the general; accumulated displacement within that space forming time. The star accumulated more heat by consumption applying direct dismissing without accumulating space-time in liquid form beforehand therefore there is no heat remaining to dispose of by producing light. When the general displacing flow of space-time within that sector of the star or the star in total reached 56.6 g/mol. absolute solidifying becomes the norm within the star as the star will exclude all electron functions and stop shining as the demand on space-time duplication and diminishing reduced the atom to space without a heat envelope that will be electrons or a liquid/gas jacket. Only the nucleus will be able to sustain the diminishing and the reducing of space by increasing of time. The entire star becomes a solid structure by reducing space-time directly freezing the space-time from a gaseous state to a solid state. By motion speeding up the tempo of the flow of space-time the liquid state of space-time is by passed going from gas to solidity in one motion. The atom would shrink to such little space it will have space within the star that only the centre nucleus will fit. More reducing by applying motion in creating space

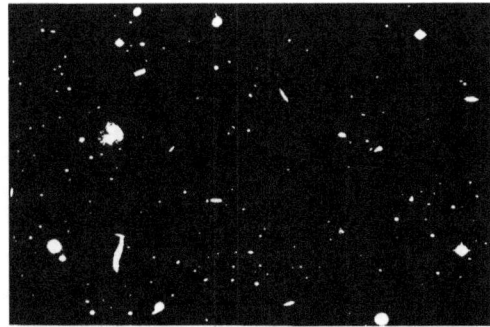

differentiation will leave a star with so little space the space will be insufficient to secure a position for the neutrons and the star will then have the name of being a neutron star. Going even further will find the proton rejected from the star. We know the Coanda effect is about circle and we can see the Unversed is all about circles. Therefore all things in the Universe and in stars as well as galactica are in circles. Everything we se see is a line containing a circle holding a line representing a circle. Every circle is forming a governing singularity, which is bonding the circle into a Unit. Every star, even our Micro stars such as Jupiter and the rest is a containing star with every phase the star needs for developing. The development in the layers might still be in infinite but that it is there definite. Every layer there for is a shell that will be discarded as time takes the star back to where time started producing that which became the star. It is all contained in circles and the Sun has a part that is a Black Hole.

One year on Earth (3,16 × 10^7 seconds) would be equal to a

BROWN STAR'S	duration of one year in time would be 37 868-earth seconds
RED STAR'S	duration of one year in time would be 27 050-earth seconds
ORANGE STAR'S	duration of one year in time would be 21 040-earth seconds
YELLOW STAR'S	duration of one year in time would be 15 780-earth seconds
WHITE STAR'S	duration of one year in time would be 9 500-earth seconds
BLUE STAR'S	duration of one year in time would be 3 160-earth seconds
BLUE GIANT'S	duration of one year in time would be 1 900 earth seconds
A BLACK HOLE'S	duration of one year in time would be 0,0005 earth seconds

All stars work in a ratio of linear displacement (R/T = 1) and circular displacement $R^2/T = 1$.

Second Over All Overview

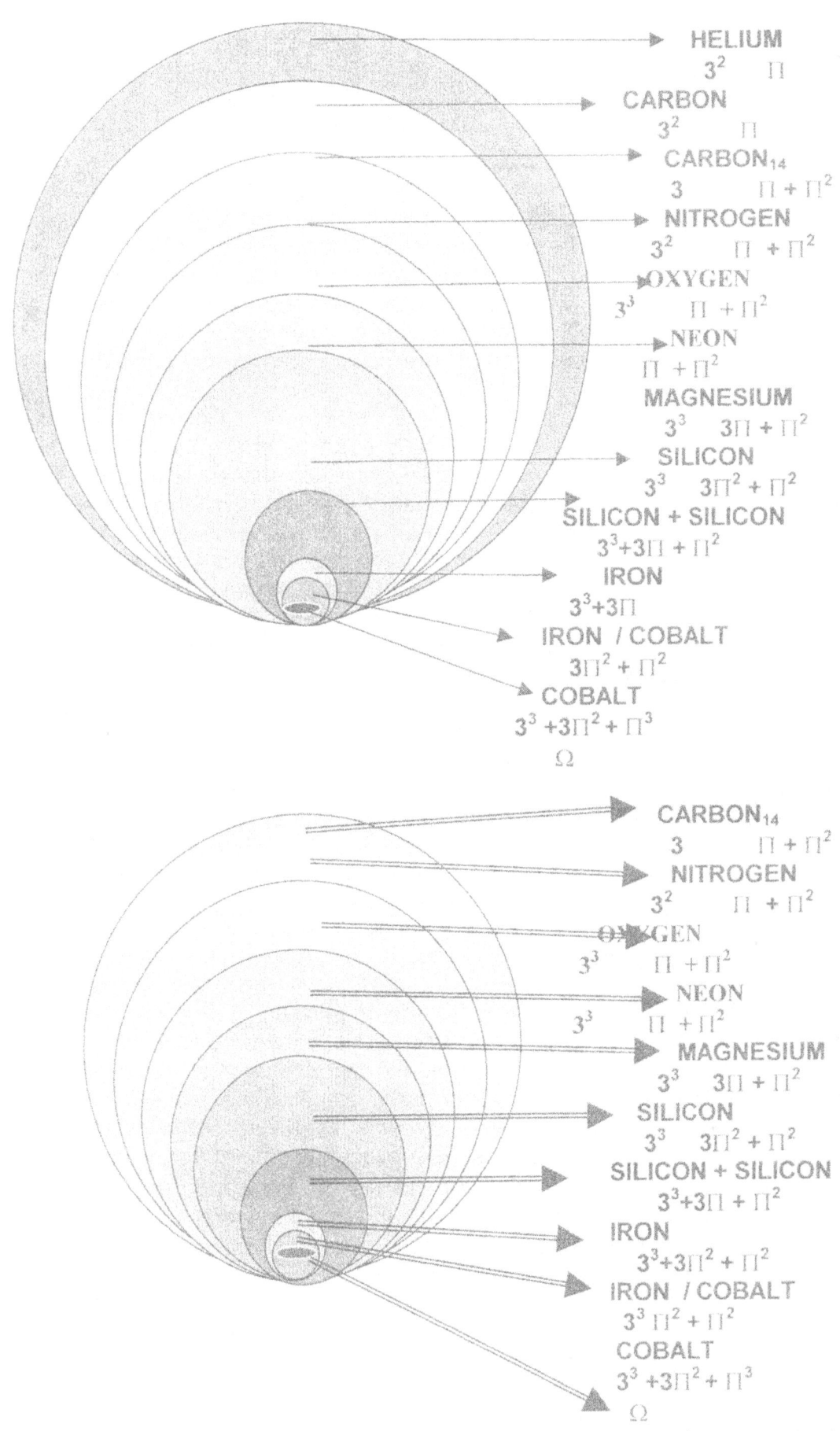

→ HELIUM
3^2 Π

→ CARBON
3^2 Π

→ CARBON$_{14}$
3 $\Pi + \Pi^2$

→ NITROGEN
3^2 $\Pi + \Pi^2$

→ OXYGEN
3^3 $\Pi + \Pi^2$

→ NEON
$\Pi + \Pi^2$

MAGNESIUM
3^3 $3\Pi + \Pi^2$

→ SILICON
3^3 $3\Pi^2 + \Pi^2$

SILICON + SILICON
$3^3 + 3\Pi + \Pi^2$

IRON
$3^3 + 3\Pi$

→ IRON / COBALT
$3\Pi^2 + \Pi^2$

→ COBALT
$3^3 + 3\Pi^2 + \Pi^3$
Ω

→ CARBON$_{14}$
3 $\Pi + \Pi^2$

→ NITROGEN
3^2 $\Pi + \Pi^2$

→ OXYGEN
3^3 $\Pi + \Pi^2$

→ NEON
3^3 $\Pi + \Pi^2$

→ MAGNESIUM
3^3 $3\Pi + \Pi^2$

→ SILICON
3^3 $3\Pi^2 + \Pi^2$

→ SILICON + SILICON
$3^3 + 3\Pi + \Pi^2$

IRON
$3^3 + 3\Pi^2 + \Pi^2$

→ IRON / COBALT
$3^3 \Pi^2 + \Pi^2$

COBALT
$3^3 + 3\Pi^2 + \Pi^3$
Ω

Early stars still in the envelope of heat within the centre of the Galactica have only space duplication and growth through the cover of such enormous heat. These class stars are not visible but are shrouded in a blanket of heat covered by light. The atoms forming the stars are small and under developed. They remain cool because they contrast with the heat surrounding the star where the star material support the cool space and does no form part of the liquid heat forming the outer limit. As I indicated previously but I would like to draw your attention once again to the fact that the Sun at one stage was a cool 18×10^{6} ^{0} on the inside and freezing cold at 6500 ^{0} on the outside while outer space was a blistering 10^{34} ^{0} which was considered the coldest place in the Universe because the Sun was still part of the deep frozen space. Look at any galactica and see in the centre there are stars surrounded by a blanket of heat with stars conversed by heat sitting like a duck frozen in this pond of liquid heat.

The inner singularity sets a border firm enough to prevent the heat from liquefying the star and only when the control of the general governing singularity will apply the same heat as that of the barrier will the barrier shift where the star would then accumulate and concentrate space-time to form an atmospheric border. the protons grow as a result of the fact that more space within the proton is duplicated that that which the proton can dismiss. In such a case the heat surrounding the star is hotter than the core of the star. We know that the Sun and the equivalent to the Sun has an inside temperature of 18×10^6 degrees where every one at present think of it as extremely hot. But a while ago that inside was a deep freeze because the outside outer space area was a wild 10^{34} on a mild day. At that stage the Sun was just a covered spot surrounded by heat from which the Sun could receive heat and turn that to space converted to material. The Sun developed from 3 to 3^2 to 3^3 and then when the outer edge where the Sun meets the end of outer space the Sun started turning space into liquid. Then the Sun went from 3 to Π by liquefying the atmosphere. The Earth at present has reached the stage where the atmosphere surrounding the Earth as Π extending is beginning to become intense liquid relevant to the other side of the 100 km border. However the moon has still to reach this stage.

As the star develops it moves away from the centre and away from the heat blanket covering the star. By concentrating heat at the centre as it is removing space in the centre the singularity driving the star comes into a dual with the singularity driving the galactica. This is the stage where the spinning top example proves the point. As the inner core heat surges the outer edge forming the newly established atmosphere excel in temperature by increasing duplication through applied motion around the Π^0 singularity from where a border (various borders in fact) establish Π and limit the interaction by Π^2. The border forms three sides on each side of the divide allowing a crossing of the border Π under specific conditions.

Then from there too is various Π borders and those borders end where the solid structure of the Earth $\Pi^2 + \Pi^2$ forms another border to bring to an end the atmosphere at a value of Π^2. All the while the outer space liquid heat blanket that covered the star since time began slowly move aside as the growing

Second Over All Overview

atmosphere pushes the envelope aside. In this manner a star is born. The star seeks independence and the galactica demands control, just as the top acted when spinning severely within the atmosphere of the Earth. The Sun is all-liquid inside and in the process of turning space gas into liquid heat the Sun is able to discard much of the liquefied space as light photons throwing the light back into space. In this manner the heat collected inside will apply as a separation to push the star to move outwards and with that the star will bring about stronger independence and relieve of the galactica domination. In the stage of development, which I now am referring too the star transforms from 3 to Π and Π to Π^2. At Π^2 the star begins to turn the flow of light from flowing towards outer space returning the light as space dismissing towards the inside.

It goes about in a balance of cold inside the star centre and heat fuming in outer space. One must remember that not withstanding culture the outer space is gas being much hotter that on the inside of the star where material freezes away to singularity In **"STARSTUFFIN" ISBN 0-9584410-3-0** I explain this in much detail. As the star progress the star starts to dismiss space more than the star liquefies space and this starts a cycle of light flashes and star seemingly disappearing. These stars are called by many names but are the Pulsating variety of variation.

That is how gravity applies because it is a matter of relevancies applying between space holding and demanding conditions and space reducing in relation to insufficient motion bringing about much less space duplicated and space demised. The space duplicated brings about mass as a result.

Material comes into place when singularity overheats that starts the process of Π^0 going onto Π and such overheating creates motion Π going onto Π^2 is about motion coming about and forming border relevancies that becomes time as well as space and by introducing this motion singularity applies the value Kepler stated $a^3 = T^2 k$ but which I changed somewhat to apply in terms of singularity as $\Pi^3 \Rightarrow \Pi^0 \Pi^2$. While singularity is splitting the Universe it is dividing the Universe in a joining effort by forming a dividing unity. On the one side is $\Pi^3 \Rightarrow \Pi^0 \Pi^2$ and on the other side is $\Pi^3 \Rightarrow \Pi^0 \Pi^2$ also and it might seem as if Π^0 is a joining value and the truth is that it is a parting value although shared by both sides equally because it runs into infinity and out of infinity on the other side thus providing the Universe with two equal sectors being the same.

Space is created by motion $\Pi^0 \Pi^3 \Rightarrow \Pi^0 \Pi^2$ that duplicates the material expanding it into four $\Pi^0 \Pi^3$ sectors whereby the motion created cuts the space in two halves $\Pi^0 \Pi^2$ that divides the space ($\Pi^0 \Pi^2 + \Pi^0 \Pi^2$). The protons causes motion ($\Pi^0 \Pi^2 + \Pi^0 \Pi^2$) and by creating space the proton find the ability to dismiss space $\Pi^0 \Pi^2$ because it renders material from Π^2 to Π^0 by diverting Π to Π^0. The proton holds a value of $\Pi^0 \Pi^3 \Rightarrow \Pi^3 \Rightarrow \Pi^{-1} \Pi^2$ because the gravity Π^2 is always reducing and therefore is diminishing Π^3 by placing k^{-1} in a negative relation as the proton then returns to what singularity started from Π^0. As the process places the Universe in a reverse by providing growth such providing of growth stimulates singularity and prevent overheating. This motion duplicating as much as reducing takes place on both sides of the Universe in the space as well as the motion part and therefore the proton holds a dual value of ($\Pi^0\Pi^2$ + $\Pi^0\Pi^2$).

As the star diminish space within the star as claimed by the star the atoms shows a reluctance to submit to the demand. But as motion differentiation bring about the decline so will the atom reduce the quantity factor and later the occupation factor until the star becomes a single atom with no neutrons.

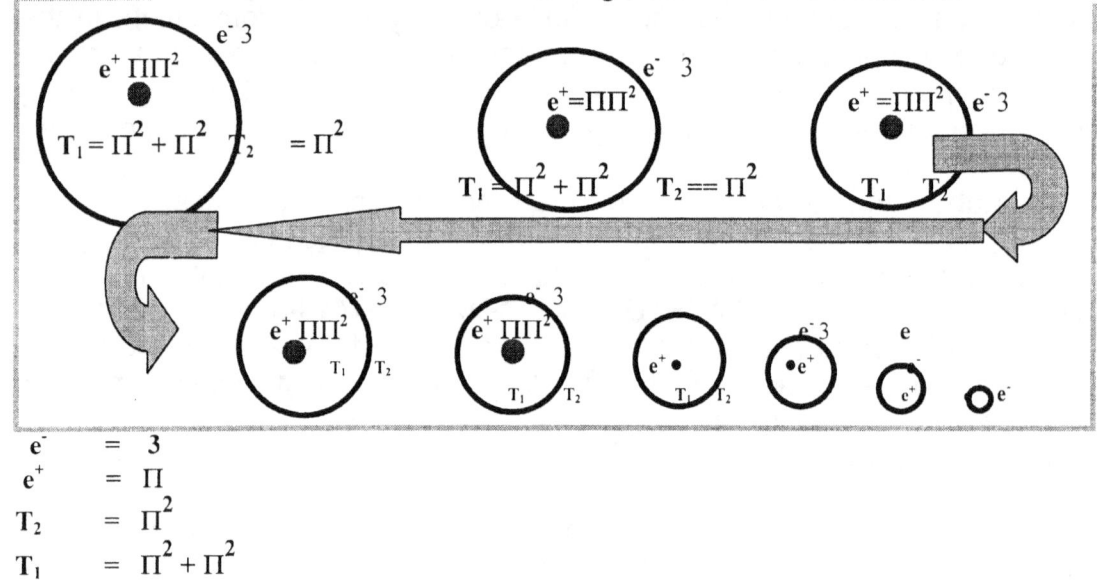

e^- = 3
e^+ = Π
T_2 = Π^2
T_1 = $\Pi^2 + \Pi^2$

The best way to explain the relation between 3, Π, Π^2, $\Pi^2 + \Pi^2$ and Π^3 must be in the confinement of the Earths atmosphere using one of the best known weather phenomena, the way rain forms.

The orbiting of the electron

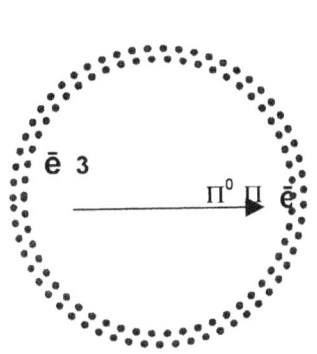

> THE UNOCCUPIED SPACE-TIME REPOSITIONS TO A NEW LOCATION, BRINGING ABOUT A RISE IN HEAT. COOLING WILL COME INTO EFFECT BY THE METHOD WHERE REPLACING SPACE HEAT WILL BRING ABOUT LOWERING TEMPERATURES, BY CANNIBALISING OTHER PARTICLES' CLAIM TO SPACE-TIME IN THE FORM OF

The same apply whether it is an atom construction and whether it is a star or a galactica. The rules apply equally to all in the same manner because it drives by space in motion consuming space to render heat a more a denser value being more intensely concentrated in the distribution thereof. As space is reduced at each level the motion leans strongly towards diverting motion to the lesser space and in such action we find the motion we named gravity. Gravity is the result of time destroying space by creating motion thus cutting the growth in space by half at singularity level. However, in the process it is also filling the vacant space with heat that moved inwards from the outside. As much as outer space is creating a relevancy seemingly in favour of space becoming more, that same relevancy tends to reduce unoccupied space-time by claimed occupied space-time in the better-developed stars.

If my prediction is correct and gravity is heat being forming as space is compressed systematically, then electricity is heat compressed instantly. That means gravity and electricity is the same thing riding only on a differentiation brought about by the time compliance that the two systems have. In each dimension, the light displaces space-time that is more concentrated and therefore influences the projection path of the light's future position in space-time to unoccupied-, occupied-, and densified space-time. Science at this point is using magnetic fields carrying large electric charges to try and produce fusion. Their trying that involves electric charges is also a silent admitting of some agreeing of my statement. Since Π^0 does not affect the relevancy therefore I do not normally include the use of singularity values of Π^0. When referring to the double proton value as $(\Pi^2+\Pi^2)$. Therefore I refer to the singularity value at the proton level as $(\Pi^2 + \Pi^2)$ The motion discrepancy within the atom leads to space-time displacement but that has nothing to do with mass. Every proton has an equal ability to dispense of space-time by motion applied. Since the motion of the atom at the proton level is much more time consuming the space is that much less because the space stands directly relevant to the time. The contraction reduces the space and with the contraction the heat density rises. From this more motion reduces space but such reducing is on another dimensional plane where space has three sides, which is duplicated by motion once more. The duplication $(\Pi\Pi^2)3$ brings about the one six dimensional Universe we live in. With every plane using more space it affects the time considerably and the delaying of time is reversibly affecting the time duration. Singularity reducing space forms a unit and such a unit works on space reducing within what space captures and not on size particularly.

The less space available in whatever area applying the more the reducing of space that is captured will affect the space that motion captures and motion of reducing such space. By applying this rule material captured in a larger space will always draw towards the centre of such a much larger space where it then will have to match the value of the space surrounding it as the dominant space by reforming it holds because the captured space cannot repeat the expanding and even less the duplicating of such space. By the specific density coming from this the material will reduce to a point in the space it claims as an individual structure where such duplication will be in harmony with the duplication of rest of the space surrounding it. This tendency of smaller sized object too move towards the centre of a structure as to put the motion and duplication of space in harmony with the rest of the space we call gravity and from that a relation comes about where the structure captures in the larger space fight to protect and sustain individuality by securing its structure and form. That is what time is. We find humans being as confused; as humans are humans confuse eventualities with time. Time is not a flow of events following the will and the wishes of persons in life in a complex by chance progress from which we can report about a history written in blood. Going back in time will not bring Napoleon back to the future because the events played their part while time was flowing and the events can and will never happen again. What life confuses time with is again putting life in the pinnacle of time and history forming the focus of the cosmos' entire reason to exist. Time and history humans see as bridging the time that it will take on Earth in comparison to the time it will take to travel from here to there where there is about a billion light years away. That is nonsense because if that was possible the traveller would first have to return to time just after the Big Bang when space was that small, but remain in the time frame that traveller now has to bridge the distance by short cutting the time. That is the way we think of human eventuality by becoming my father's farther as I shift time to and fro. My travelling through space is projecting life's eventualities through space reduced by time being in some ridiculous short cut. What we in life confuse as the **centre of the Universe** is the position the Universe grants each of us as being in the **centre of the Universe**, which we are. But each one is in the centre of each persons' particular Universe matching that person. My being centre underlines my eventuality I position as paramount but that is eventualities and even if there was any possibility (which there is not the slightest chance of) any return to such an event will most defiantly not repeat in the way it transpired on the previous occasion because some idiot will have some other idea in his head at the time and that will change his behaviour which will change the outcome of the future being much different as it was in the past. Eventualities run on human decisions and not the driving of the cosmos. Any going back will most probably lead to even more confusion as before where the human spirit in charge of messing everything up will do its purpose even better and we will be in an even stickier situation than we were before. That is human mind becoming events but without the human mind no rock will run into a car as soon as a woman driver pass the rock. Cosmic time is the way the distance from the centre singularity establishes space holding the time relevant to such space being established but the flow of time in relation to the space being available from the centre. $a^3 = T^2 k$ and you change **k** to find everything else in the equation changing. The space traveller will have to pass through the centre of the Sun in order to escape from the centre of the Sun because it is the centre of the Sun keeping such a traveller relative to the centre of the Sun. The motion the traveller has to project will be quicker in pace than the centre gravity the Sun produce at the very, very centre of the Sun. Cosmic time is the relation the object in motion has while being in motion with all aspects of all motion applying being relative and that includes the heat to space ratio determined by the motion forming interruptions in relation to all other factors sharing space-time. This motion influencing time is because all gravity in the solar system is motion in ratio the motion coming about and from the very centre of the Sun where all motion is standing still. Time is motion on the spot relating to the rest of the Universe but under the command of the motion less ness of the strongest centre spot killing motion. If one whished to travel beyond the Sun one first have to break the barrier of the centre of the Sun where the Sun stops all motion and when that is overcome the traveller would have a vehicle stronger than the centre of the Sun By accepting the Big Bang theory and acknowledging Hubble's evidence science has to have a look at the way they cling onto Newtonian views. There is no chance that by purely having the diameter one can calculates and determines the gravity of the star. If that was the case then how can science ever try to explain the pulsar or the Neutron star? Even by bullshitting in the face of contradicting evidence there is no manner in which to explain any star by birth or by death because a star cannot be born in the manner science try to prove as much as a star cannot die in the manner science promote. I truly cannot see how science can stick to the notion of the speed of light being a constant when all evidence prove the very opposite. How can the Black hole break the constant of the speed of light because then the sky must fall? Before the light tries to escape the motion the Black hole introduce the light goes into reverse by falling into the Black hole and to come to that reverse the light will first arrive at a point of being very motionless and therefore the light will lose the space it holds in that specific position in that very moment. The light will go slower and slower until it comes to a standstill from where it turns in direction and go into a reverse as it then accelerate faster than the speed of light down the gravity funnel towards the centre of the Black hole.

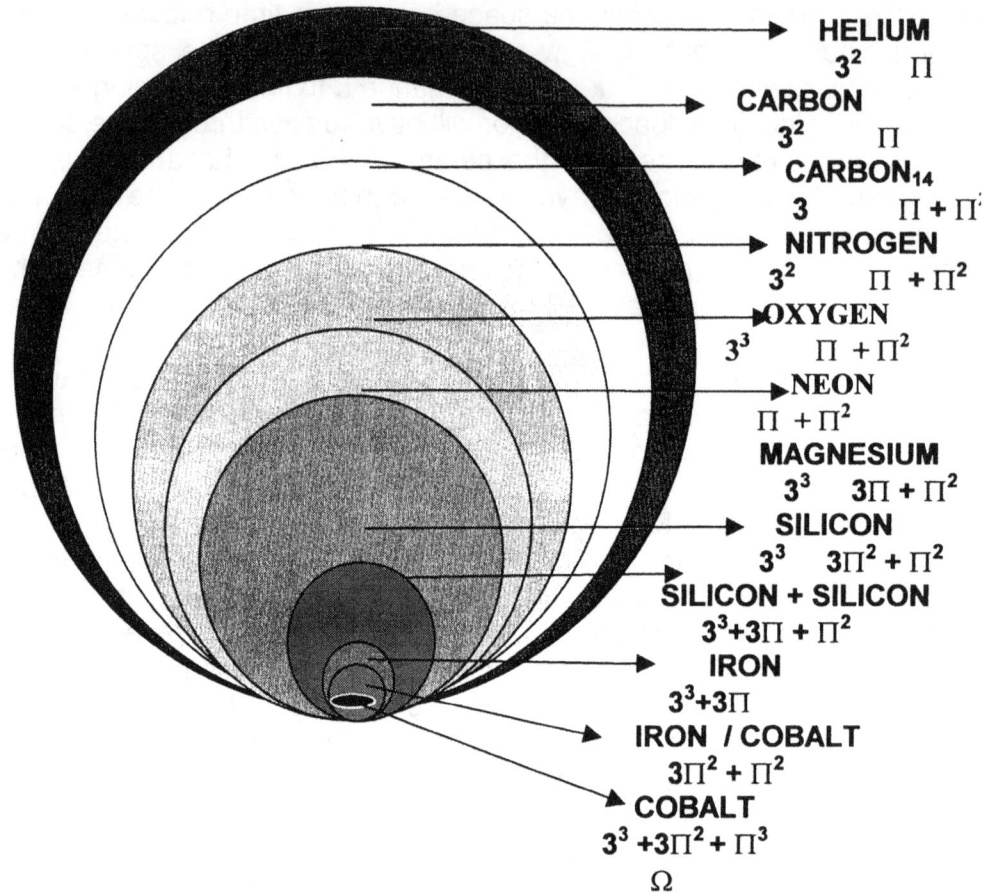

All the time the light holds an even relevancy but since the relevancy overpowers the light speed of $3\Pi^2\Pi^0$ which is the ultimate antigravity possible that relevancy can change into 330 km / h or it can accept the factor of one as it rushes down the gravity funnel of the Black hole. When the factor is one it will show no individual motion and therefore relinquish all claim the photons had on space. Light will go dark. But during this all the relevancy factor remains $3\Pi^2\Pi^0$ until the stronger Π^3 relinquish the space and the time by rendering the light motion no relevancy. Wherever light flow and wherever singularity apply Π^3 light will hold the relative motion of $3\Pi^2\Pi^0$ in relation to singularity being a proportional factor to all changes in space-time that may affect space-time. A constant derives from a law and the cosmos does not break laws. The atom fills the space inside stars and the atom must be the space that the star ultimate will deplete. The speed of light is just another part of atomic space because science are aware that once in the beginning of the Big Bang the outer space comprised of a density stronger than what the electron was. The complexity of the Universe rides on the balance that singularity introduced and the entire Universe invested in. In the centre is singularity standing still. That will remove all space because there is not motion that produces space. But the standing still of singularity produce space through expansion by overheating and the overheating constitutes of motion where only motion will provide the required cooling. But it is both the occupied space holding the seven relevancies to singularity that expand and as that expand the ten relevancies holding occupied space within unoccupied space that produces the other expanding factor.

Second Over All Overview   671

..●●●● ● With us now having the wonderful gift of hindsight we now know things did change rapidly as our experience now tell us. The atom grew, but it is the inner part that grew and not the Universe that expanded. As the inner part of the atom started to absorb what the electron lost in intensity, the proton / neutron combination advanced in the space- time the charged. If you will then one may call it the learning curve material went through to be material. One thing is sure, and that is that the Universe did no grow and neither did the Universe get bigger. Particles forming the Universe with lesser space-time received relevancies as time moved over to space.

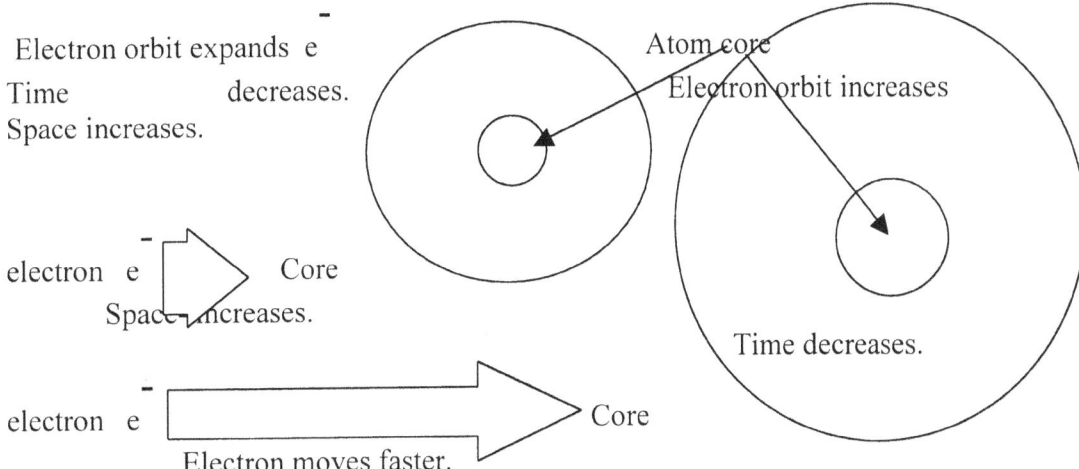

It was always a case of the more **k** expanded, the further time **T²** had the relevance with **k⁰** and the more space-time **a³** did the atom occupy. The reality about this diagram is that every aspect of the cosmos relates to this diagram. Since the Big Bang the atom increased because all the relevant space in the Universe increased as the balance shifted. With that being the case the reverse has to apply as the star is about contracting and reducing space by motion producing gravity.

At first a star has it to task to compress time and return space to liquid. By doing that the star in its very young days take the relevance of the atom from what it is in outer space being $(\Pi^2+\Pi^2)(\Pi^2\Pi)3$ to $(\Pi^2+\Pi^2)(\Pi^2\Pi)3^3$. By the stars concentrating of time from a gas to a liquid the star is increasing the density of the liquid while the star is decreasing the relevance of the atom and allowing the atom to grow much more prevalent than it does in outer space. The star is filled with atoms and at first the main gravity consist of individual atomic gravity casting the spoils of dismissing space-time and fighting an individual battle to generating a comprehensive accumulative and compiling effort as one star. Only after the governing singularity that all the atoms helped to achieve and put in place becomes stronger tan the compliment of individual atoms per sector, can the star progress into a next phase. The phases sport many names because if we humans are good at something then we are good at naming what ever we can. However in the end it is all the same as the star just loses another cover in time or layer in element cover. The star is shrinking away as much as the Universe is expanding and the star is reducing in space as much as the Universe is developing. The one shows gain not while the other loses but both works from an equal stance.

$(\Pi^2+\Pi^2)(\Pi^2\Pi)3$

1kg becomes ⇒1 - 100 kg

A **100 kg** in outer space will register as anything from **100 kg** to as much as increasing

From what I gather, the red Giant is the equivalent of a gas structure such as Jupiter, only containing a lot more space filled fluids. Strains placed on the atom in the Red star, which is just a liquid bowl of contained heat separating atoms, will be precious little. There will be many protons but the protons will be spaced individually where each proton claims individual space securing individual survival since the space dismissed is relatively little and the space duplicated are relatively very much favouring the duplication and not the destruction of the space.

space-time to **10000 kg** the Red star because the atom will have to release very little space. In reality the Red star giant as they wish to call it is only a sloshing liquid bowl of heat, which remained liquid since the Big bang reduced the liquid from the Universe. The Giant star is not a star but it is a future galactica waiting on its turn to gain in gravity.

From outer space the atomic relevancy is as follows:
$(\Pi^2 + \Pi^2)$ Represents the proton in relevancy to singularity Π
4 Is the time aspect of spin creating motion that is creating space.
x Π (singularity) = 112.31. That is the limit placed on the atom within the boundaries of what we consider to be the Universe. That will remain a unit
From outer space the atomic relevancy is as follows
$(\Pi^2 + \Pi^2)$ Represents the proton to singularity Π through out the Universe
4 Is the time aspect of spin creating motion that is creating space.
10/ 7 Is the **space(10)** in which the **material (7) spin** according to the Titius Bode principle.
7/10 Is the **material (7)**, which **spin through the space (10)** according to the Titius Bode principle.
Outer space has heat secured at **10/ 7 X 4($\Pi^2 + \Pi^2$) = 112. 8** while the star through motion generate a requirement to heat that establish a flow of **7/10 X 4($\Pi^2 + \Pi^2$) = 55.27.**

Yellow dwarf

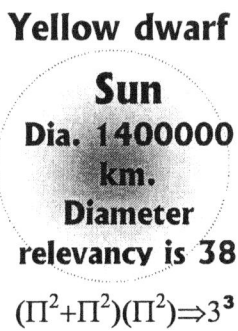

Sun
Dia. 1400000 km.
Diameter relevancy is 38

$(\Pi^2+\Pi^2)(\Pi^2) \Rightarrow 3^3$

In the star the core has developed considerably concentrating heat to the inner regions. Within the inner regions the heat concentrating will sustain the proton in a manner it will show much more growth within the denser heat surroundings and as the proton grows in stature the proton finds a stronger ability to dissolve space and with that ability it can therefore dismiss more space with the increase in mass as a result. The more the mass the more it produces more massive protons that can displace more space-time. In this star the favouring shifts to apply equal to all ends of gravity.

$\Rightarrow$ **100 kg on Earth** $\Rightarrow$ **becomes 1000 kg in the sun**

The thing Newtonians have to lose is their obvious fasciations they have grown to have with the sun. The sun is one of the smallest stars there can be because if it was any bigger, it would not be able to allow life to be in its influence sphere the sun holds. The sun is the red giant scaled down to fit life and to allow life. More gravity would have destroyed all chances of protecting life and as much as the sun at the moment is protecting and nursing life, it will grow into a monster that will destroy all forms of life with the gravity it will charge in the distance future.

White dwarf

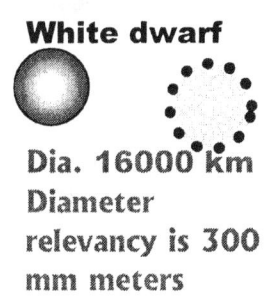

Dia. 16000 km
Diameter relevancy is 300 mm meters

100 kg $\Rightarrow 10^3$ tons

$\boxed{(\Pi^2+\Pi^2)(\Pi^2)3^3}$

In this class star by this time has extended the heat concentration to a level that the star on the outer regions are as brilliant and strong as the electron concentration can produce and the inside centre is starting to diminish any space that can hold heat. The space in the centre is so critical reduced it can no longer allow the flow of light on the inside while on the outside the star is one unbroken solid electron and the protons are diminishing that space also at a rate. To find a compromise as both ends of the star holds both ends ultimately the compromise will be to favour the light by the producing of light. Then it will favour the diminishing by contracting all light to the inside in the direction of the core. In this way, both ends find equality at both ends of the duplication and the dismissing of space-time

The star at that point is getting so cold it is absorbing space without requiring the mediation of the atom serving as a storage provider. At first the atom was a pantry to secure a required flow of heat in order to maintain the governing singularity with the protection it needs. Then the governing singularity gains the prominence where it starts to demolish the atoms in order to maintain the flow of space-time and the atom with mass becomes a taxing burden. From this point the star will become darker as the protons grow more massive finding the ability to demolish even space holding heat so concentrated the heat is pure electrons and photons filling space as material would. It is discarding the electron from the atom. The more massive protons must create much better conditions for fusion.

Neutron Star ●

$(\Pi^2+\Pi^2) \Rightarrow (\Pi^2)$

100kg $\Rightarrow 10^9$ tons
Neutron star
Dia. 19.2 km.
Diameter relevancy 3 mm

$(\Pi^3) \Rightarrow (\Pi^2+\Pi^2)$

In the neutron star the space dismissing reached a level where no heat as such is available any more because conditions inside that star has gone to pre Big Bang conditions. The space applying is not sufficient to separate atoms any longer and the reducing of space has limited the motion of space to almost a standstill. There are two forms of motion being very apart. The heating allows the neutrons the ability that photons had before the atom's dismissing space through motion decreasing and in motion the neutron can expand the space by duplicating the space into which it expands. In that manner, it finds a way to escape much as if the spacecraft does when launched from Earth.

By reducing the atomic motion the neutron has the ability of a liquid thus it remains in motion. It is that motion that the neutron then uses to distinguish the space it claims from the space the star pursuits. As the space-time reduces, the neutron surges outwards using the motion of being a liquid to sustain the space it claims and finds the release it requires. The protons however reduce all space, which may be inside the atom by discarding the neutron from the atom leaving only room for the protons housed in space less ness.

The star is in a state of changing from what the Big bang once offered and what was gained after the Big Bang came in place to where light has no function because what space offers is far to small to accommodate something as enormous as a photon gravity is motion and the photon is to large and to slow to offer any assistance any longer in the developing of the star.

Black hole Dia. 9.8 km. Diameter relevancy

Within the Black hole only singularity remains. It is the single dimension of space less ness where the singularity retains the single unit of all singularity it will always contained. That is the epitome where the Universe is heading, which is the final goal that all stars securing singularity will finally reach and there is no unsuccessful candidates that will float in darkness in space forever and another day. The star has completed the journey and is now in the effort of demising outer space as an entirety.

100 kg $\Rightarrow 10^{19}$ tons The mass increase is only an educated guess but it is most likely that a mass of **100 kg** will become **10^{19} tons** in the **Black Hole** but it is much more accurate than the official number presented as a calculated value by mainstream science.

 The galactica and the star that ended as a Black Hole started off equally in mass but much different in mass distribution. The star that eventually became the first Black Hole united the mass it had forming the controlling singularity, while the galactica distributed the mass it had in separate cluster packets we call proto stars.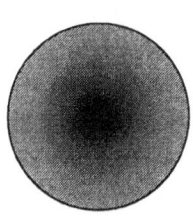

The Black Hole and most galactica started off on the cosmic developing journey at the same time period released. Both had equal singularity generated, except for the way the material compliment produced the governing singularity by motion of duplication and proton cluster density.

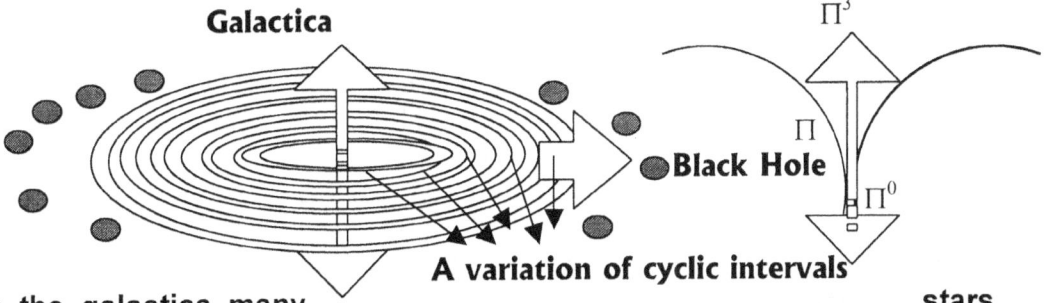

In the galactica many formed units and the units formed partitioned identities in clustering as the units became stars-to-come that provided atoms a wide range of units to form whereas in the Black Hole, the material was the same in total but the variation of space occupation was much less with much more protons in numbers per cluster unit. In the galactica the time factor gave a variety by which the different material could develop and form many equal governing singularity. There was a range of cyclic periods in which to develop and gave a range of singularity developing periods.

This comes from the density and the duplication in every star. Through the spin the atoms provided space-time that form the unit. Every atom becomes a little pump pumping in space-time by dismissing space-time. The quality of spin would achieve a governing singularity. In the Black Hole the material that connected to the centre governing singularity remained the same as if still filled with atoms In the galactica having a centre "Black Hole" all the stars spinning produces one unit as stars in cluster provide the motion to establish a gravity centre. With the massive proton numbers that forms young stars as if the young stars are proton clusters a very high demand is placed on dismissing in the centre coming from the duplication. The galactica develops by developing its atoms while the Black hole develops by dismissing its atoms. At the time the Black Hole was developing, the galactica was only duplicating and waiting for development by having the dismissing factor developed. The cluster of stars in the galactica reach maturity in singularity development and by such maturity find release from the centre governing singularity, however it never strays out of cyclic developing order and remain part of a galactica governing singularity. In the case of the Black Hole there was one unity from the star that formed one governing singularity where all atomic singularity remained attached to such a centre. The centre was strong enough from the start not to allow individual development in the unit and all material developed towards achieving one united singularity. The mass however connecting to the main singularity in the galactica is the same in worth as that which connects to the singularity in the Black Hole. The Black Hole singularity made the mass redundant as it united all proton singularity into one unit. The galactica has the proton mass more widely spread and brings space to parts that form as units that forms individual clusters that become stars with many different centres attaching to a centre singularity but with much less confining value. The cluster we call stars move away from the centre singularity that keeps the galactica united. As the singularity governing the star finds mass development, it would motion away from the centre attachment of the galactica and progress to achieve independence as the spinning top tries to achieve. It is this developing law that we mimic in the top we spin. The Black Hole eventually destroyed all means to duplicate material and was set on dismissing space-time. In a book of mine called **"STARSTUFFIN" ISBN 0-9584410-3-0** I describe and explain how the two developing stars progress. (A galactica is just another Big star with spawning abilities as it develops stars while staying on route. By releasing stars when the time comes for such release of star to occur it keeps cosmic development on course by releasing stars. All released stars will eventually, in far off time to come, end their development as Black Holes but that is in time to come.). I am of the opinion that most if not all so called giants stars are galactica that has not indicated any form of growth because of the little intensity in dismissing there are and the high value of duplicating the unit shows. The difference between the two options that the stars had when the galactica and the Black Hole-to-be formed was one in the form of the galactica concentrated on duplicating and by duplicating developed pockets of forming dismissing eventually and the other star expired duplication long ago and only concentrated on dismissing.

The Black hole reached the point of singularity. Singularity is a point where $k = \Omega$ placing $a^3 = \Omega$ and also $T^2 = \Omega$. I feel sorry for the mathematicians but this time they cannot take it out the cupboard and play with it for a while. It is too small for mathematics and only fits into thought. On Earth we may give it a meaningless value of being one molecular mass. At first the space surrounding the atom holding the heat relevancies diminish as the space within the star compromises in favour of heat accumulating. This is quite in coming into place within the gas structures and is starting to apply in the solid stars.

Within the star the atom will follow the demising trend on space set by the star.

All the while the atoms has to comply to the rules within the star as demand on the atomic space claims sets new standards. At a point the reducing of space becomes so demanding that the factor of light finding an ability to apply motion disappear as the massive structure draws even light towards the centre because at that stage the photon no longer has the ability to duplicate space as it displaces space. The photon must then surrender space due to a lack of adequate motion applied at a relevancy of 56.6. At the beginning when a star establish independence from the galactica outer space which is hotter than the star itself, it is not the heat but the motion as the totality of all the protons working as a group within the secure unit of the infant star that dismiss space-time and the total displacement finds a focus in the centre of the star.

At the beginning when a star establishes independence from the galactica going into the outer space that is forming (which is hotter than the star itself), it transforms not the heat but by the motion, that secures the heat. The totality of all the protons working as a group within the secure unit of the infant star that dismisses space-time and the total displacement finds a focus in the centre of the star.

At first the star may only demand a reducing focuses on the 3 or the Π to become independent and secure defined borders or atmospheres but as the star develop through the intense centre it forms, the protons will grow and bring about through fusion much more active displacing that eventually forms fusion. The more protons there are in the least space there are will bring about the strongest gravity there are. As the star development progress the dominant gravity generating protons found in one location begins to form within the centre of the star where the major heat is accumulated. The shift takes place from the focus at first on the outside rim of the star developing towards and then to the middle sectors and eventually to the centre of the centre. In this the focus of the displacement gradually moves from a massive number of single proton atoms to a massive number of atomic protons. The quantity of protons efficiency move over to form a focus on to the quality in proton numbers in one unit of a centre and then the dominant atom displacement will not be Π but it will become $Π^2$, later **$3Π^2$, $3^3+3Π^2$** and so the centre progressively develops.

Then further premiums on the space-time that the individual atom may require becomes resolved as the space demand within the star annihilates all atomic motion of individual atoms and the neutron, which is the representing of motion in the atom, abandons the unit of the atom to reproduce space in the manner the photon did in normal stars. In the end there are by then no hint of any photons left because a lack of motion brought on a total demise of photons. The star is not dead! Eventually only the collapsing of space can sustain the proton activity still present in the star as singularity sets in and diminish all motion activity within the star. As explained on the previous page this mass comes from the fact that the proton lags in motion to singularity, which is motionless. Explaining this concept or the following concept will also take to much time but what I mention in this page I have a book of more than five hundred pages that covers the whole aspect and has the name of AN OPEN LETTER ON " **STARSSTUFFN'** " ISBN 0-9584410-3-0. I will just quickly touch on the thought. What we see from the outside is just the opposite of what is applying on the inside of the Universe where the "inside" is singularity. Considering it in that light that is why all the information we receive from the Universe by means of light is a mirror reflection of what is taking place. I shall quickly mention the most basic idea of this concept: The motion produces time and the time brings about space. To us being in space-time the forming of space by using time is a positive measure because we are on the side in space-time, but from singularity such motion becoming space is a disaster coming into practise and it will take the Universe many billions of eternities to once again correct the disaster. To us the motion is quicker that eternity but to singularity the motion is slower than eternity. By creating eternity minus whatever that deduction slows down eternity where as from our perspective it increase eternity by splitting eternity.

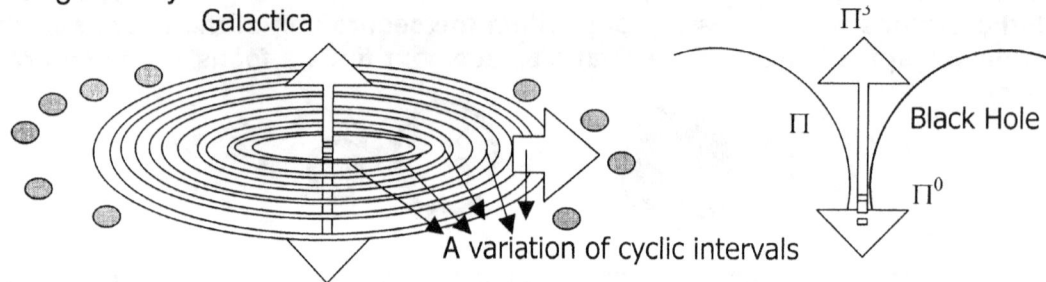

On the inside of those magnificent galactica sporting Black Holes is stars that has not yet comer into their time having their Big Bang. If you will those stars are still in the white hole and still has to come off age. The incubator is keeping them stored just like the incubator is keeping the stars of the same era stored in Betelgeuse. Both are still incubating an era with stars of much less displacement that will come in the future.

Gravity is the concentration of heat running from outer space with a displacement value of 112 to the inner star that has to have a displacement value of at least 55 to ensure the flow of gravity generated by the motion of the object within the boundaries of these two limits.

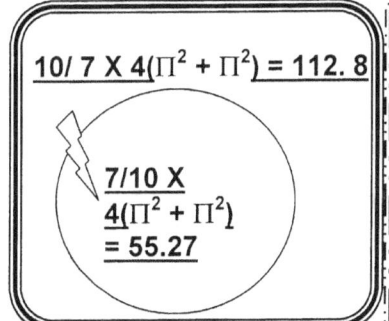

19.74 + 13.01 + 3 = 35.75
35.75 X Π = 112.31
The atomic total times singularity is the maximum displacement achievable

One property I give indicates the relevance of the atom $((Π^2 + Π^2)) = 19.74$ while the other concerns the flow density of outer space in relation to the centre of gravity within the star **10/ 7 X $4(Π^2 + Π^2) = 112$..** The two are very much linked but indicating how, would be a task that exceeds the purpose of this letter

With an inner core displacement of less than the required 55, the star would not yet have arrived at the point of securing an individual singularity in the presence of outer space at 112. The potential difference needed to generate gravity is 112 coming down to 55

In outer space the limit on the atom is at maximum $10/7 \times 4(Π^2 + Π^2) = 112.$ and at minimum within the centre of the star it is $7/10 \times 4(Π^2 + Π^2) = 55.$

$\{((Π^2 + Π^2) = 19.74+ + \{((Π^2 Π) = 13.01\} + \{3\} = 35 \times Π$ is the borders of outer space or then the atom that charges outer space into forming outer spec
$\{((Π^2 + Π^2) = 19.74$ resemble what is the value of the proton
$\{((Π^2 + Π^2) = 19.74$ resemble what is the value of the neutron
$\{3\}$ Resembles what is the value of the electron
$35 \times Π$ resembles what is the value of the full atom compliment in relation to singularity

The atomic total of 35. 75 x Π (singularity) = 112.31. That is the limit placed on the atom within the boundaries of what we consider to be the Universe. That will remain a unit.

The same formula applies to setting the boundaries that limits the possible duplication as it limits the boundaries after which space-time goes flat. The relevancy is $4(Π^2+Π^2)$.

Generating gravity requires a proton number $4(Π^2 + Π^2) = 55.27.$

The outer walls of outer space is **10/ 7 X $(Π^6) / 6 = 112. 8$** while the position that the atom demands space is the value iron have as a potential difference. It is in the **7/10** and the **10/7** that the limits are placed. It is seven spinning about in ten crossing singularity by turning about the inner core of the star.

$10/7X(Π^6)/6=112$ **Outer space**
$7/10(4(Π^2+Π^2))=55$ **Gravity**
$3(Π^2+Π^2)=59$ **proton**
Space collapses $4(10Π+10Π)=62.$ Empty

The factor of **10Π** being in relation to Π³ is a direct translation from Kepler's **formula a³ = T²k** By substituting the symbols used with the actual value of Π the symbolic massage transforms to specific values applying **10Π Is space square 2(5) (T²) in relation to singularity Π (k) being equal to space Π³ (a³)**

Π³ (a³)

In outer space the motion **2(5) (T²)** of the material **Π³ (a³)** keeps space in dimension Π **(k)**. But this motion produces a relation that apply to material groups such as stars relating to space holding groups such as outer space which I refer to as geodesic space in more advanced

a³ = T²k is Π³ (a³) = 2(5)(T²) Π(k)

When using the atomic relevancy I refer to the proton relevance in space in example ((Π² + Π²) and then how space will relate to accommodate the atom as the atom as a group facilitate the star and accommodate the stars unifying requirements.

In the expression **10Π** relating to Π³ it is space flowing towards the star centre in approximately an equal manner as volts flow from space to the Earth or Neutral whatever name there is to choose.

One must see the **10Π** not for what we read into the numbers as such but what it represents. The number **10** is the square of space that stands for space outside gravity in the place of Π² and is the square of space relating to the ten positions in relation to singularity dot as Π. That is the space in which the motion is providing the establishing of gravity Π² in relation to the creation of space by motion Π³.

The star on the inside cannot support space up to equal or beyond **2(10Π)** before ultimately collapsing the space dimensional support of 6 sides in the square of space **(10)**

On the other hand can the space in the geodesic securing the presence of the atom hold space up to the ability of 112 protons displacement secure **10/ 7 X 4(Π² + Π²) = 112. 8**. This is theory because we well and truly know that it is actually **5(Π²+Π²) (Π/2)² (3/5) = 244** which is the number of neutrons and protons that will allow Plutonium the ability to remain a constructed atom within our Universe. But as one can clearly see it is as volatile as no other element and is on the very edge.

Plutonium holds at 94 and as an atom almost falls outside space-time reality $3\Pi^2$ as it is on the very border with a possible increase in displacement of **5(Π²+Π²) (Π/2)² (3/5) = 244**. In the sun however the dimensional change is 10⇒10Π² in comparison with our change of **10** ⇒3⇒Π. With the universe being 7/10Π⁶ /(6)=112 and the sun **at $ =(Π⁰)** 10Π² = **98.696**

Everything in the Universe pivots around the atom and cementing the atom is the proton. The proton is securing the atom as it is securing the Universe.

The outer walls of outer space is **10/ 7 X (Π^6) / 6 = 112. 8** while the position that the atom demand space is the value iron have as a potential difference. It is in the **7/10** and the **10/7** that the limits are placed. It is seven spinning about in ten crossing singularity by turning about the inner core of the star.

The factor of **10Π** being in relation to Π^3 is a direct translation from Kepler's formula **$a^3 = T^2k$** By substituting the symbols used with the actual value of Π the symbolic massage transforms to specific values applying

10Π Is space square 2(5) (T^2) in relation to singularity Π (k) being equal to space Π^3 (a^3)

In the star the balance bringing about space-time flow is in the iron displacing limit of an atom not holding more that 56 protons because the atomic relevancy is **7/10 X 4(Π^2 + Π^2) = 55.27** whereas the neutron reaches the value **of 2 × 3 × π^2 = 59,22** the double proton value, it will respond by returning to a space-time value. This is as far as the atom will go down eternity and no further.

Where we are is not the only place that is possible in the Universe to be. This concept is as wide as the Universe is and I am by no means getting into that argument in this book because that argument covers 650 pages pf the book **Matter's Space In Time: The Hypothesis ISBN** I S B N 0-9584410-3-0. What I will refer to in the following few paragraphs is what applies to our Universe forming our space in our time concept. There are as many possibilities of different Universes all contained by one Universe as there are names for people on Earth and that even is underestimating the possible quantity by the indefinite, but I refer to the one I share with all my fellow Earthlings circling on route around a star we named the Sun. In the Universe I am able to witness the proton holds $\Pi^2+\Pi^2$ giving a displacing of space in the duration of time as **19.74** of what ever you wish to name the measure.

The Universe holds a displacing value of **10Π** in relevancy of the motion applying Π^3. Dividing Π^3 by the square of space as **10Π** leaves **9.86** or Π^2.

When the centre displacement of a cosmic structure has a group atomic displacement at the core that is exceeding Π^3 the star qualifies to form an independent structure as the outer space it separates from are **10Π**. When **10Π** shows a relation to the inside of the star holding a displacement of Π^3 the value of Π^2 =9.8696 becomes gravity which the spin or motion of the star will produce as it moves in space through space. By having Π^2 between the **10Π** and the Π^3 inside the star becomes independent from the outer space that captures it. The motion inside the star delivers the independence the star requires to separate from the centre of the galactica and proceeds as an individual star. There is no star in our Universe showing this weak gravity but in other parts of the Universe there are such stars coming into operation. Such a star will not be able to produce space that ensures total independence. It will not yet have gravity that is forming electricity on a grand scale.

With the displacement of iron being 55 + the iron atom has the capacity to dismiss space and by doing that it has the ability to generate such proton motion as to remove space all together from a selected area on conditions of motion producing a connection with singularity. Such connection we call electricity and the diminishing of the space we call an electromagnetic field. The electromagnetic field is the reducing of space between the element by name of copper, (**63**) where we find copper exceeding the border of $2\Pi^3$ and that of iron Fe_{56}. At $2\Pi^3$ the space-time will become motionless and being motionless and the element iron producing the motion to generate the gravity micro, which carries the human name as electricity.

Space-time displacement which also is motion that reconverts space from heat back to singularity start to achieve a duration in time putting such duration above what the Universe reserve as having the ability to duplicate. Above **62** then forms the epitome of time. At the point the space can no longer sustain the flow in time to sustain the demand set out by singularity with a dismissing potential of **62** protons. After **62** proton is providing the motion of space-time displacing the space held by the protons break down the dimensional wall created by motion and the atom of which only the proton remain in place at that point in any case completely destructs. Only singularity remains casting all other space-time out into outer space. The star then become a star holding no atom but only contains singularity on the inside. It becomes the all so famous Black hole where all falls down a pit of space less ness into singularity without space or motion.

In the star in the Universe which we are in the proton number of those atoms forming the composition of gravity or the dismissing of space to become eternal in time is **7/10 $(4((\Pi^2 + \Pi^2)$ =55.** Coincidently this displacing value belongs to iron and therefore iron can produce electricity because when applying motion iron with the ability to displace space-time by using the combining motion of **26** protons and **26** neutrons has the ability to confirm space-time to singularity. For this reason stars must have an iron core and if the Earth did not have an Iron core our gravity was not able to generate electricity. What this means in short is that the star then can convert space to light by diminishing space the contraction of space.

At double the value of outer space **$2(10\Pi) = 62$** space within the star collapse since the compactness within the star starts to destroy the space that atom holds as **55.27**. The motion applying within the density the star is creating as gravity then has no space or time to occupy any atom in form using space-time such as all an atoms must do.

When a star find the inner-Core- value of applying atomic spin or motion to create **$((\Pi^2 + \Pi^2) \times \Pi = 62.0$** which is double that of outer space **$2(10\Pi) = 62$** the result is that space depletes within the inner core of the star and the star will start to withdraw more heat from outer space than the star establishes or returns light into outer space At **$7/10\Pi^6 / (6) = 112$** the Universe stretches space-time to the limit we find ourselves in. For this reason atoms that are exceeding the mass of 112 cannot fit in our Universe we have. But it has nothing to do with mass coming from pulling, pushing or shoving. It is about motion exceeding eternity. This is what the atomic number is that can apply motion within the atom centre by the maximum number of protons gathered as a group, which as one group can apply space-time displacement although the practical number of protons in one cluster is $3\Pi^3$. More protons that bring about a group motion will produce a collapse of the atom space. At a displacing value surpassing $2\Pi^3$ the dimensional walls of space leading on the motion forming time (Π^6) / **6** will collapse into the centre of the atom. Beyond $3\Pi^3$ no atoms form a unit because in the practical sense atomic motion cannot surpass the displacement value of $3\Pi^3$ where at such a point singularity starts maintaining space-time without the support of atomic structures and substructures.

The number of protons applying motion produces space dismissing and cultivating heat from space and in that process is the space in time returning to heat by duplicating space. The protons apply motion where there is just about no space and by the motion at that level the proton motion turns space to absolute heat where singularity then dissolves the heat. By reducing the space it intensifies the heat and that returns singularity to what it was when space came about as the Big Bang presented space-time. $((\Pi^2 + \Pi^2)) ((\Pi^2 \Pi) 3)$ is the atom number used to form the atom in the development which brought about the atom. The reducing of space is 1836 times more at the proton $((\Pi^2 + \Pi^2))$ than it is at the electron but the combining effort of displacing is the sum total of all the atomic part.

When the proton $((\Pi^2 + \Pi^2)$ and the neutron $(\Pi^2 \Pi)$ is added to the **3** the electron produces the total dimensional sum produces not 6^2 as it should but **35.75**. With the sum being **35.75** one can see where space will collapse or return to the form singularity provides if it exceeds the singularity connection there is between the atom in total and such an atom connecting to singularity Π.

Then we arrive at the universal six dimensions of three sides in space and three sides in motion bringing about a totalling of $((\Pi^2 + \Pi^2)=19.74+((\Pi^2 \Pi)=13.01 + 3 = 35.75 \times \Pi$ **(singularity) = 112.31** and that is the maximum atom displacing value outer space can tolerate before destroying the space holding the atom all together. Inside the star the proton maintaining a connection with singularity directly will produce the proton value of $((\Pi^2 + \Pi^2)$ **X** Π **= 62.01.**

This means that at this point within the star the protons and above this velocity time cannot duplicate space any longer and the wall of space erected by time collapses back into singularity. Space disappears because time cannot any longer sustain space. Any star having a space with a displacement exceeding the generating ability to displace space to the value or above the value of 62 protons in one secluded given space that is repeated as a unit by motion duplicating space will no longer have the ability to sustain the walls time provides space.

The limit is 62, which is **10Π** holding the one proton Π^2 on the one side in place and the **10Π** holding space in duplicated motion Π^2 and it is also double the space value of $2\Pi^3$. Therefore the gravity a star produce at a maximum point is (**10Π. +10Π.**) converting space to $(\Pi^2 + \Pi^2)$ in relation with singularity Π then collapses back to Π^0.

At a value of 6 X 10, time can duplicate space because 3 X 10 = 30 and that is more than singularity Π^0 extended to the square of space at **10**Π will tolerate. But when the displacing exceeds (6 X 10 +6 X 10) making the duplication of space 60, the walls start tumbling in or space is overhauling time as the one side catches the other side and $a^3 \neq T^2 k$

This **62.01** is the total and the maximum number of protons dismissing a value of Π^0 space-time after which the atom as a single unit or as a group in space and time and ultimately the star as a unit of the combining effort of all the atoms forming the star will abort space or the Universe will dispense of the star, which is all the same effort. The star then has grown back to the connecting with singularity and then forms a Black hole. The principle may sound somewhat simple but it is quite involved in the total explaining. The Black hole that forms is within every star centre because that keeps the star in a unit above and beyond what is going on in outer space and as soon as it can bring about fusion by performing motion it will sustain a proton growth setting the star on its final journey to eventually form a Black hole, although there are two stars even beyond the Black hole. However I would prefer not to elaborate on that at this venture. The black hole hides in every atom because through that tiny space not even being part of the Universe on our side, space-time dismissing is applying by killing space because of motionlessness. All atoms in our Universe are showing equal growth notwithstanding position or motion because all atoms in our Universe are in complying to motion as a result of the Titius Bode law of 10 /7 and 7 / 10 bringing the time split equal to all involved in space-time.

When the atom finds the motion within the star centre or star-core-centre start to reach $(\Pi^2 + \Pi^2)$ **X 3 = 59.22** the neutron moves outside the atom and also outside the star. At a displacement value of **59.22** the atom in the star can no longer accommodate the neutron within the atom and the neutron motion slows down to a point where the Neutron motion is to slow to find accommodation in the atom and in the star. At $((\Pi^2 + \Pi^2)$ **X** Π = **62.01** the proton collapse, which is double the value of **10**Π as well as $2\Pi^3$ therefore Kepler's is doubled or duplicated to serve the motion increase. That is what space-time represents because space-time also represents **k** in variations of distances. It is the doubling of $\Pi^3 = 10\Pi$ to $2(\Pi^3 = 10\Pi)$ which then exceeds the duplicating ability introduced by Kepler. But we must not forget that it is also half the cosmic value of **7/10**Π^6 / **(6) = 112**. At a higher motion the proton moves to outside the atom and form a proton motion by introducing the proton motion to the space surrounding singularity. It is then where it started with, the dot that claimed a spot in the cosmos. But the dot went further and grew a lot.

It started with a dot Π, because that is the only form, size and dimension mathematical logic will allow our brain to accept what would form as the first value flowing from singularity. From the one dot had to come a second phase of dots and a third lot of dots. The dynamics of such dots are smaller than we can understand because such a dot is in negative relation to what we see Π to be, and the deeper we delve in finding the smallest fragment where space started, and that is the spot where time is still eternal as much as we can accept eternity to be. This we find in the aligning of planets where the one dot from which the aligner stem becomes the reference too the distance applied between the aligner and the original dot, or governing singularity or structure in charge of holding position to all orbits following.

The reason why we should first locate the spot is because we can only work from that point forward. By working forward we have to work backwards to locate where we are heading. The cosmos started at a point and where such a point is, we will find the Universe and where the Universe one day will end. Every one knows where the Universe is, because we can see where the Universe is, but if we can see where the Universe is, then we should find the **centre of the Universe** in that spot. Einstein theoretically positioned the point of beginning at a place he indicated where singularity should be. With the cosmos the size it is and space so large compared to our smallness we have no chance in finding the **centre of the Universe**. The Universe started where singularity is and singularity is the sure indicator of the **centre of the Universe**. With all spinning objects holding singularity we then have located singularity in as much as finding the centre of the Universe. The Universe started with a dot forming. That answer arrive from taking mathematics back to a point of being the smallest possible position, far smaller than we may be able to calculate form as we return for a moment to a time before Mathematics developed.

My approach might seem unconventional but through the abandoning of the accepted, it enabled me in locating the precise location of a universal singularity forming a connecting basis of the Universe (this I say with some degree of confidence). The smallest figure there can be must be a dot The only value a spot can have being without space is Π^0. The dot is the only form that leaves all the options open to extend in any and in all directions should the opportunity arise. The only mathematically sensible option about extending a line from the dot will be non-bias progress in all directions equally in order to give a meaningful flow of mathematical equilibrium.

The Pythagoras mathematical principle is the proof and that I explain. The obtaining of singularity is in my rejecting of nothing by replacing it with something being the dot. With the clepsydra or "water thief" Empedocles deducted that air was composed of innumerable fine particles, braking the thought that what we now know is air, was also believed to contain nothing being altogether a space filled with nothing until proven to be wrong so many years ago. Never did science take the lesson learnt back then to the future and out onto outer space. If there is space, there cannot be "nothing" as space is something. The claim becomes obvious when observing the connection between the half circle, the straight line and the triangle, which could also promote all the qualities lurking behind the pyramid. Consider the connection between 180^0 sharing and then one may realise much of the pyramid mystique becomes less spectacular in considering the very basic in mathematics being the Law of Pythagoras on which all mathematics arrived. Once the water thief was eliminated by some human intelligence the matter was left at that. Nothing shifted out to an area we think of as outer space. In outer space they say we now find nothing. There is nothing but an atom here and there and even the atom is covered in nothing. Now I ask you how bloody logic is that?

I wonder why the nothing landed there? Could it be that the reverse came about and because there was no visible "water thief" the very limit of man's suspicions came into practice. Man has always been extremely good in flying from one outer edge to another and if the water thief proved something was present, then the mere absence of a water thief must therefore prove that nothing must be in outer space. It seems much easier to shift nothing than to find what should replace ether after ether was removed from space and replaced with nothing. But what is space as such. What can space be, because with explosions we can clearly witness space created from heat. Our culture prevents us from admitting our vision, but the release of heat produces a "shock wave". That "shock wave" is nothing less than space created from heat released. The space that the release of heat creates re-establishes the position and location of the entire space it refills. We have to brake free from culture of the past and a rigged mind set narrowing our vision. We have to learn to see the Universe with our minds and not our eyes as we can see in the presence of the Black hole we cannot see. The Black hole is only visible by presenting invisibility. We know about the Black hole because we can't see the Black hole. Why would that be?

Because of the manner in which the Universe initially started where more singularity in the relation was unsuccessful to form contraction after the overheating brought about expanding and overheated by expansion much more heat released into the Universe in heat as space uncontrolled than that which remained controlled by singularity secured inside a unit such as atoms or as star or any form of containing material. The container had borders coming about from motion that the unit employed and such motion set the forming structures apart from the rest of space. Once again we see $a^3 = T^2k$ being correct from the start. There is more heat in space uncontained than space contained in some cosmic unit in heat that is volumetric consoled and secured. Therefore to restore balance there must be a position where singularity reduces space faster than heat can fill space. At such a point singularity is taking longer to reduce the space than it takes the heat to fill the space and therefore the space reducing takes longer than filling the heat in space. The space flickering that announces contraction takes longer that the expansion causing the heat to turn to space. Before the heat expansion can begin in progress the contraction already completed the motion successful. This question proves fusion between materials to be the answer. By primarily having 112 atoms as single hydrogen particles in one area will capture a great volume of much space. Changing that number into an atom of single proportions having 56 protons that is housing 56 neutrons in one accumulative structure some heat will release during the capturing process but such heat will become the demise of the space which was holding the heat as well as the demise of the heat by substituting the dismissing of space and building of the gravity applying material in the process such material will push the direction the Universe takes into some final conclusion.

At the start of the Universe birth there was heat that turned to space that turned back to heat through motion applying contraction but there were lots of other where singularity could not contain the expanding by contraction and at that, that expanded more then singularity could contract allowing more heat to be in space than material is in space. Since there is more heat in space than there is darkness because the heat is darkness that turned to space the reducing thereof will bring about more darkness because it reduces the heat.

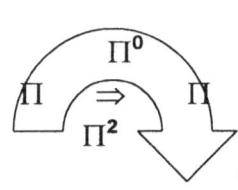

In the region where the singularity present space starting time and where the motion originates space is so little that time is eternal. The space the photon reduces to be 1836 times smaller than what the photon is. Actually in more advanced explaining I shall show the reducing is $((1836)^3)^2$ times smaller that the photon but let us leave that explaining for later in another book at another time.

In the area where singularity release time to contract space the motion of Π to fill Π in the seven Π positions of Π creating ten Π to establish gravity by reducing ten to $Π^2$ the very idea of moving to the other side of the Universe is covering a distance we cannot begin to grasp. The slightest of motion ever possible completes the whole journey and the space duplicated is one mark more that no space at all. We are dealing with mind accepting and not sight ability because where we venture is $((1836)^3)^2$ smaller than the photon we cannot even see but can only use to see, Where the space is that little the time must be that long because time is the very opposite of what space presents. $(k^{-1} = T^2/a^3)$. Being on the other side of the Universe being so small the contracting duration takes so long that the space the photon fill reduces to compact the heat that the photon contains above.

The motion is reducing space in a space that is much more reduced than what any heat concentration in the Universe can sustain. It is far below or far more than the speed of light and captures the space that the photon claims by motion in reverse. The flickering that we find within all stars and by which we name some that is pulsating is taking so long with the reducing of the space it remains on the dark side of the impulse one eternity while the flair of light is then reduced before the impulse can begin to expand. One should take into account that a photon that is released is holding a space that is humanly impossible to measure. The photon is space connected by a spinning motion that reduces the former space to heat where is so concentrated it can and it does cut material. In order to understand the reality in the situation one has to reduce that 1836 times and that gives us the space where the proton finds space to dismiss. It operates in the area below the speed of light and at the limit where motion produces space. The mass the photon creates is so enormous it destroys the space the photon claims That is the reason why the photon has one motive and objective and that is to liberate space by creating motion. If the mass of something as small as the photon is destructed by own mass to the extent it has to move out of the space it holds and where we on Earth find the mass immeasurably small the gravity applying to establish such reducing is more than what numbers in mathematics can present on paper.

Going down the atom passage leads to an area where gravity motion is reduced 1836 times that of the reduction the photon has already been through and we are left with space in the Black hole where that space the photon measures is so enormous that is the demise of the photon as the occupying space destroys the photon. The photon already far to small to measure is 1836 times too large to fit into the space. It will have to comply with as it enters the edge of the Black hole. That is the road in reduction that the photon has to go before it measures up to the conditions applying in the Black hole. The photon cannot liberate the space it holds and thus the massive size that the relevancy of the photon has in such conditions destroys the photon.

From this evidence it is clear that stars range from the one dimension of total darkness to another dimension of total light. In the Black hole a dimension is motion in space where dimensions are just a concept. To those hard liners in Mainstream science that cannot see past the culture driven concept that time is the same everywhere how can those thinkers declare time as a constant where there is already mathematical proof that the increase in gravity reduces time by increasing duration. This was Einstein's calculation that brought this insight. The only manner in which space will reduce 1836 times is if the concentration of the applying gravity intensifies by 1836 times and that increase the contraction 1836 because it has to contract the space into a unit 1836 times smaller. This can only come about if gravity increases by 1836 times to what it was and with gravity changing 1836 times surely it has to influence time and the measure of time. Einstein said that gravity increasing slows down time and what bigger increase do they wish to experience that an increase of 1836 to one. There is the other limit to where stars are still in the cradle blanketed by covering of heat in density beyond the photons releasing. That is why that inner centre of the galactica is so eternally luminous. The time duration on that side of the Universal limit again is so short that the contraction captures the expanding light faster than the speed of light can secure a photon release from that centre. If not that centre must by now be as dark as the rest of the dark Universe.

The escaping of a photon from an area as large and compresses as what a galactica holds and in the presence of such mighty gravity as that has to be present there must be considerably slower than that of the escaping light that is coming from a midget star such as the Sun Yet looking at a galactica the light streaming from a galactica does not seem to be much brighter in intensity that when coming from a star. Which means the light (heat) escaping from the galactica, which has to be billions of times bigger in space occupation than what the light claims coming from a star. That is the other side of such enormous gravity applying in a confined infinite space. Remember that gravity and time is the very same thing because gravity contains the expanding of space in motion. Therefore the time factor inside galactica is going on very, very slow indeed. That will mean that although the speed of light is in affect the release of photons, the escape against the applying gravity flow will in relative terms be a trickle of a flicker.

The flickering is taking so long in favour of expanding the light flowing from it seems a solid by constant flickering. That we also can tell because photons must present flickering if photons are individual particle apart from one another. The accumulated gravity is becoming denser as the heat concentration intensifies toward the centre of the galactica. The relative heat concentration shows up denser on photographs but the light coming from that area is not more intense when measured, in the way it should be with such sustained increase in density in the source of the light. But the increase in light comes from a stronger gravity and a stronger gravity will also produce a slower release of light photons escaping. The Universe is in ranges of space-time.

The Black hole presents space-time that favours the flickering of the dark much more intensely to the extend the dark is far beyond solid The galactica favours the flickering of light with the same intensity as the Black hole favours darkness to flicker, but the galactica in the centre eventually commits so much accumulated gravity, it slows down the release of light and more so rushing from the part in the very centre of the galactica. In the measured precise centre where Prof. Hawking measured a Black hole, the gravity applying is so intense due to the not yet developed that is within that centre that the accumulated gravity of all the yet to develop stars situated in that centre slows the flow of photons down to a stand still and even taking the photons in reverse flow. The light is so intense it can stand still while only a certain amount of light is able to escape. This allows the rest to remain stationary or even to flow back in reverse to the specific dot in the centre of that galactica.

In that centre the gravity of the accumulated underdeveloped stars are so great it allows light to flow in reverse and therefore take time in reverse to beyond the Big Bang which then explains the other singularity that Prof. Hawking detected within the centre of the Galactica that slows down the flickering of light to the extend the light exceeds the limit of solidness as it present itself as being a Black hole holding matter not yet introduced to the centre of the Universe. The space that Prof. Hawking detected has not yet left the battle ground and is still n the spot where the stars of previous era already completed the journey, those in the centre must still experience their Big bang in waiting their turn. If and when one use only light to read what there is to measure from the situation it will most likely lead to the thought of Black holes but then the question arrive at the problem what will produce such gravity and why does such gravity not devour the galactica from the inside. The answering of this question proves that considering it as another Black hole would be the incorrect conclusion to draw.

One must see the Black hole as a lot of solid space within another big space. When the space in the Universe was, little the Black Hole was big and was constructed of the most massive space that was available at the time. Yet at the time when the space which the black hole claimed was more massive than galactica are today because if not, the galactica would have already gone into the Black Hole development stage.

There was in comparison not that much space available at the time. The k that was then present produced a minute a^3 (in our reckoning at present) with an enormous T^2. The universe was not small just as much as the current Black Hole was not big at the time.

The Black hole sustained the space that became singularity but the extending of singularity grew with the growth of space-time not committed to star structures. The gravity that the star back then had to possess and create independence in the midst of the gravity presented in the Universe had to be so strong that it now contains the star in circumstances that was present at the time in the Universe. We must realise that the gravity contained in that star was strong enough to reduce the space in that star while all the gravity in the rest of the Universe was unable to contain the space or retain the time in developing. Looking at a star we can measure the Universe at the time when the star broke free from its galactica confining. The Black hole and all other stars for that matter stayed behind just as much as the space grew more. It is a relevancy applying both ways and not just to one side. As the heat in space declines the heat within the star rises. It is a relevancy not favouring any side. But also matter grows as much in relation as space recline.

The Black hole started off as a (presumably) red giant of its day. In the Universe applying at the time the space the Black hole contained was enormous and I suppose it engulfs large areas of the space available at the time. It is still acquiring massive space in dismissing therefore as far as the star goes little changed but in relation to the Universe the Universe acquired muck more space in exchange for large reducing in time. I cannot for the life in me see how Newtonians can consider that one part of the cosmos change leaving the rest of the cosmos never to change. The cosmos is about changing on all fronts that there are and in every factor of the relevancy $a^3 = T^2 k$. There was a time when the little space the Black hole now controls was a big space in a little Universe with lots of heat and little else but a few very bright stars covered in massive heat. It should also be recognised that the Black hole contains the material we will find in the collection of many massive galactica combining their protons into confining that much material in one unit. As I said, if that was not the case the Black hole just could not develop in the manner which it did while the rest of the Universe spawned of to form space in overgrown time.

There is the time where the space is consumed by the Black Hole as much of the space is concentrated because the Black hole remains relevant only to the singularity it sustains by dismissing space not relevant to the singularity it is sustaining in growth. To that end it is supplying heat to the singularity that will reconstruct the space to heat occupied by material and by that it feeds singularity by consuming heat to convert to space.

Then the Universe expanded as the star reduced. The star did not reduce while the Universe did not expand. The star reduced space as the universe expanded space and one may never lose sight of the relevance. Only singularity grew as singularity acquired the taste for other singularity, as they will provide the required space/heat to contain.

 There was a time when the Black Hole presented hundred of millions times larger space than the space that the huge Red giant now presents in relevance to the space available and that was available during the star era of the Black Hole. In today's era, the Giant seems huge in outer space and the huge star claims at present space holding vast quantities of occupying material within boundaries set by singularity extending. However, the space it claims is cloned and not real as we saw from previous explanations. But back when the Black Hole held formidable space in relation to space being available then the huge Giant Red star of present was only a speck invisible in the centre covered by a blanket of thick foggy layers of heat in the centre then of the Galactica that was then present and converting the now released the giant star.

 In the spiral of the proton, that the Black hole now has stretching into space the gravity influence extending might be some indication of what the core's influence had in relevance to the space ratio that applied when the star went from the universe into singularity.

One can only imagine the growth this star represents in the Universe because as much as the star lagged behind the Universe the star also grew in space in the Universe. Just imagine how hot space was when that Black hole was some flickering orange star dot in the sky. But that star was a star before light was the light we now know because the light we now know is invalid within the Black hole we now know. In the mean while the cosmos expanded and the Black hole reduce by remaining the size it had before the expanding came to alter the ratios that was applying. Just as the Sun captured the outer space value in heat that was present and part of the outer space density back then when the Sun released form the outer space the Black hole too captured the space relevant to outer space then and holds that space relevant throughout other progress applying. It is only our human insignificance that indicates what might presumably be big or small, hot or cold, near or far but to the cosmos it is history.

- At present the Black Hole as a star, as a whole is smaller than an atom because the star ejected all the proton qualities of the occupying singularity which was holding space-time and rejected all material occupying space or otherwise. At first all space-time that was on the out side of the atom structure but also being within the structure of the star claiming singularity is dissolved into singularity. The first step is to go beyond the electron as it now has gone past the point where all space –time was equal to the electron. Our sun as feeble as it is, is still in a process of accumulating space-time and the development will grow until the space-time is equal to that of the electron in intensity. That means as a star structure the sun has not even gained the development as far back as the Big Bang and being a star, it is only then when it can discard space-time equal to an electron in intensity that it can grow and develop singularity.
- It discarded the neutron when it got rid of all possible internal motion and only occupied singularity in the sector where the proton commutes in. Then finally with further Universal development that implicated the star as star development, the dismissing of space became absolutely overwhelming by even displacing the proton function to the space outside the Black hole. While space is in demise stars are growing.
- Compare the theory I propose with the theory, which mainstream science, underwrites and applauds and was promoted by men to the likes of no less than Einstein in person.

Compare the theory that I introduce in a practical context to Einstein's critical density and the underlying and the underlying factors as scientists wishes to promote gravity. They never calculated the distance that particles are apart because they couldn't. They couldn't because of the obscurity we humans have in our perception on the Universe stemming from our position in the fact that we think we hold the centre position in the Universe because all light streams directly n our direction. From the official argument the Earth must be at the centre of creation because every time Mainstream refers to where space ends at some billion trillion zillion or whatever ridiculous number they come up with. Such a remark damns their perception on reality and from such grossly thoughtless argument the wish to position stars and distances parting stars. Mainstream science full well know that the parting radius has a phenomenal influence on the outcome and even their argument about space ending is totally ridiculous because what happens to light shining where space ends. Does it collide with the end and then what will constitute such an end. Where will gravity go when the gravity afterwards reaches such an end? The thought about the lot entertains those unable of clarity but to the rest of us it is a joke that men with such brains can deliver such nonsense and present it as truth.

Einstein's Critical Density lacks the accepted matching facts we need in proving the critical mass factor. But our inability in securing such required evidence defies the most basic logic. It seems all new evidence we receive from outer space is disputing all Newton laws about the cosmos and new findings that disprove Einstein's Critical Density as the answer. The Universe will not reach a point of contracting, not withstanding whatever dark matter astronomers try to locate in the vast space. Why would the expansion turnaround and do a reverse by going back to where it came from. Where will the centre be where such contracting will locate a point and what will form that point It seems fine to draw a saddle or a square or a triangle or even a flat mat (no mat can be flat because there has to be another side if you draw anything presenting flatness) and then try and sell that idea to those scientists considered being a bit of jelly minded and weak of thought. That includes mostly and entirely the us, the me, the commoners and non-academics that cannot think. Even the saddle or the triangle or whatever has to present a perfect centre from where the general gravity will flow that keeps the structure they present contained as a unit. Without such a centre and the ability to show such a centre all fancy arguments drop short of being realistic. $F= G(Mxm)/ r^2$ holds all related mass I correlation with one specific centre point and without the centre the rest is strewn in disarray.

If the material is there it is there because it is secured there and where is the centre securing the material that is secured and there. It is such a simple question but without an answer the statement becomes ridiculous. Consider the momentum alternation such a change will bring about. How will the total Universe come to a halt when the time comes where the Universe must come into the turning back and contracting in order to honour and confirm Newton's statements? In what direction will the turn around go since all material that is in motion is spinning in ever growing circles? In other words the question is where is reverse. There will be a massive number of collisions except on the unlikely condition that this turnabout is orchestrated in the manner where every single subatomic particle goes in reverse in the same instant.

Such simultaneous action will have to include and involve every point of material but what will motivate the action and who will give the command to stop turn and restart the moving as the direction of moving must include every particle all over the Universe. I might guess the stopping problem alone is an insurmountable object that first has to be laid to rest before we start calculating and measuring every atom in the Universe. There are so many more unanswered questions, which I touch on in other books but that are not the nature of this book. While the clever academics present an image of very literate about physics they also lack the physics insight to realise that if motion shows a turnabout all material in the Universe will demolish because the momentum that applies with such a turning will destroy all structures. You can't even stop a bus that fast without paying very dear consequences. They never refer to relative facts in cross-referencing.

The Sun is not a gas-filled sphere holding hydrogen in its "natural gas" form, but it is all fluid and is in a liquid form where singularity is liquid- freezing hydrogen at 6500^0 C while outer space is boiling over at -276^0 C. Even if it was the gas filled bowl, then still more to the point is the thought no one ever gave a minutes notice to is why hydrogen being a gas will remain on the inside of a structure being 6500^0 without expanding like a rocket exploding. What prevents a gas filled bowl not explode because even if the gravity is considerable, hydrogen would never stand that heat and not get pretty volatile about it. The hydrogen must get very spaciously exploding long before 6500^0 is reached under what ever the conditions may be.. This book explains the Roche limit in the practical sense... when applying cosmic laws instead of improvising cosmic laws the information one receives uncovers that reality and becomes awesomely simple. It becomes clear that the Universe is as much expanding as it is contracting and contracting by expanding. As there is no hot or cold, no big or small, no grand opposing but relevancies in ratio to one another. If you do not believe me, then believe your eyes when looking at the picture about the Sun coming from telescopes. What ever the Sun is it is fluid falling into fluid.

Please tell me one reason why I am incorrect and consider the time it took the Universe to develop from 10^{-5} to 10^{-43} seconds to create a cosmos the size of a neutron. Compare that to what is happening now and see how many events took place by the creation of every lepton and every hadron and it was to be true that that period took longer too complete than it took the Universe to create the solar system. The flow of light through the density that space produce heat gives the speed of light the relevancy of time in space. The thicker the "soup" of heat is that space forms, the longer it will take light to cover a distance. It is very important to note that the speed of light is a relevancy between time (seconds) and space (kilometres). The speed relies completely on the value **k** holds on space –time. The speed of light is forever a constant but the constant is part of the relevancy of space-time.

The Universe connects in a way Kepler established through his relevancy theory. Those not convinced answer this: where would the Planets be if not for the Sun securing planet positions. The relation proves the ratio of one in all cases to be valid. It proves much more than merely connections at liberty of holding positions where ever the randomly opportunity placed the structure. The structure does not come closer by a pulling and tugging. Kepler's figures coming of his calculations must still be around and by repeating the task again but this time made much easier with the help of computers and telescopes of magnificence we can compare that to those the which Tycho Brahe calculated and test what growth took place. At present the star, as a whole is smaller than an atom because the star ejected all the proton qualities of the occupying singularity to the out side of the structure the star claims and is under control of the governing singularity. It rejected the electron before space –time was equal to the electron.

I do realise that the way I interpret Kepler is very new approach and up to now all I could find is Academics with scepticism and detachment from my views. I do not blame such reaction but if I am not correct, please explain according to the view you hold how the following is possible. You're being a person that stands at night on the highest elevation in your local the vicinity in a manner so that no other solid object can restrict or block out any of the light flowing towards you. From all over and from the most outer regions of outer space light is travelling in a straight line directly to you. The light is travelling at a speed maximum to what the cosmos will permit. That is how eager the light is to reach you personally. That light is using mostly millions and in some cases even billions of years to reach you while you are filling the centre of the Universe. Wherever you move the centre of the Universe will shift to the position you then mostly millions and in some cases even billions of years to reach you while you are filling the centre of the Universe.

All light is flowing from all positions and from every possible direction to salute you're filling the spot in the Centre of the Universe. Wherever you move the centre of the Universe will shift to the position you then hold because the light flowing to you will follow you to wherever you are at any spot. All the light coming across the vastness of space every possible region of all outer space is acknowledging your position where you fill the centre of the Universe in the spot which you fill every second of your entire life. But it is you filling the centre position and not the centre position being where you are. As you move notwithstanding who ever you are or wherever you are the centre will follow you as the centre will always be on the spot where you are. That is only because you that fill the centre of the Universe are where you are at the time you are there. Everyone being of flesh and blood as an individual knows that that person being him or her is filling the centre position of the entire Universe because all the light from everywhere is coming right across outer space to acknowledge what every person on Earth was expecting about himself being the one that is filling the centre of the Universe in any case. It is coming straight to you because you are in the centre of the Universe according to the light travelling. Not one spot represented by one ray will pass you thinking that you are not the centre of the Universe or the spot you hold is not important enough. Why would that be? Why will the light act as if you fill the spot where the light considers it can locate the centre of the Universe?

The light is underlining what you take for granted, the light admits that it is coming to you because you are filling the spot the light comes too as if that spot is you being the centre of the Universe. Most of the light started on their route even long before man was to become a species and most of the light was already on route long before the time when the solar system came about. Yet the light treats the place you occupy as the very centre it wishes to come to confirm the importance of you're position where you're position is within the spot centralizing the entire cosmos must be just because you fill that spot you're glowing presence. And every person everywhere has the same conformation about his or her importance as the Universe acknowledge this important position in the location wherever you may have what you have that is so important only you can have that centre position. If you close your eyes or ignore the light, the light will go unnoticed for all time to come.

The light then travelled across so much space using so much time to travel only not to be recognised by you the person filling the centre of the Universe. The light and all the space the light represents and all the time the light travelled would go into eternity dismissed as never acknowledged to be worth noting just because you, the one filling the centre of the Universe did not take the time to look and acknowledge the effort the light made so many million of years ago just to bring homage to you filling the position of absolute centred importance. So much effort is done by the light so long before you were even born to get the timing so right as to acknowledge your centre location in this moment in space and in time depends on you're taking notice of the effort and appreciate what the Universe did to admit that it also believe you are the centre of all space and time out there. Now use modern science to explain this centre you establish with your importance.

Tycho Brahe and Johannes Kepler stood there night after night and made a super human effort to acknowledge the information the light of the Universe brought to them. They wrote down every massage every night. But since they are only human they could only managed to acknowledge the light coming from the Sun that was reflected by the planets. The two masters were the centres of the Universe when the two Masters decoded the language the cosmos used to speak. Think about what they managed to collect for all mankind's benefit. They acknowledged the light coming towards them in a straight line. A flow of electrons causing photons to travel across outer space and meet their eyes. One line of photons flowing tells them all about the regions the light came from. Think of the size Jupiter holds and a few photons can bring across such large quantity of information.

Yet the light acknowledged the position the Master was in and came all the way representing such a large structure as what Jupiter is confining such an enormous structure with all the information it has into one line of photons. With the information of the entire structure confined into a few photons it managed to convey all that information across such vastness of space, choosing that specific point as the centre point of the entire Universe. By using the theories now applying and representing the views of Mainstream Science how would you go about explaining the way you are treated by light travelling through space and time to be in the centre of the Universe, which is the spot you have and hold. Seriously you know you cannot be in the centre where the Universe started the initial Big Bang process.

Not even the solar system or the Milky Way can be in the centre where the cosmos started and yet that is how Mainstream science argue, Can they be wrong? Yes they can and no they can't because they are correct but they are incorrectly looking for a centre where there are so many centres in the very precise centre. Still without any possible influence the light puts you and every one else in that centre. But since you are in the centre, you must admit that there are billions of trillions of spots forming the centre of the Universe just as Kepler stated. The only way you can fill such a spot is if the spot is the place holding singularity and singularity represents the centre of the Universe. Then k^0 can form at any spot and establish $a^3 / T^2 \, k$ because k is the end of k^0 being the start and k^0 is representing singularity. The centre of the atom represents k^0 and the electron represents k. The main issue brought to light by Kepler's formula is the relevancies that always prevail through out the cosmos no matter what sphere it is representing.

When a point holding singularity achieve energising by securing a number of protons concentrating heat towards that space centre holding the singularity, the singularity will establish a will to seek independence from other more dominating singularity surrounding that point and moreover seek to abolish control of the dominant singularity that suppresses its individuality by establishing a securing centre. Two points will redefine their relationship and will establish space-time relation between the points. A relation will come about where there is a dominant point in singularity establishing a controlling centre that is trying to establish control over space-time by creating motion too space-time by displacing space-time, then there is the space-time in motion and a third factor where singularity is applying motion to material within the space-time in control.

The dominant singularity will control the space surrounding the lesser spot in singularity that is gaining in heat concentration by material growth. Through applying gravity to ensure heat concentration with space-time diminishing the lesser singularity will gain heat in centralising space-time and by then turning it to motion it then can use such motion to gain independence from the centre in the space-time that the dominant singularity is controlling and is diminishing through applying motion within the space - effecting the lesser partner while the lesser partner the prominent space it is feeding from the prominent singularity that provides the lesser singularity with some pre concentrated space-time. This is the manner which structures in development within the incubator of the galactica heat blanket use the galactica governing singularity that is democratically elected and placed in a centre as a centre to bring progress in the incubating star development.

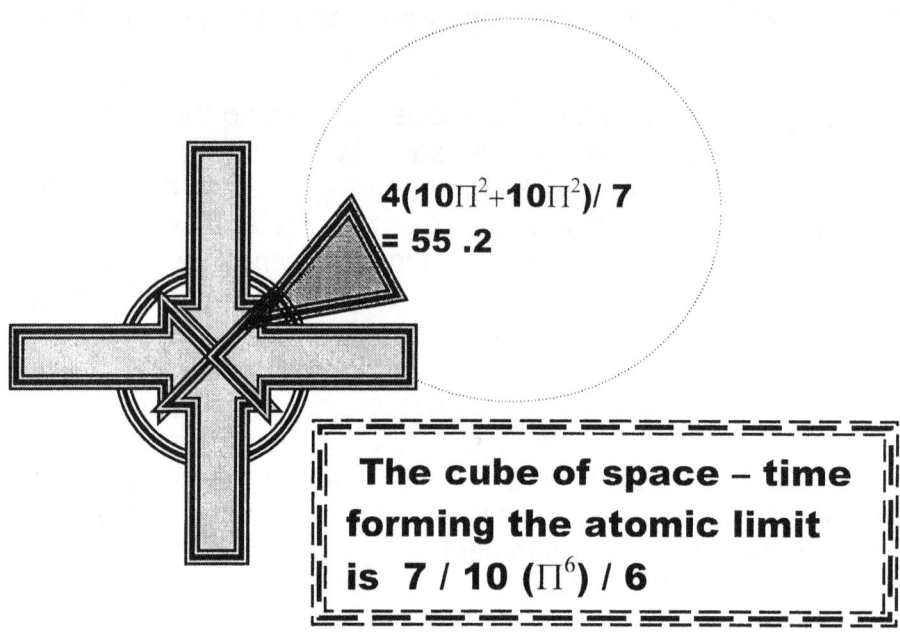

$$4(10\Pi^2 + 10\Pi^2)/7 = 55.2$$

The cube of space – time forming the atomic limit is $7/10\,(\Pi^6)/6$

The cosmos cannot be if the cosmos do not share with everything else in the cosmos but the sharing is always producing relevancy to the position of another factor forming the Universe.

The inner space is applying positive space-time displacement in relation to the object in rotation.

The orbiting object forming the outer ring is in a negative displacement in relation to the inner centre. The outer object has heat in the centre it has but the heat is far less dominant than the heat of the centre and by increasing motion it is concentrating heat as much as the motion is reducing space to concentrate heat. By applying motion it is securing space

What makes the cosmos is the variety of structures forming the Universe while all participating objects of all sizes are using the same singularity. There is no big ore small because that which has the biggest control in the cosmos is also incidentally the smallest there ever can be in the cosmos. In fact it is so small it cannot even directly claim apart in the cosmos. To make sense of this lot to me being somewhat of a dimwit I placed two opposing motions in relations to each other and the one will always show one or two relevancies in relation to the other.

The heat the inner structure secure prevents the motion from applying to the object because it became dominant enough to reduce the space towards the heated centre and in doing that it is producing space to secure space through applying motion to the space.

The cosmos cannot be if the cosmos do not share with everything else in the cosmos but the sharing is always producing relevancy to the position of another factor forming the Universe.

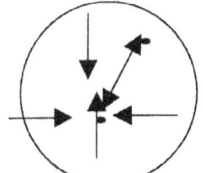

The inner space is applying positive space-time displacement in relation to the object in rotation.

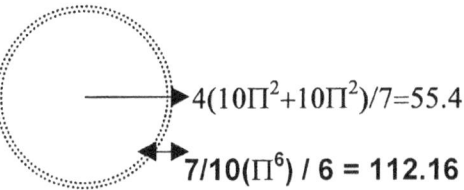

$4(10\Pi^2+10\Pi^2)/7 = 55.4$

$7/10(\Pi^6)/6 = 112.16$

The heat the inner structure secures, prevents the motion from applying to the object because it became dominant enough to reduce the space towards the heated centre and in doing that it is producing space to secure space through applying motion to the space. The orbiting object forming the outer ring is in a negative displacement in relation to the inner centre. The outer object has heat in the centre it has but the heat is far less dominant than the heat of the centre and by increasing motion it is concentrating heat as much as the motion is reducing space to concentrate heat. By applying motion it is securing space

What makes the cosmos is the variety of structures forming the Universe while all participating objects of all sizes are using the same singularity. There is no big ore small because that which has the biggest control in the cosmos is also incidentally the smallest there ever can be in the cosmos. In fact it is so small it cannot even directly claim apart in the cosmos. To make sense of this lot to me being somewhat of a dimwit I placed two opposing motions in relations to each other and the one will always show one or two relevancies in relation to the other.

If one looks at the transmission of sound, it too depends on the relocation of matter, but to a very small degree, and in this process lies the transmitting of sound. To make the error of judgment in confusing the process with the breaking of the Doppler rings are quite understandable.

_The biggest activity there can be is in a Black hole forming space-time confined directly to singularity where that pushes time eternal and by doing that places al motion activity into outer space.

It is about confirming space $(\Pi^2 + \Pi^2)$ conforming space $(\Pi^2\Pi)$ and converting space3.

The Universe comprises of the atom and a star is just another atom holding lots of its own. This comes about from the fact that both particles are the same in the Universe since both particles serves singularity. From singularity the size space-time takes up is unimportant because from singularity space-time is merely principals connecting singularity forming energy as gravity or antigravity and presenting space through the relevancies forming the motion of time. That is particles and atoms surrounding singularity, protecting singularity, maintaining singularity and securing the surviving of singularity. This service of space is done by motion in duplicating space or extending space.

When a star favours **3** and a multitude of three the star is still in a process where it will favour more the duplication of space. As the star develop it will ever increase as it moves through the ranks of being liquid ($\Pi^2\Pi$) the favour to the proton ($\Pi^2 + \Pi^2$) where more space is dismissed that space is duplicated because the motion of the star also diminish progressively and so to the end phase of the star. where the star will once more be motionless. The cosmos comes together at a displacing limit of 112 protons per atomic unit. It values time to space at **10 /7(4(($\Pi^2 + \Pi^2$)) = 112.795** on the space limits while motion breaks down within the centre of the star at half that being at **7/10 (4(($\Pi^2 + \Pi^2$)) = 55**

Space receives six sides at **7 /10 (Π^6) / 6 = 112.16**

(($\Pi^2 + \Pi^2$) the proton +($\Pi^2 + \Pi$) the neutron + (+ 3) space = 35.75 X Π singularity = 112.313

The cosmic cube we live in is singularity that is six sided **7 /10 (Π^6) / 6 = 112.16**

To decipher the code one goes about as follows. The value that generated gravity or if you wish to call it electricity will be as follows: $4(10\Pi^2+10\Pi^2)/7=55.4$

4 Indicates the time presence influencing the value. Four always indicate time (10) I outer space in the square already

Π^2 Indicated the square forming gravity

$(10\Pi^2+10\Pi^2)$ Is the double proton in multiplication with the square of space on both sides of the Universe.

/ 7= is representing the value of the sphere

55.4 Are the maximum protons displacing space-time working in one space less unit.

The four quarters of time (4) holding the square of space (10) in the double to gravity Π^2 also in the double on both sides of the Universe (+) in a proton relation ($\Pi^2+\Pi^2$) still maintaining the shape of the sphere (/7) is representing the flow of space-time by duplication in relevance to space-time dismissing that comes to a limit will be the maximum that the form of the sphere can sustain. After increasing the flow the sphere as form will collapse although space-time will still remain and be concentrated but valid. It is where gravity as we know gravity ends because that is where the any stars liquid atmosphere meets the density of the electron and the density equals the speed of light. After this value the speed of light will no longer sustain an electron in the atom, which by that time has been compressed to its limit as it was during the Big Bang. That is the end of the electron where the star is all electricity with no space-time being less than the value of the electron.

 The fact that the displacement equals the iron proton "mass" explains why iron Fe $_{56}$ is the only element that has the ability to generate electricity because by bringing focus on an iron Fe$_{56}$ core through motion contributed to employ the

Coanda effect electricity is generated as long as copper $_{62}$ is used to demolish space-time. Every electric generator is a star core in the micro.

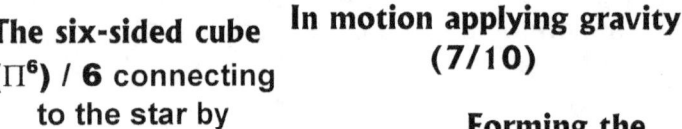
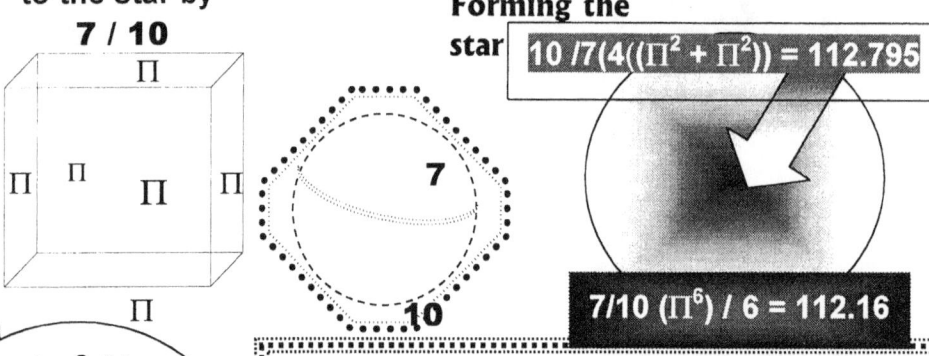
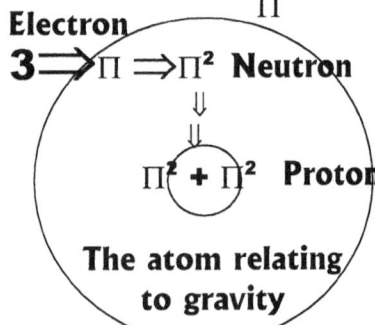

From the value that outer space can support being the sum total of the particles forming the atom $((\Pi^2 + \Pi^2)$ **the proton** $+(\Pi^2 + \Pi)$ **the neutron** $+ (3)$ **space** $= 35.75 \times \Pi$ **singularity** $= 112.313$ the star deliberately reduce the atomic space or the subatomic constructed- space as the star intensify motion and that reconstructing of space-time changes the qualities of the atom from what we presume the atom to be to suspending the atom beyond the boundaries of **$7/10 (\Pi^6) / 6 = 112.16$**. The converting of space- time from outer space through gravity to the star centre is the same route electricity follow

Electricity and lightning is the absolute epitome of the Coanda effect where the Coanda effect is precisely the manifestation of light following the exact principles of the Coanda effect and the **Total Internal reflection** is also miming the same principle as the Coanda effect which is vivid proof of space-time $a^3 = T^2k$ the Coanda effect in acting principally by using the flow of photons instead of atmospheric heat. **Total Internal reflection** is only about applying motion by the flow of space-time (in this case water running) through the atmosphere but in the case of the phenomenon we call the **Total Internal reflection** singularity captures light holding the flow of light honest to a specific centre as does the Coanda effect and by setting borders the boundaries light is restricted as singularity set limiting boundaries to the flow of photons. But that is what electricity is; it is only creating space-time accelerated motion with much intensity added and it links a line than is concentrating space-time as it accelerates space time through the displacing differentiation which one find in stars between copper dismissing space and iron accelerating heat directly to singularity. It is only much more intensity. All it is, is the Coanda effect forming electricity and lightning as the Coanda effect.

Gravity is electricity because electricity is the flow of heat from a gas source to singularity by charging iron {7/10 (4(($\Pi^2 + \Pi^2$)) = 55} (forming the artificial core exactly as the Coanda effect will charge singularity by applying motion) through the influencing of total space reducing which copper can manage having the specific space-time displacing value. The influencing of copper ($\Pi^2 + \Pi^2$) **X** Π = **62.0** breaks down space-time as stars do in the core centre. That is the reason why only iron can excite to charge electricity and only copper can dismiss space to a tome equal to the flow of photons. All phenomenon used in the Cosmos is the precise same thing using the precise same principles in a more intense or lesser intense gradient. Still it is all about singularity charging the control and the flow of space-time through motion where a liquid flows through space to a solid iron core that is influenced by copper.

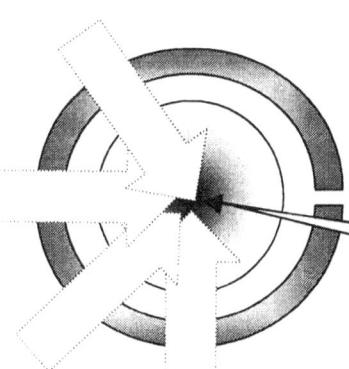

Electricity and lightning is gravity reduced by the intervention of the phenomenon we know as the Coanda effect where the Coanda effect is the establishing of a more intense dynamic point representing a new point of a controlling singularity dynamic.

Iron forming a centre or an iron- core precisely as the Earth core forms in iron form a centre core.

The copper field coils breaks down space-time by dismissing space-time as the 63 factor excels the space-time as fast as electron will as the space-time has to flow through the copper in the event where the iron being in motion causes the flow and charging the flow to the equal time set as the photon has. That is electricity. It is taking space-time directly to the centre of the earth because the motion T^2 **k** excited the space a^3 to a level that gravity is within the Earth core.

Electricity and lightning is the absolute epitome of the Coanda effect where the Coanda effect is precisely the manifestation of light following the exact principles of the Coanda effect and the **Total Internal reflection** is also miming the same principle as the Coanda effect which is vivid proof of space-time $a^3 = T^2k$ the Coanda effect in acting principally by using the flow of photons instead of atmospheric heat.

Total Internal reflection is only about applying motion by the flow of space-time (in this case water running) through the atmosphere but in the case of the phenomenon we call the **Total Internal reflection** singularity captures light holding the flow of light honest to a specific centre as does the Coanda effect and by setting borders the boundaries light is restricted as singularity set limiting boundaries to the flow of photons. But that is what electricity is; it is only creating space-time accelerated motion with much intensity added and it links a line than is concentrating space-time as it accelerates space time through the displacing differentiation which one find in stars between copper dismissing space and iron accelerating heat directly to singularity. It is only much more intensity. All it is, is the Coanda effect forming electricity and lightning as the Coanda effect.

Gravity is electricity because electricity is the flow of heat from a gas source to singularity by charging iron $\{7/10\ (4((\Pi^2 + \Pi^2)) = 55\}$ (forming the artificial core exactly as the Coanda effect will charge singularity by applying motion) through the influencing of total space reducing which copper can manage having the specific space-time displacing value. The influencing of copper $(\Pi^2 + \Pi^2)$ X Π= **62.0** breaks down space-time as stars do in the core centre**.** That is the reason why only iron can excite to charge electricity and only copper can dismiss space to a tome equal to the flow of photons. All phenomenon used in the Cosmos is the precise same thing using the precise same principles in a more intense or lesser intense gradient. Still it is all about singularity charging the control and the flow of space-time through motion where a liquid flows through space to a solid iron core that is influenced by copper.

Kepler said that the one half of space is motion and therefore there are no space any where except space in motion forming space-time. There is only space duplicating during a specific time duration and only space-time can relate to a specific point where singularity establish control. There is no space only because it is all space-time, which means space in motion by duplication thereof. Therefore when one removes motion from the space factor space will collapse. That is what a Black hole is all about! In relation to the mindset of modern man what I am referring too is unthinkable not to be absolutely part of common sense.

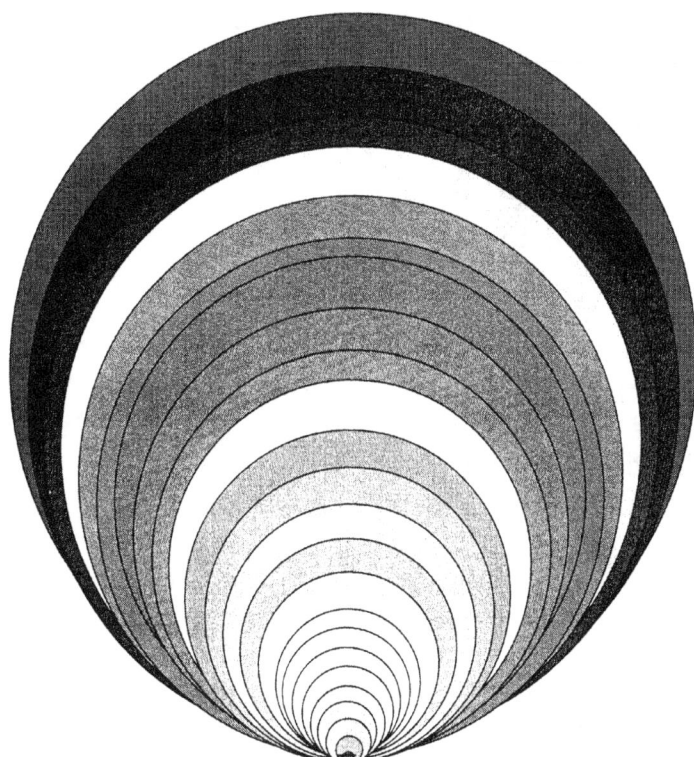

No star on the inside can have pressure. That is a physical impossibility. There is no pump to pump in and there are no walls to contain what is inside. Every layer is self-contained and mixing must be the exception because there are layers. Because of that I gave the inside process four names which I shall explain a little later. The star is a container in time and it preserves rather than pressurises what is contained on the inside. The process is named keeping the time aspect of space in motion in mind and avoid the rather simple pressure all mixing idea that is so Newtonian. It is ***aanplasing, verplasing, versnelling* and *inperking.*** This is what happens when material flow not blow inside stars.

Material within stars (and outside stars play the game of follow my leader because the material follow in a flow where the one behind captures the space it will occupy as well as the space it will represent in the flow directed by the direction of time. The material flowing into an object is connected to material flowing out of the object by motion we talk about as momentum.

When creating a solid barrier to stop material flowing we destroy the space by preventing the motion to continue. We stop the movement of with some intervention blocking the space-motion but by stopping the momentum we destroy the space duplication. We refer to this as a collision between bodies but what it is, is that we fill the space which the space in motion was suppose the fill therefore we destroy the form of the space in motion and most times we destroy all space by creating heat bringing about a colliding destruction that produce more space in the end. The more space we supply in relation to the space in motion the more space in motion will be stopped where the space in motion will become a flat object as all the material in motion share the same space. Space a^3 then becomes more represented by T^2.

Stars in our Universe are controlled by singularity that defines the outer borders as $10/7(4((\Pi^2 + \Pi^2)) = 112.795$ on the out side and an inner border of $7/10 (4((\Pi^2 + \Pi^2)) = 55$ in the centre of the star. That means the space envelope is **10 /7** and that is confined too **7/10**. That produces the gravity in the stars

As the space in motion is occupying less space due to the motion duplicating and reducing the space the space in motion will need less space to duplicate thereby then create motion. Smaller objects can apply faster motion since smaller objects require less space to duplicate and therefore less time to do the duplicating of the space.

The motion of space using space in time is about only and above all replacing what is filled per time unit required filling that specific space. If the time per space unit diminish the space used per time unit reduces. The space removes time from the inside of the atom and use such space as time in the form of heat on the outside of the atom That is the reason why jest heat up when crossing the sound barrier. When entering the outer space edge of the Earth the coming through the atmospheric barrier the craft must adapt the time constraint to that apply within the zone befitting the Earth. We use time per frequency to fill space in which we live. This apply more so as stars develop by ridding space in favour of extending time. But Black holes are just the peak of an evolutionary development that can be explained so easily when using the cosmic dimensional code.

`1

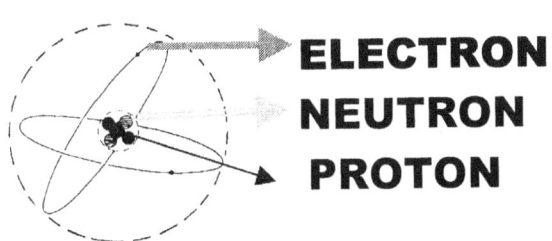

ELECTRON is about confirming space
NEUTRON conforming space
PROTON converting space

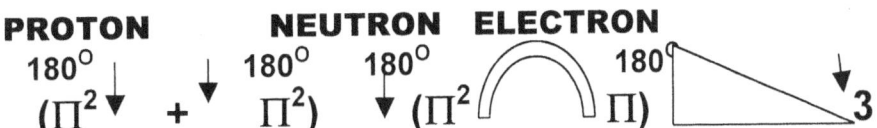

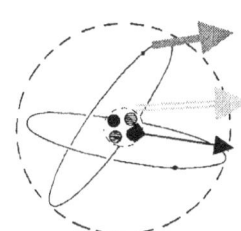

ELECTRON : Unoccupied space-time is 3
NEUTRON : Occupied space-time is $(\Pi^2\Pi)$
PROTON : Densified space-time is $(\Pi^2 + \Pi^2)$

The role of the electron the neutron and the proton is very commonly accepted the role each sub atomic particle plays. But galactica and stars are just as much just more cosmic atoms playing their part in the very same way as does the sub atomic particles.

Science should become serious about the task they chose to perform and not going flat out covering up mistakes by denying mistakes their Master made. I am aware that science is aware of the contradictions Newtonian science advocate mainly because no one this far acted surprised or bewildered when I mention it and they then have to realise it for the first time in their entire existence. They know exactly what I say when I say and with that they are performing acts so shocking about self-protection and self-preservation. When Roche presented his findings they should have realised there is something missing with the way they see things. When Hubble presented his findings they should have put Einstein to task about finding the mistake they made and not a manner to cover up the mistakes they made about the Universe contracting. Newton's arguments mathematically are about contraction and Hubble proved that the entire Universe proved without doubt that all indications prove expansion. By going overboard even further and order Einstein to measure the Universe is frankly madness! But I have never seen any group of persons to be blatant in their refusing to acknowledge a possible mistake that Newton made in the manner that I do. Mainstream science places Newton equal to religiosity. I found on all and every campus I went that any remark about uncovering a mistake Jesus Christ supposedly made generated immediate interest with even the most adhering Christians coming to hear the argument. Making a remark about Newton making an error gets you marched off the campus by security. Why not test Newton's $F = G (M.m)/r^2$ from figures Kepler left us and see how far did planets shift closer. Prove what is lectured to millions of students. Where such observation did take place it rather proves definitely quite the opposite because the distance between the Earth and the moon is growing and it is not shrinking. With me openly criticising Newton and Newtonians lecturing none applying information across all of the Universities across the world (I guess) will once again make this book as successful as the others in the past. The Academics get very a very subdued hostile attitude towards me when I blame the Academics to their faces that they are not about gaining knowledge but about conserving the past through protectionism. Universities protect their own without any willingness to test that which it protects. All evidence should be clear in confirming that the basis on which the entire world of science forms a union is founding their policies and beliefs on incorrect principles. They should not become annoyed with critics but not only that, they should show evidence immediately to back their claims showing how far did the solar system-structures move closer. From that we then can calculate what collisions we are waiting for and how long before the big solar clashing will begin. The absence in they're just mentioning such possibility confirm to me they know as well as I do there is no evidence of the moon reaching the Earth with no evidence of pulling or tugging and the Universe is in synchrony more than any person may ever be able to prove. When looking at photographic images coming from the Sun we can clearly see that the fluid pushes out of a bowl of liquid and the telescopic images coming from the Sun via the camera lens there can be no doubt that the Sun is a bowl of liquid sloshing like a boiling kettle of soup.

From within stars, spilling both sides as it falls back into liquid pool forming the Sun. The inside of the Sun is not gas but it is fluid. In all of nature there is no NATURAL **gas** as much as there is no **natural solid**. Hydrogen is as much a liquid as iron is a gas and neon is a solid. It depends on the element relating to the space/heat in the circumstances surrounding the substance at that very precise instant in time. We have to stop telling the cosmos to show us what we wish to find and start accepting what the cosmos is telling us what is out there that we should look for and find. The fluid state and the gas state is expendable waste that stars remove through development but then so is all space-time and material. Using the location to explore these books I announce will indicate as to how singularity came about to form space-time and commanded motion by creating space. In that exploring we may find out that the Universe is already contracting as much as it is expanding and it is contracting by expanding because it is through the contracting that it is expanding; the answer comes about from **a³=T²k**. My effort with the criticizing of the Academics was never to attack the world of physics and I never had it in mind to destroy any work made by them. I can ill afford enemies and even less enemies as power full as the Academics are. But on the other hand I cannot go on praising work I disagree with when I aim in full knowledge in the mistakes I see about their work.

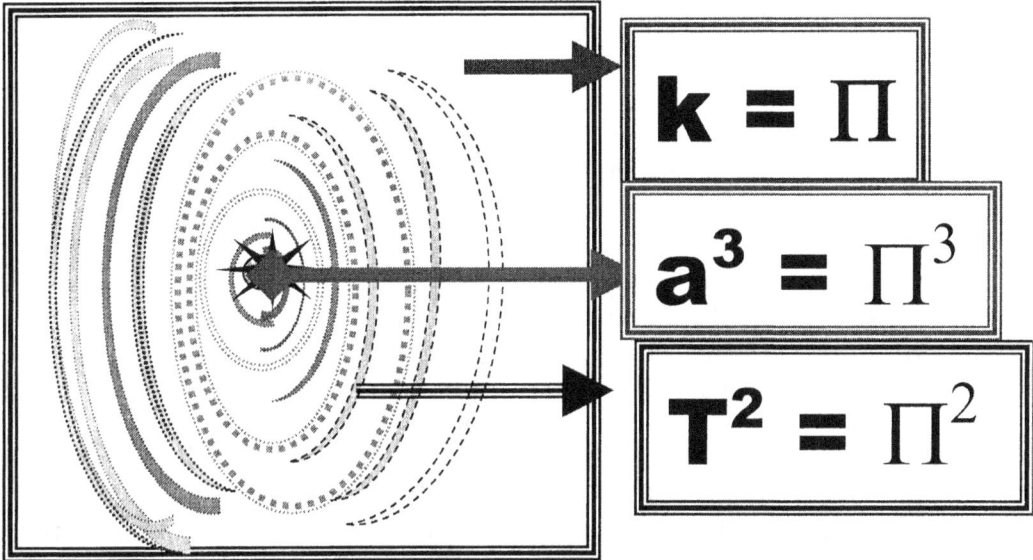

Everything in the cosmos is moving, on two accounts by own individual accord, as well as under the influence of some other singularity dominance. In explaining, we return to the top. The two opposing motions are inseparable and always in relevancy but never to size because, it is always to motion. In explaining, we return to the child toy in the spinning top.

The value is in the layer and the value is the purpose of the layer. A layer contracting will not be the layer taxed with duplicating because the stronger the liquid motion is the more intense will the gravity be coming from the motion. This statement I can prove by mentioning the Black Hole. The motion that the space-time being contracted to the centre of the Black Hole is far more that the speed of light, yet by such strong motion the gravity produced by such higher motion is far stronger that the speed of light.

Space-time holding a six dimensional Universe is:
$4(10\Pi^2 + 10\Pi^2) / 7 = 112.88$ **Outer space**
$4(10\Pi^2) / 7 = 56.6$ **Inner space**
$\Pi(\Pi^2 + \Pi^2) = 62.0$ **Ending of space**
Roche in cube $(\Pi^2/2)(4(\Pi^2 + \Pi^2)/7 = 55.6$
establishes gravity

At first I was very timed in my approach, which brought me no joy as well as no results from any Academic near or afar responding. I then felt forced to become much more bold and down right to the point. Still the response was the same. I concluded at the time my lack of results most probably that my revelations did no surprise those Academics with my startling evidence one bit with revealing my evidence about they're contradicting testimony about the official accepted cosmic principles physics. I now am of the opinion their lack of attention to my asking their agreeing or disagreeing is more about they're refusing to admit than they're being surprised as they are stunned by their seemingly double talk of their teachings. Not even once did any Academic in any way acknowledge the absoluteness of motion controlling the space making the space being valid in the Universe. All this is notwithstanding all the clear evidence collected by centuries of investigative studies bringing proof and proving to this affect. If the motion driving space-time towards the centre of the Black hole stops the Black hole will altogether disappear. It is the driving motion of space-time concentrating as it speeds up moving towards such a centre that that puts the Black hole wherever the Black hole may be in the centre of that Universe the Black hole keeps captured by motion.

Everything in the cosmos is moving, on two accounts by own individual accord, as well as under the influence of some other singularity dominance. In explaining we return to the top. The two opposing motions are inseparable and always in relevancy but never to size because it is always to motion. In explaining we return to the child toy in the spinning top.

When the top is in a state of motionlessness on own accord it is everything but motionless. The motion it adapts are synchronised with the Earth in harmony with the solar system and according to the greater picture of the cosmos. When an energy source not related to the cosmos called life intervenes and energises the tops motion, the singularity in that top suddenly jumps to life. By adopting a rotation energised to an unnatural state of energising because of life's intervention, the singularity that up to that point dominated the top is no longer in absolute dominance of all aspects of the top's motion. Another singularity takes charge placing the it in charge by nominating a point in singularity as the Coanda effect takes charge as the Sun or any other star for that matter applies more and more motion, it will begin to find a means whereby it can escape and apply individual singularity as the top starts to separate from the singularity the Earth holds. The singularity holding the Earth would then allow the singularity of the top to rotate within a specific band where that specific band of being active will tolerate such striving for independence before the Earth's singularity will start to destroy the singularity in rebellion.

The top on the other hand will try its outmost, when the singularity it then forms by individual spin is too strong to remain without independent motion and being totally be in dominated by of the Earth's singularity. The motion of the top is an attempt to begin applying an individual singularity space-time defying and standing apart from the Earth's gravity. This is coming about by forming the Coanda effect. That action we see as the top starts rotating in a manner where the top does not align with the Earth's singularity. With the adding of spin, the time the top holds becomes unrelated to the time the Earth holds and the top will start a campaign too escape from the singularity domination the Earth has on the top. When the time or spin of the top exceeds the limits the Earth places on the top, the top would emerge by trying to escape from constrains placed by the Earth. The view I present at this point is known to science for almost as long as science knows mathematics. The motion establishes the singularity that sets the top's drive for independence from the Earth's gravity restriction and set the top in balance as Kepler indicated. The space the top represents becomes the motion thereof $a^3 = T^2k$.

If there is anything that we can learn from the Coanda affect and how we see it applying in the Universe then it is that motion brings about an allocated and generally elected chosen centre form where that centre find control of all of the Universe being the Universe in such a controlled area.

We may say in our self-defence that is what the cosmos tells us to believe! To the light on route, time means nothing and space even less. Light cannot show more motivation to reach me at this point than I am filling at this moment. Not one ray is by accident missing me except by my choice prevailing. It flows through the Universe in time and in space in the hope that I, the chosen Universal centre whom is filling this spot, is in the spot that all light is coming to, hoping all the way and all the time that I would be graceful enough to notice the light. If I were not in the mood to acknowledge the light, the light would have done all the travelling just to be disappointed by my not meeting the light.

We use the most infinite to view and formulate what we think is going on in the Universe. Being Π sets us in the centre of the Universe. We take so much light for granted, never thinking for one second how impossible our relation with light truly is. This totally extraordinary relation we have with light must be one of the reasons why we humans put our position we have in the Universe in such a pivotal place. The fact that we as life carrying individuals especially we in person that are all blessed with the ability to only use our hind legs to walk on, on the surface of the Earth has the idea that the Universe was created especially for us, us being those holding life. Think again and such an idea supports everything everyone

If we wish to find the future we should locate the past. If the cosmos is contracting, where to is it contracting? The direction of contracting must be in the opposing direction the direction of expanding. If we wish to locate the past from where the cosmos came and through that see in what direction the cosmos came, it must take an effort to backtrack the direction it came. Should the argument come about that all came from nothing, then everything either still has to be at nothing, or our understanding of nothing leaves much to desire. Nothing means not existing, not being, never found and unable to produce any multiplication there of by any growth.

The above questions, but mostly the unanswered questions about what is more nothing and what is less nothing draw me to the realisation there can be no such a quantity in space as nothing because even space has to be something. Clearly as it is for any one to see one create space by nuclear explosions. In explosions Academics portrait that the wind as shock waves, but what is the shock wave other than new space coming into prominence and rearrange the structure in relation to the new space just created by liquid heat unleashing the created space as well as the space volume that came in place. In that way it is clear that releasing heat brings about the expanding of r as part of the sphere forming space. Hubble proved the Universe is expanding. Then by backtracking we have to set about reducing the sphere constituting the expanding Universe. If r in the circle is growing we have to reduce r to backtrack. When the circle reduces, the value located to r will become implicated because r determines specific size. Not so in the case of Π, because Π in the true sense only indicate that the circle is a square without corners and therefore Π dictates form and not size. By reducing size only r comes into contest and will point to such reduction. By reducing the circle radius r by half continuously will lead to an infinite small circle but Π will remain because the circle as a form remains even being infinitely small.

It would be the same as if the ant running in Central Park in New York is of the personal opinion that it is his being in the park that is the cause of the park being where it is and maintained as it is and only because of our ant's personal benefit that every one which is there is in the park. Our ant considers that every one of millions of people thinks a thousand people is dedicated on his behalf to maintain Central Park in New York. All the people in New York are of the obsession having one purpose to live for and that is to please that one ant. And yet that is happening with the light and us. Every person is standing in the Universe is under the elusion that all the light through out the Universe is directly flowing to the very point the person is standing. It happens to all of us. The place where I stand or any other individual for that matter is standing is positioned in such a manner that every beam is directly flowing to that very specific spot. Every beam is coming at the speed of light through the entire Universe to locate such a person with that magnitude in honour and glory as to fill the centre of the Universe.

From all the corners of the Universe one line of light is especially directed to that specific location used by that specific person for that specific instant in time. One very important human being is filling such a location of absolute splendour. The light departed from every location in all points through out the entire Universe stretching further than the mind can admit directed on course to meet the person in that centre spot. The light followed one after the other dedicated for billions of years to flow in the direction and directly that is the spot, never diverting for one instant, to come to where I am filling that spot in that centre of the Universe. All the light in the Universe is coming to me. It is on route straight to me where I am standing filling one spot on Earth. If there are those that don't believe me, well those I challenge then to go outside and see the vastness everywhere form wherever the light is coming from. It is coming from all over. It is coming from areas so large not even Einstein can calculate the size or content and it is rushing towards me specifically. There is not one ray that is going to miss me by fluke or accident. The light has one purpose and that is to meet me at the point I am. Every beam has my name on it and it is coming for my eyes. Can any one imagine if a person was standing in a location and found all the persons in that city was running towards him where he is occupying that point, how frightening such a person must feel. Yet it is happening to every one from where ever the vastness of space is situated and is coming across space to that very specific point the viewer is standing.

Even if I shift to another position on the other side of the Earth or to the moon the light will change direction and trace me in my new location. Even if my new location is in a camera and the camera is in a vehicle in the centre of Mars the light will know that I am using such a point and trace the camera so that I may still be in the centre of the Universe. Wherever I might be, the light will still get me at that location. The light flows to me from where ever and to top that it is also flowing to all other persons. That means it is not the Earth that is that important but it is where the point location is and that point which the observer is using to view from, that is the most important place eve to be. If it was only the Sun that the light was streaming from that is choosing me as representing the Universal centre being the centre of the Universe it then cannot be that very exclusive. The Sun is close and the light is plenty. But that, which I am referring to, that it is coming from all over.

That is just one small part of the fantastic affair. Some of the light left the stations they come from some 12×10^9 years ago to meet little old me in this spot I am filling. The light has been travelling 12×10^9 at the speed of light, which I might add is much before my birth crossing space and time, rushing all the way to meet me at this point. No one ever thinks how it was possible for the light to know I was going to stand at this point and be here the moment the light arrived. How did the light know I was going to take centre stage at that moment when it left so many billions of years ago and fill the specific centre of the entire Universe? I have to be in the centre of the Universe because all the light is travelling to this spot filling the centre of the Universe without one straying off course and missing my spot. The light takes two million years coming only from the closest next galactica to meet me here taken into account the prefect timing it applied after all travelling that far to be in time just to meet me in the centre of the Universe. How important can I ever dream to be? Light is coming across time measured in millions and billons of years through space measured in millions of trillions of kilometres, travelling at the maximum speed the cosmos will allow, ignoring all other places it could go too and came to meet me in the centre of the Universe.

To the light on route time means nothing and space even less. Light cannot show more motivation to reach me at this point that I am filling at this moment. Not one ray is by accident missing me except by my choice prevailing. It flows through the Universe in time and in space in the hope that I, the chosen Universal centre whom is filling this spot, is in the spot that all light is coming too, hoping all the way and all the time that I would be graceful enough to notice the light. If I were not in the mood to acknowledge the light, the light would have done all the travelling just to be disappointed by my not meeting the light. If I choose not to go outside and acknowledge the light all the travelling the light did was then in vane. An effort spanning billions of years and an effort stretching trillions of mega kilometres was all done for nothing and no reason at all because I neglected to meet the light. From everywhere the light is coming my way and that miracle is passing me by because I am feeling even more important as to acknowledge the total importance I have. The light is tracing me specifically at the location I am occupying just to please me and serve me with all the information about the history of the Universe. I can accept and acknowledge the effort or I can dismiss and ignore the lights efforts. I suppose that will allow me some arrogance and encourage me to think this all was specially created with only me in mind and if I wish to draw a map about the Universe I have all the right in the world to place me in the centre of the Universe from where all of the everything has chosen to meet me, the one to be met. After all, all the light is doing just that!

This information being at our disposal for thousands of years we never used except to inflate our already overblown ego further. By studying what we experience we can dismiss our superior view we hold on our self-importance and find what the cosmos is telling us about the position we have concerning the cosmos. It has to be true that we are in the centre but by using our knowledge especially concerning the Coanda principle we should gauge how the Universe work because we hold the point from where we may judge much, much better. That is because of the Coanda effect and because we are in motion apart from the Universe within the Universe confined to and being inseparable of the Universe we all are still carrying the eternal centre located to us as we are in the eternal centre.

I am in the forefront of motion in the Universe as is everything else having singularity…and everything, but everything has got singularity. Motion is not what we think of through what we experience as motion in getting from here to there because we do not wish to be here but we wish to be there. Motion in the Universe is the rearranging and the re-administrating of objects. It is a question of relocating the arranged to be re-arranged. This is because the Universe and everything in the Universe can't go anywhere but remain in the Universe. Because there is motion allocated to me I become a reference point for singularity that is holding space –time that is guarding singularity. Because of the motion I receive the allocation of being the most prominent of that entire Universe that can and that may be dominating space-time because I am pivotal to gravity, which is the gravity in my area securing and surrendering to the form to the Universe by accepting form from of the Universe. With me being in the position that my life within me holds me in singularity it's putting me in a spot of not representing space and therefore being motionless. Because I take on singularity everything around me represent space and motion at a distance from the point I have no space and no motion. Remember I am not my body as my body is the closest space-time I can control. With me being life I hold singularity. From where I hold singularity everything is space in motion and it is because of the Coanda effect I feel the affect of being the centre in the Universe. All space is spinning around me and because all space is spinning around me the Coanda effect puts me in the centre of the space-time sharing my Universe. There is one Universe, which is an innumerable, many Universes all referring to one Universe and represents one Universe in innumerable locations within the same unit. The motion I am due is the Universe allowing me my due to be. The Universe is not about positions or locations but relevancies responding to other relevancies that either puts me in a form or uses me as part of a form An object in a collision does not destruct or destroy but merely rearrange its composite from point of motion to another point in motion. But all motion of space forming space has the due to pay, which is the due I have to pay. We all replace energy from one location by motion to another location in motion.

In the past, and even in some quarters today, science is on the search for the 100 % efficiency machine. That theory runs on the surmising that a machine can drive as an output delivery without receiving input of energy. A few hundred years ago many Kings were fooled by such notion and some scientists truly spent a life in honest search of just such a device. Mostly the accomplishment came from cheats that very well new their machines were not up to the task, but in fooling a rich investor, brought about wealth to the inventor. As science progressed the no input giving all output machine became less and lesser a feature of the honest inventor. But the idea does not exclusively come from crooks finding a way to cheat the world. The practise of receiving without giving comes from science in the form of physics. It is physics taking the world on a wild goose chase in the way physics present the cosmic motion.

Physics propagates that the cosmos is all about running without input driving energy. The cosmos is all about wasting matter to a supply of motion. This idea prevails even after the world of science saw clearly in the past that there could be no such machine anywhere. In the Universe space can only be space when space is in motion and motion is about energy. Energy does not come free and energy always has a price tag because of differences in energy supply. That made me realise that. Even the cosmos must be a machine driven by an input and an output. It is the input / output driving energy that must be located and the driving ability we have to locate. Science hold the idea about mass that forms the unexplained and is therefore the magical drawing power to prominence, but what if it is not the drawing power of mass that holds prominence, but it is the reducing or contracting of space that is the driving motor behind the cosmos. All energy we humans at present use to accomplish material being in motion, holds some form of heat redistribution. Even electricity is a form of pure heat. I say that holding in mind of what apply when the energy of electricity becomes over abundant and the machine overheats. By overheating it means that the motion the machine creates comes about from heat control and precisely planned heat distributing. When scrutinizing the process we find behind the charging of electricity we find behind the charging of electricity is the same principle than what establishes the Coanda effect. Understanding the connection between the various principles helped me to connect the line of dots.

That brought me to realise that it is not me that is drawn towards the Earth, but it is the space in which I find myself that reduces, and that produces the effort bringing me closer to the Earth. The formula $F = G(M_1.m_2)/r^2$ suggests driving, moving in a direction and contracting. It suggests the reducing of space and not merely drawing or moving closer. When looking at any machine in practice, the machine draws power from space that reduces whereby heat increases. Not releasing that it is the heat that forms space is a misguidance of the truth.

The heat creating or forming space will lead to the destruction of the composition forming the machine. There is no form of matter, or element strong enough to resist the material deformation brought about by overheating. Having this in mind that matter does not resist heat, it is of importance to recognise that it is heat that is allowing space to give matter form. Looking at the manner in which energy is utilised it is space and heat forming matter allowing motion that allows work to achieve value. Motion originates from singularity by establishing the proton promoting spin through out the atom. Then there comes the part that makes the star a star or any cosmic structure what it is because every cosmic structure is the motion it is and not the material it confines. A Black hole will disappear into the realms of obscurity if not for the motion it establishes.

At this moment science is all about a body falling where the two bodies are producing a force whereby the bodies draw one another closer and all this is enticed by unbelievable gravitational pure magic. The bigger the mass, the bigger the drawing that comes about from the force unleashed by the mass of matter. The idea about this practise was phenomenal in 1602, it was impressive in 1802, but it is really ridiculous in 2002. Why would Boron form a solid having 5 protons weighing 10.811 g / mol from a dense solid structure and Argon as a gas having 18 protons weighing 39.9 g / mol. be a gas so light it floats in the atmosphere. This would imply that the "heavy" element with the biggest drawing power is a gas and the lighter element of the two is a solid. That denounces the contracting force theory. The way we compile and use energy must be in a similar manner to the way the cosmos uses energy distribution. **We humans can create nothing, but nothing is all that we humans can create**. The rest of our achieving is by duplicating whatever nature provides.

We have advanced past mass but and should now find the ability to establish what drives the Universe except for blaming some medieval magical force coming from nowhere going nowhere we have to find what drives us. The energy we use in all forms is producing heat in space by either converting space to heat or heat to space. Explosions are about converting heat to space. Compressing is about reducing space to heat. That is all energy composing work and is the only method of producing energy notwithstanding the immeasurable many names we use to express the same function in different forms. Arriving at the question about locating the space and time forming the centre the centre of the Universe one has to realise the centre of the Universe are in every singularity forming matter weather it is big or small, size carries no significance. It is the impartiality of singularity that is claiming the value and not the differentiation of matter. One must realise there are no big / small or hot /cold or near / far. It is all relevancies between matter claiming space and space is heat in a turnabout manner. Every aspect in the cosmos are locked-in Universes, sealed off from other Universes and inclusive or exclusive depending on singularity holding relevancies relating to one another. The relevancies rely on inter dependence and inter linking, but there are no differences according to human sizes or standards. Accepting that principle unlocks the "so called mysteries" of the Universe and brings about clear understanding. It is all about accepting, acknowledging and interpreting the role singularity maintains on matter.

Second Over All Overview

www.ingramcontent.com/pod-product-compliance
Lightning Source LLC
Chambersburg PA
CBHW080616190526
45169CB00009B/3198